T0134938

Human–Computer Interaction Series

Editors-in-Chief

Desney Tan
Microsoft Research, Redmond, WA, USA

Jean Vanderdonckt
Louvain School of Management, Université catholique de Louvain,
Louvain-La-Neuve, Belgium

The Human–Computer Interaction Series, launched in 2004, publishes books that advance the science and technology of developing systems which are effective and satisfying for people in a wide variety of contexts. Titles focus on theoretical perspectives (such as formal approaches drawn from a variety of behavioural sciences), practical approaches (such as techniques for effectively integrating user needs in system development), and social issues (such as the determinants of utility, usability and acceptability).

HCI is a multidisciplinary field and focuses on the human aspects in the development of computer technology. As technology becomes increasingly more pervasive the need to take a human-centred approach in the design and development of computer-based systems becomes ever more important.

Titles published within the Human–Computer Interaction Series are included in Thomson Reuters' Book Citation Index, The DBLP Computer Science Bibliography and The HCI Bibliography.

More information about this series at http://www.springer.com/series/6033

Oliver Korn
Editor

Social Robots: Technological, Societal and Ethical Aspects of Human-Robot Interaction

 Springer

Editor
Oliver Korn
Offenburg University of Applied Sciences
Offenburg, Germany

ISSN 1571-5035   ISSN 2524-4477   (electronic)
Human–Computer Interaction Series
ISBN 978-3-030-17109-4   ISBN 978-3-030-17107-0   (eBook)
https://doi.org/10.1007/978-3-030-17107-0

This Springer imprint is published by the registered company Springer Nature Switzerland AG
The registered company address is: Gewerbestrasse 11, 6330 Cham, Switzerland

Introduction to Social Robots

Social robots are robots which cannot only do services for us but also communicate—thus, they could come very close, into our homes, into our private lives. Do we want them to help elderly people? Do we want them to help us when we are old ourselves? Do we want them to just clean and keep things orderly—or would we accept them helping us to go to the toilet, or even feed us if we suffer from debility or even Parkinson's disease?

The answers to these questions differ from person to person. They depend on cultural background and personal experiences—but probably most of all on the robot in question. While the industry is continuously developing humanoid robots, my team's research in the Affective & Cognitive Institute (ACI) at Offenburg University of Applied Sciences, Germany, indicates that many people feel distanced towards human-looking machines. When I gave a talk at an event called "Children's University" in summer 2018, a little boy brought it to the point: "I always want to be sure if I talk to a machine or a human". For children, but also for many adults, small robots which resemble animals or cartoon-like robots might well be the better design solution.

The design space. For social robots. What a big and potentially overcharging task. This book approaches it by providing a compendium, grasping the phenomenon of social robots by looking at their different aspects. The basic idea was to actively include different research disciplines. This interdisciplinary approach goes back to an idea competition I won at AGYA, the Arab-German Young Academy for Sciences and Humanities. Since this academy is open for researchers from all disciplines, I had to find a topic where everyone could contribute: mathematicians, medical doctors, philosophers, linguists, sociologists, political scientists, law experts and several others. Social robots are a topic, which indeed requires input from all these areas. To me, the successful design of social robots is one of the biggest challenges of this century, as it combines artificial intelligence with machines which have the power to act in the real world—and thus can potentially create physical harm.

For over two years, we discussed the design and future application perspectives of social robots at workshops at the Active and Assisted Living (AAL) Forums in Coimbra 2017 and in Bilbao 2018, as well as during a dedicated workshop in Corfu

2018 at the conference ACM PETRA (PErvasive Technologies Related to Assistive Environments). We invited renowned scientists from the International Conference on Social Robotics (ICSR) to work with us and talked with dozens of experts all over the world. We created a film to show some of today's social robots and their broad range of forms and functions (https://affective-lab.org/videos/)—and a comic featuring a crime story about social robots, showing how emotional ethical debates on this topic can get (https://affective-lab.org/comics/). Many of these activities and findings are reflected in this book. To give you an overview, I will briefly introduce each of the thirteen chapters.

My friend from AGYA, the historian Christian Fron from the University of Heidelberg in Germany, was so kind to work together with me and compile Chap. 1. It shows how the concept of robots or "automata" dates back over two millennia— and interestingly provoked similar reactions throughout history.

While we analysed the past and drew lines to the present, the engineer Tarek Mokhtar from the Alfaisal University, Riyadh, Saudi Arabia, goes beyond that, providing an outlook into the future by Chap. 2. Those who agree with my assessment that humanoid robots are a problematic design approach should have a look at this chapter.

Five other authors who aim to look into the future are H. P. Chapa Sirithunge and colleagues from the University of Moratuwa, Sri Lanka. Indeed, it is a central design aspect if social robots do just wait and listen for commands or are allowed to act proactively, asking their owners questions and even present recommendations. The authors evaluate this topic in Chap. 3.

Dimitra Anastasiou and her colleagues from LIST in Luxembourg also want robots to act naturally. However, in their case the robot is an avatar that "mirrors" the actions, gaze and gestures of a distant person. In this area of "mobile robotic telepresence", social robots with their potentials regarding embodiment are key. If you wonder if the current state of the art already suffices for the vision, or if you are interested in telepresence altogether, have a look at Chap. 4.

My friend Gerald Bieber and his colleagues from Fraunhofer IGG in Rostock, Germany, consequently go a step further: if a robot may initiate (Ravindu et al.) or imitate (Anastasiou et al.) communication, it requires more information about its human counterpart. Bieber et al. suggest that "robots should use unobtrusive vital data assessment to recognize the emotional state of the human". Clearly, this is a controversial topic: emotions lie at the core of human self-conception. For the little boy I mentioned above, an emotion-aware robot surely would be something to take in. On the other hand, we readily complain about the "coldness" and lacking empathy of robots. If you are interested in how emotions can become a part of our interaction with social robots, read Chap. 5.

Sarah L. Müller-Abdelrazeq and her colleagues from RWTH and Cybernetics Lab in Aachen, Germany, also wonder how a positive attitude towards robots can be created. However, their field of application is not hypothetic and futuristic but very real, as they investigate industrial contexts. Starting their long way as dumb and highly dangerous machines in the eighties, industrial robots are now ready for collaborative work. In an attempt to combine "robots' power, consistency and accuracy with humans' creativity and flexibility", they share their experiences in Chap. 6.

Marketta Niemelä and her colleagues from the VTT Technical Research Centre, Tampere, Finland, take social robots into the public space: they describe how customers and staff react to a social robot in a shopping mall. The pepper-based robot was generally well received. The authors share their findings on co-designing a robot both "entertaining and useful" in Chap. 7.

Saida Mussakhojayeva and Anara Sandygulova from Nazarbayev University, Astana, Kazakhstan, also investigate social robots in public spaces, and again a shopping mall is selected for the field study. However, in contrast to their colleagues from Finland, their chapter focuses on multiparty interaction, namely when a robot interacts with children and adults: Chap. 8. If the robot detects an adult and a child: should it adapt its behaviours to the adult or to the child? Interestingly, the authors also investigate if the answers to such questions are culturally dependent by comparing participants from the USA and from Kazakhstan.

The relationship between social robots and children also is the main research area in the chapter by David Silvera-Tawil from the e-Health Research Centre of the Commonwealth Scientific and Industrial Research Organisation (CSIRO) in Sydney and Andrew Brown from the department Art and Design at the University of New South Wales (UNSW), also in Sydney. They investigate how a socially assistive robot called KASPAR can support therapy with autistic children and share their "lessons learned" in Chap. 9.

The four chapters in the last third of this book focus on the probably most prominent areas of application for social robots: **health and care for elderly persons**. In many industrialized countries, the demographic change leads to an age imbalance. According to the United Nations World Population Prospects, even on a global scale the population aged 60 or over ("the elderly") is the fastest growing. In the more developed regions, the ratio of the elderly is increasing at 1.0% per year and is expected to increase by 45% by the middle of this century. Thus, the "old-age support ratio" (i.e. the number of persons aged 15–64 per person aged 65 or over) is continually going down: In Germany, Italy, Japan and Sweden, there are only three working-age individuals for each older person, in Japan only 2.5. It is no surprise that such countries invest heavily into developing automated solutions like social robots which support the human staff in care and health.

As pointed out above, I believe in the benefits of user-centred design: while asking the future users may not result in disruptive technologies, it helps avoiding unaccepted and thus unused technology. Jaana Parviainen from the Research Centre for Knowledge, Science, Technology and Innovation Studies (TaSTI) at University of Tampere, Finland, takes this route by asking several hundred healthcare professionals where they really want help from social robots, from bathing over physiotherapy to moving assistance. The findings are described in Chap. 10.

In the chapter by Kimmo J. Vänni and Sirpa E. Salin from the University of Tampere, Finland, similar questions are investigated: based on survey with 220 nurses and several substudies, they take a look at Chap. 11. While also looking at general activities like "activator and motivator" or "material logistics", they additionally present in-depth findings covering areas like robotics training for healthcare students.

After learning a lot about the perspective of healthcare professionals in the previous two chapters, it is time to take the perspective of the person who is in care or the patient. Neziha Akalin and her colleagues from the School of Science and Technology at Örebro University, Sweden, do just that by Chap. 12. However, they also present helpful models and tests and try to weigh central factors of safety and security. Thus, this chapter is especially interesting for readers who plan to conduct studies with elderly persons.

Last but not least, Lucas Paletta from JOANNEUM RESEARCH, Graz, Austria, and his nine co-authors present a field study on a social robot working with elderly persons: Chap. 13. Focusing on the needs of users with mild cognitive impairment (MCI), the chapter describes a solution based on the pepper and a tablet encouraging "robot-based coaching for playful training". It is one of the few studies, where longer periods of interaction (here: one week per household) are investigated.

Thank you! A large book project like this one creates tons of work and requires a team to succeed. I thank my research group from the Affective & Cognitive Institute (ACI). Project assistant Andrea Küntzler kept the threads together, especially when communicating with the authors and working on the index. She was supported by our three student research assistants, Annika Sabrina Schulz, Franziska Schulz (who are not related) and Lara-Sophie Bornholdt. Special thanks go to Rúben Gouveia, our Ph.D.-student and soon-to-be postdoc from Madeira, who helped a lot with the reviews of both the initial submissions and the revisions. Thank you, Lea Buchweitz, for working your way through the "antique" literature in the history chapter—for computer scientists, references dating back a few hundred years are a rare challenge.

This book could not have been realized in this quality without the support from AGYA, the Arab-German Young Academy for Sciences and Humanities. The German Federal Ministry for Education and Research (BMBF) supports this network for young scientists, which allowed us to apply a professional scientific proof-reading process as well as funding the best paper award and a workshop. Special thanks go to Sabine Dorpmüller, Peter Nassif and Dominik Ceballos Contreras.

Finally, I want to thank the Springer team: Martin Boerger for supporting the initial idea, Alfred Hofmann for his direct and frank feedback on the first concept, and last but not least Helen Desmond for working with us with great patience on both administration and editing issues.

Offenburg, Germany   Oliver Korn
February 2019

Contents

Contributors

H. M. Harsha S. Abeykoon Intelligent Service Robotics Group, Department of Electrical Engineering, University of Moratuwa, Katubedda, Sri Lanka

Neziha Akalin School of Science and Technology, Örebro University, Örebro, Sweden

Dimitra Anastasiou Luxembourg Institute of Science and Technology, Esch/Alzette, Luxembourg

Niklas Antony Fraunhofer-Institut für Graphische Datenverarbeitung, Rostock, Germany

H. M. Ravindu T. Bandara Intelligent Service Robotics Group, Department of Electrical Engineering, University of Moratuwa, Katubedda, Sri Lanka

Gerald Bieber Fraunhofer-Institut für Graphische Datenverarbeitung, Rostock, Germany

Scott Andrew Brown Art and Design, University of New South Wales (UNSW), Sydney, Australia

Sebastian Brunsch Humanizing Technologies GmbH, Vienna, Austria

D. P. Chandima Intelligent Service Robotics Group, Department of Electrical Engineering, University of Moratuwa, Katubedda, Sri Lanka

H. P. Chapa Sirithunge Intelligent Service Robotics Group, Department of Electrical Engineering, University of Moratuwa, Katubedda, Sri Lanka

Maria Fellner JOANNEUM RESEARCH Forschungsgesellschaft mbH, Graz, Austria

Christian Fron University of Heidelberg, Heidelberg, Germany

Max Haberstroh RWTH Aachen University, Aachen, Germany

Marian Haescher Fraunhofer-Institut für Graphische Datenverarbeitung, Rostock, Germany

Frank Hees RWTH Aachen University, Aachen, Germany

Päivi Heikkilä VTT Technical Research Center of Finland, Tampere, Finland

Florian Hoepfner Fraunhofer-Institut für Graphische Datenverarbeitung, Rostock, Germany

A. G. Buddhika P. Jayasekara Intelligent Service Robotics Group, Department of Electrical Engineering, University of Moratuwa, Katubedda, Sri Lanka

Oliver Korn Affective & Cognitive Institute (ACI), Offenburg University of Applied Sciences, Offenburg, Germany

Silvio Krause Institut für Informatik, Universität Rostock, Rostock, Germany

Annica Kristoffersson School of Science and Technology, Örebro University, Örebro, Sweden;
School of Innovation, Design and Engineering, Mälardalen University, Västerås, Sweden

Lara Lammer Humanizing Technologies GmbH, Vienna, Austria

Hanna Lammi VTT Technical Research Center of Finland, Tampere, Finland

Thibaud Latour Luxembourg Institute of Science and Technology, Esch/Alzette, Luxembourg

Amy Loutfi School of Science and Technology, Örebro University, Örebro, Sweden

Gerald Lodron JOANNEUM RESEARCH Forschungsgesellschaft mbH, Graz, Austria

Tarek H. Mokhtar College of Engineering, Alfaisal University, Riyadh, Kingdom of Saudi Arabia

Sarah L. Müller-Abdelrazeq RWTH Aachen University, Aachen, Germany

Saida Mussakhojayeva Nazarbayev University, Nur-Sultan, Kazakhstan

Marketta Niemelä VTT Technical Research Center of Finland, Tampere, Finland

Virpi Oksman VTT Technical Research Center of Finland, Tampere, Finland

Lucas Paletta JOANNEUM RESEARCH Forschungsgesellschaft mbH, Graz, Austria

Sandra Pansy-Resch Sozialverein Deutschlandsberg, Deutschlandsberg, Austria

Jaana Parviainen Faculty of Social Sciences, Research Centre for Knowledge, Science, Technology and Innovation Studies (TaSTI), Tampere University, Tampere, Finland

Dimitrios Prodromou Humanizing Technologies GmbH, Vienna, Austria

Sirpa E. Salin Tampere University of Applied Sciences, Tampere, Finland

Anara Sandygulova Nazarbayev University, Nur-Sultan, Kazakhstan

Kathrin Schönefeld RWTH Aachen University, Aachen, Germany

Sandra Schüssler Medical University of Graz, Graz, Austria

David Silvera-Tawil Australian e-Health Research Centre, Commonwealth Scientific and Industrial Research Organisation (CSIRO), Sydney, Australia

Christoph Stahl Luxembourg Institute of Science and Technology, Esch/Alzette, Luxembourg

Josef Steiner Sozialverein Deutschlandsberg, Deutschlandsberg, Austria

Tuuli Turja Faculty of Social Sciences, Tampere University, Tampere, Finland

Lina Van Aerschot Department of Social Sciences and Philosophy, University of Jyväskylä, Jyväskylä, Finland

Kimmo J. Vänni Tampere University of Applied Sciences, Tampere, Finland

Julia Zuschnegg Medical University of Graz, Graz, Austria

Chapter 1
A Short History of the Perception of Robots and Automata from Antiquity to Modern Times

Christian Fron and Oliver Korn

Abstract Robots and automata are key elements of every vision and forecast of life in the near and distant future. However, robots and automata also have a long history, which reaches back into antiquity. Today most historians think that one of the key roles of robots and automata was to amaze or even terrify the audience: They were designed to express something mythical, magical, and not explainable. Moreover, the visions of robots and their envisioned fields of application reflect the different societies. Therefore, this short history of robotics and (especially) anthropomorphic automata aims to give an overview of several historical periods and their perspective on the topic. In a second step, this work aims to encourage readers to reflect on the recent discussion about fields of application as well as the role of robotics today and in the future.

Keywords History of robotics · Robots in antiquity, medieval, and modern times · Robots in past visions of the future · Robots as machines in the realm between magic and reality · Technology acceptance

1.1 Introduction

The steam engine, one of the basic conditions for the industrial revolution, had its predecessor in antiquity. The engineer Hero of Alexandria (1st cent. AD) writes in his *Pneumatics* (§50) about the aeolipile ("the ball of [the wind god] Aeolus"), a steam turbine device. In contrast to the British steam engines of the eighteenth century, the

C. Fron (✉)
University of Heidelberg, Heidelberg, Germany
e-mail: christian.fron@zaw.uni-heidelberg.de

O. Korn
Affective & Cognitive Institute (ACI), Offenburg University of Applied Sciences, Offenburg, Germany
e-mail: oliver.korn@acm.org

ancient steam turbine was not constructed for mechanical work, but merely as an interesting, also delighting experiment.

Like these engines, the idea of robots and artificial life has a historical background dating back to ancient times. Moreover, there are some concepts and actual implementations of anthropomorphic automata and "proto-robots." This raises some questions regarding the perception of robots and automata as well as their envisaged area of application throughout different historical as well as cultural contexts. In the following lines, we will focus on three historical periods: the antiquity, the medieval times, and finally early modern perceptions of robots and automata.

1.2 Ancient Perceptions

"For if every instrument could perform its own work when ordered, or by seeing what to do in advance, like the statues of Daedalus in the story or the tripods of Hephaestus which the poet says 'enter self-moved the company divine,'—if thus shuttles wove and quills played harps of themselves, mastercraftsmen would have no need of assistants and masters no need of slaves" (Aristotle, 1932). Reading these lines, one might directly be reminded of recent debates about whether or not robots are or will be threatening jobs (Arnold, 2018; Elliott, 2018; Sawhney, 2018). However, these words were written by the famous Philosopher Aristotle in the fourth century BC. Moreover, these words do not reflect Aristotle's fear for losing his "job": He is more concerned with the changes to society the construction of robots might cause. Especially he thinks about the effects on those at the "lower end of the job market," namely the slaves (Devecka, 2013). The only possible escape from an economic system based on slavery for Aristotle was a technological one; and since there were no technological developments ready to replace slaves as workers, his argument meant to confirm the current system (Devecka, 2013).

Similar reflections on technological developments (but with greater concerns about the working individuals of lower society) are not only found in ancient Greek literature—according to the biographer Sueton they are also addressed by the emperor Vespasian in the first century AD: "To a mechanical engineer, who promised to transport some heavy columns to the Capitol at small expense, he gave no mean reward for his invention, but refused to make use of it, saying: 'You must let me feed my poor commons'" (Rolfe, 1914).

In overall, robots and automata surely were present in the ancient utopian and fictional visions as well as in literature. However, from the ancient perspective looking for utopian visions means not looking into the future but rather into the always evolving and adapting glorified past (Hesiod, n.d.; Ovid, n.d.; Stroh, 2016). This way, new ideas and visions were presented to the reader as events back in the glorious days of the past, including historical as well as mythical times.

Mythical descriptions of robots have recently been analyzed by Adrienne Mayor (Mayor, 2018). In the following, we will focus on some representative examples that grant deeper insight into the mentality behind the ancient perception as well as the envisioned applications of robots and anthropomorphic automata.

Fig. 1.1 Pygmalion and Galatea, sculpture by Auguste Rodin, modeled 1889, carved around 1908, at "The Met Fifth Avenue," New York

One of these early cases of "social robots" in its broadest meaning is given by the ancient author Ovid (43 BC–17 AD) in his *Metamorphoses* (Transformations). At the same time, it reflects on the human desire to recreate an ideal model of a non-vital lookalike to match personal desires and necessities. Right after describing the transformation of the ruthless daughters of Propoitos into stone as part of divine penalty in Book 10, the story of the life of Pygmalion of Cyprus is told. Pygmalion, in disgust of women's imperfection, focused all his passion and time to the creation of an ivory statue of his female ideal (Fig. 1.1). He then fell in love with his own creation. He treated it like an idealized female companion, offering gifts and even sharing his bed with it. The function of the statue itself in some way is comparable to today's sex robots like "Synthea Amatus." It is only thanks to the help and mercy of the Cyprian goddess Venus that Pygmalion's statue becomes alive and is even able to bear children. In modern times, she has been named Galatea. In contrast to the later reception of the story, where elements of tragedy are added, and the perspective is changed (Dinter, 1979) the original story features a "happy ending": The birth of a common child named Paphos. Nevertheless, the story's moral is quite clear: Human creativity and invention is limited to the idealized figural and external imitation of Life while the creation of Life itself depends on divine impetus.

In the same way Talos, Crete's mythological iron guardian protecting the island, had to be given to King Minos by the god of handcraft, Hephaestus, or in other story versions to Europe by Zeus (Papadopoulos, 1994). The mythical craftsman and hero Daedalus was known to have created statues that looked like they would be capable of movement (Heron Alexandrius, 1899, Chap. 16). This slowly led to new story versions where some of his statues were actually capable of moving (Diodorus Siculus, 1935; Euripides, 1882). However, Daedalus as a craftsman belonged to these mythical good old days of the past, when gods and men stood in direct contact.

One more comment outside the classical field of myth, which might be of interest to modern robotics: In the classical attic comedy, even the idea of self-moving furniture could be expressed: "Each article of furniture will come when he calls it. Place yourself here, table! You there, get yourself ready! Knead, oh kneading trough. Fill up, ladle! Where's the cup? Go wash yourself" (Devecka, 2013; Homer, 1990).

Throughout classical antiquity, there is a strong connection between the creation of "artificial beings" or automata through mechanics and the divine. Statues resembling deities and ideals of human beauty were in some cases provided with complex mechanics.

- The sculptor and iron caster Canachos of Sikyon added a mechanically movable deer to his bronze statue of Apollo in Brachnidai (esp. Pliny, natural history 34, 75) (Strocka, 2002).
- It also seems highly probable that the bronze statue of Diana on one of the luxurious Nemi ships, built for the emperor Caligula, was standing on a rotating platform (Wolfmayr, 2009).
- There is the story of a tax collector, a "robot" in the shape of a woman, who had iron hooks on her hands and on her chest. Like Talos, she put her arms around the victim to (in this special case) enforce the paying of taxes (Polyb. 13,7) (Devecka, 2013; Veyne, 2002).
- The statue of Nysa in Hellenistic Egypt was described to be able to stand up and pour milk from a golden phial (Athenaeus of Naucratis, 1960).

Indeed, surprising and almost "godly" special effects were a common method to enrich the antique theaters: A crane-like stage machinery was used to allow "deities" to suddenly appear up in the air on stage (Cancik, Schneider, & Landfester, 2006). Until today, the term "deus ex machina" reflects this practice.

Heron of Alexandria describes several other automatic machines in the context of ancient religious life:

- Two statues automatically giving libations, whenever there is an incense offering (Heron Alexandrius, 1899, Chap. 12).
- Automatically opening doors of temples (Heron Alexandrius, 1899, Chap. 38).
- A statue of Hercules, shooting a snake on a tree with an arrow (Heron Alexandrius, 1899, Chap. 41) and many more.

All these automatic machines were designed to amaze the audience and demonstrate "divine power." Moreover, some of these automats, presented by Hero of Alexandria and Philon from Byzantium were clearly designed for feasts, symposia and banquets (Amedick, 2003), this means for representative contexts. This use is continued in the following historical periods.

1.3   Medieval Perceptions

The continuation of antique ideas and knowledge into medieval times was especially granted by the Byzantine Empire and Arab culture. Liudprand, the bishop of Cremona, describes some of the *thaumata*, things as well as creatures to admire and be amazed of, he exhibited in a ceremonial throne room in the imperial palace of Constantinople in 949 AD:

"In front of the emperor's throne there stood a certain tree of gilt bronze, whose branches, similarly gilt bronze, were filled with birds of different sizes, which emitted the songs of the different birds corresponding to their species. The throne of the emperor was built with skill in such a way that at one instant it was low, then higher, and quickly it appeared most lofty; and lions of immense size (though it was unclear if they were of wood or brass, they certainly were coated with gold) seemed to guard him, and, striking the ground with their tails, they emitted a roar with mouths open and tongues flickering. Leaning on the shoulders of two eunuchs, I was led into this space, before the emperor's presence. And when, upon my entry, the lions emitted their roar and the birds called out, each according to its species, I was not filled with special fear or admiration, since I had been told about all these things by one of those who knew them well. Thus, prostrated for a third time in adoration before the emperor, I lifted my head, and the person whom earlier I had seen sitting elevated to a modest degree above the ground, I suddenly spied wearing different clothes and sitting almost level with the ceiling of the mansion. I could not understand how he did this, unless perchance he was lifted up there by a pulley of the kind by which tree trunks are lifted. Then, however, he did not speak at all for himself, since, even if he wished to, the great space between us would render it unseemly, so he asked about the life of Berengar and his safety through a minister" (Brett, 1954; Constantine VII Porphyrogenitus, Emperor of the East, & Reiske, 1829; Luidprand, 2007).

In this text, the reader gets a detailed description of the scenery in a proto-mythical mechanical paradise, in which the emperor exposes himself to the western ambassadors thanks to a lifting mechanism in his throne. At the same time, this difference of height does not allow any direct conversation anymore, without shouting at each other. What was originally meant to amaze visitors is often located in the fields of magic and diabolic tricks by western medieval authors (Canavas, 2003). In the western European world, mechanical wisdom was lost until in the fourteenth century and henceforth. Technical progress as a whole was viewed as something that stood in the context of magic and wizardry (Campanella, 2007; Truitt, 2016). Thus, it is by no means surprising that a *diavolo meccanico* was developed in the early fifteenth century, which actual purpose was to scare and terrify the viewers (Frieß & Steiner, 2003). However, magic was not just something to fear, but could also be used in representative contexts. Therefore, in the twelfth century, these automats could for instance also symbolize the dominance and exploit of the ruler over nature, granting the aura of divine power (Friedrich, 2003).

Medieval "robots" have recently been analyzed in more detail by Truitt (2016). He identifies different fields of application like romance (kissing lovers), defense (golden archers/copper knights) and the afterlife (corpses perfectly preserved by human art) (Truitt, 2016). For instance, the priest-king Johannes created some iron warriors, which, thanks to bellows, could spit fire (Scharfenberg & Wolf, 1952).

At the same time, the works of the ancient engineers Hero of Alexandria and Philon from Byzantium were copied, translated and influenced the Arab world since the ninth century (Al-Hassan, 1992). There has always been a great interest in sophisticated mechanical devices in the Arab world—thus it is no surprise that the modern humanoid robot Sophia was granted citizenship in Saudi Arabia. Some of these illustrious Arab medieval engineers are Abu 'Abd Allah al-Khuwarizmi (tenth century AD) with a technical encyclopedia, the Banu Musa's (ninth century AD) *Book of Ingenious Devices* or Ibn al-Razzāz al-Jazarī's Book of Knowledge of Ingenious Mechanical Devices (twelfth/thirteenth century AD) (Al-Hassan, 1992; Hill, 1974). These books contain sketches and descriptions of complex mechanisms like a humanoid machine pouring wine (Hill, 1974). Another humanoid mechanism, depicting a slave, had the task of pouring water over the king's hands while he was performing ritual washings (Hill, 1974).

In Western Europe, there are other elaborated automated figures again in early modern times, like the moving mechanical monk (Fig. 1.2) from sixteenth century at the German Museum at Munich (Frieß & Steiner, 2003) or a moving figurine of

Fig. 1.2 Moving mechanical monk from sixteenth century at the German Museum in Munich

Jesus Christ at the monastery church of Dießen, used in the context of sacral theater, or a lute-player from the early sixteenth century (Berns, 2003).

1.4   Modern Perceptions

Early modern society's fascination with automated machinery is well exemplified by the late eighteenth century "mechanical Turk" (Fig. 1.3) or "automaton chess player." The machine was fake: A human chess master hid inside, operating the machine. However, the audience did not know that, since the chess master was well hidden inside the mechanism. Everyone was invited to look inside the machine itself, before the chess game started. This autopsy [in its original Greek meaning: The authentication and verification of something, which might be beyond the common believe, by the authors and visitors' own sight] was of great importance for the success of the mechanism, since the chess masters' abilities were beyond the spectrum of contemporary robotics and seemed unbelievable to the attendants of his performances. And indeed, the Chess Turk itself was a huge worldwide success, getting full attention in every city the machine visited. The machine's creator Wolfgang von Kempelen toured with it throughout Europe and the USA, showing it to nobility and political leaders like Napoleon Bonaparte and Benjamin Franklin (Standage, 2003).

What makes this automaton especially interesting is on the one hand its historical context (the beginning industrial revolution) and on the other hand its impact for the ongoing perception and the imagined field of application of robots. This "robot" directly challenged the human intellect, since he for instance was able to solve the chess riddle of the "knights' tour." Moreover, the Automaton later on was capable to communicate to the audience using a letter board (Standage, 2003). The Turk could nod twice threatening the opponents queen and thrice, when bringing the opponents King in check. Moreover, the chess automaton could identify any illegitimate moves and react to them by moving back the wrongly moved chess piece and starting his turn right away.

While the robots before had merely focused on mechanical imitations, this automaton was not only imitating human movement but seemed to have human intelligence. This is the core fear behind every criticism of modern robotics and the foundations of it are to be found in the eighteenth century. It comes to no surprise, that Mary Shelley's novel *Frankenstein* was published in 1818, a time at which the Chess Turk was still on tour. One famous visitor of an exhibitions, where the chess Turk was presented, was Edgar Allen Poe. His highly elaborate and extensive reaction was published in the Southern Literary Messenger in 1836:

"Perhaps no exhibition of the kind has ever elicited so general attention as the Chess-Player of Maelzel. Wherever seen it has been an object of intense curiosity, to all persons who think. Yet the question of its *modus operandi* is still undetermined. Nothing has been written on this topic which can be considered as decisive—and accordingly we find everywhere men of mechanical genius, of great general acuteness, and discriminative understanding, who make no scruple in pronouncing the

Fig. 1.3 Although the Mechanical Turk was fake, it fascinated a European and American audience for decades. Copper engraving from 1783

Automaton a *pure machine,* unconnected with human agency in its movements, and consequently, beyond all comparison, the most astonishing of the inventions of mankind. And such it would undoubtedly be, were they right in their supposition. […] The Turk plays with his left hand. All the movements of the arm are at right angles. In this manner, the hand (which is gloved and bent in a natural way,) being brought directly above the piece to be moved, descends finally upon it, the fingers receiving it, in most cases, without difficulty. Occasionally, however, when the piece is not precisely in its proper situation, the Automaton fails in his attempt at seizing it. When this occurs, no second effort is made, but the arm continues its movement in the direction originally intended, precisely as if the piece were in the fingers. Having thus designated the spot whither the move should have been made, the arm returns to its cushion, and Maelzel performs the evolution which the Automaton pointed out. At every movement of the figure machinery is heard in motion. During the progress of the game, the figure now and then rolls its eyes, as if surveying the board, moves its head, and pronounces the word *echec* (check) when necessary If a false move be made by his antagonist, he raps briskly on the box with the fingers of his right hand, shakes his head roughly, and replacing the piece falsely moved, in its former situation, assumes the next move himself. Upon beating the game, he waves his head with an air of triumph, looks round complacently upon the spectators, and drawing his left arm farther back than usual, suffers his fingers alone to rest upon the cushion. In general, the Turk is victorious—once or twice he has been beaten" (Poe, 1836).

Computers are the modern successors of the automat. But this time without any trickery or magic. The chess computer *Deep Blue* has already proven, that a machine is capable of winning against a human world chess champion (Garry Kasparov) and Google's AI Program *AlphaGo* beat the Chinese GoMaster Ke Jie (Mozur, 2018).

Moreover, social robots are already found in many different aspects of everyday life, beginning with robotic lawnmower or cleaning robots. In the elder care, robots like Paro and Robear are already in use, but not broadly received yet (McGlynn, Snook, Kemple, Mitzner, & Rogers, 2014). Robots have even found their way into sex industry, where Synthea Amatus can be seen as a modern version of Pygmalion's statue. Nevertheless, it lacks the divine assistance granted to Pygmalion and cannot bear any children.

Moreover, there still is some "trickery" or at least very intense specialization used in robotics. For instance, *AlphaGo* is not capable of beating Kasparov at chess. The robot Sophia can imitate human expressions and human dialogue but require pre-programmed answer alternatives (Sigalos, 2018). Nevertheless, there is constant progress in robotics, and the public's response to these new achievements usually is amazement, excitement, and fear. Interestingly, these reactions form a continuum throughout history (CNBC, 2016; Collins, 2018; Pettit, 2018). Also, these emotional reactions seem to be an integral part of human progress. International ethical guidelines and laws should help us distinguishing between what we can do and what we should do in the fields of robotics as the individual approach seems to depend strongly on both society and individual exposure to robots in everyday life (Bartneck, Nomura, & Kanda, 2005).

1.5   Conclusion and Outlook

Within three sections, we discussed different perceptions and fields of applications of robotics and automata in antiquity, the medieval times, and modern times. Engineers in antiquity already had the means to build machines and robots based on water pressure as well as steam. Therefore, a first version of a steam engine was already developed in antiquity. Yet, this did not lead to an industrial revolution. While the answer why this did not happen is rather complex, one important partial response seems to be that technical developments were seen as something endangering the social system, based on slavery and cheap labor. Therefore, many of the imagined applications of robots in antique literature seemed to have the purpose to replace daily tasks of slaves.

Nevertheless, the main function of robots and automata actually implemented was to amaze, thrill and sometimes frighten the audience. In most cases, there remained a strong connection to pagan religion and the ancient rite. Whenever a door seemed to open by itself or statues gave libations or were moving by themselves, the presence of divine spirits became more real to the worshippers. Therefore, the case of Pygmalion is very illuminating, since it states a clear division line between what is possible for humans and what is reserved for gods.

In medieval times, in Western Europe the main hindrance to new technical developments was the religious belief that such developments were magical and potentially even demonic. Therefore, European automata were mainly built in late medieval times. Yet, in the Byzantine and Arab World, antique robots and automata kept being realized since the antiquity, with slight improvements. Moreover, the social systems kept being based on cheap and virtually slave labor with no urgent need for technical improvements. Robots and machines were in many cases imagined as refined replacements of typical servants' duties.

In modern times, the industrial revolution and the spread of democracy and capitalism slowly improved the individual conditions and lead to technical innovations. However, the intentions of machines like the chess Turk remained to amaze and thrill the audience. Nevertheless, the chess Turk did something new: It was not limited to mechanical tasks but was believed to be highly intelligent. As was already stated in antiquity, intelligence is considered the characteristic feature of humankind. Thus, the development of artificial intelligence challenging the human mind is one of the basic fears. However, today complex robots are no trickery anymore. They will evolve further and slowly change our society, in some fields for the better, in others for the worse.

In any case, the dream of robots is not just a recent one, as this discussion has shown. Yet, in most historical contexts, they were either used to delight or fear humans—or to partially replace humans in fields of works, which typically were done by servants or slaves. In this way, robots were always imagined as something that belonged to the rich and not to the poor. Their realization on a bigger scale always seemed to be surreal. However, in the not too distant future, we will be able to build robots in mass production, which potentially does no longer limit the use of robots to the rich. Society will surely accept robots in areas with less favorable or very specific tasks like lawn mowing or lifting. However, acceptance will fall when it comes to demanding and well-paid jobs and. As long as work is considered crucial to individual self-perception and social status, robots will be considered a threat to human society.

Some consider the upcoming "robot revolution" no different than the industrial revolution and technical progress itself throughout history: Many professions like millers, charcoal burners, or telephonists vanished and other, new jobs established. However, others insist that the combination of artificial intelligence and robots marks a turning point in human history, a new step of human evolution or "Life 3.0" (Tegmark, 2017)—or, alternatively, the end of human history. The next decades will tell, in which way modern society adapts to the challenges of AI and robotics. As stated above, only international ethical guidelines and laws can help us distinguishing between what we can do and what we should do in the fields of robotics.

References

Al-Hassan, A. Y. (1992). *Islamic technology: An illustrated history* (1st ed.). Cambridge; New York; Paris: Cambridge University Press.
Amedick, R. (2003). Wasserspiele, Uhren und Automaten mit Figuren in der Antike. In *Automaten in Kunst und Literatur des Mittelalters und der Frühen Neuzeit* (pp. 9–48).
Aristotle. (1932). Aristotle, Politics (Loeb Classical Library 264; H. Rackham, Trans.). Retrieved from https://www.loebclassics.com/view/LCL264/1932/volume.xml.
Arnold, A. (2018, March). *Why robots will not take over human jobs.* Retrieved January 31, 2019, from https://www.forbes.com/sites/andrewarnold/2018/03/27/why-robots-will-not-take-over-human-jobs/.
Athenaeus of Naucratis. (1960). *Deipnosophistae* (Charles Burton Gulick, Trans.) (Vol. 5). Harvard University Press.
Bartneck, C., Nomura, T., & Kanda, T. (2005). A cross-cultural study on attitudes towards robots. *ResearchGate.* https://doi.org/10.13140/RG.2.2.35929.11367.
Berns, J. J. (2003). Sakralautomaten, Automatisierungstendenzen in der mittelalterlichen und frühneuzeitlichen Frömmigkeitskultur. In *Automaten in Kunst und Literatur des Mittelalters und der Frühen Neuzeit* (pp. 197–222). Wiesbaden.
Brett, G. (1954). The automata in the Byzantine "Throne of Solomon". *Speculum, 29*(3), 477–487. https://doi.org/10.2307/2846790.
Campanella, T. (2007). *Del senso delle cose e della magia.* Retrieved from https://www.laterza.it/index.php?option=com_laterza&task=schedalibro&isbn=9788842083979.
Canavas, C. (2003). Automaten in Byzanz. Der Thron von Magnaura. In *Automaten in Kunst und Literatur des Mittelalters und der Frühen Neuzeit* (pp. 49–72). Wiesbaden. Retrieved from https://historische-bibliographie.degruyter.com/hbo.php?F=titel&T=HB&ID=23011496&target=_self.
Cancik, H., Schneider, H., & Landfester, M. (2006). Deus Ex Machina. In *Der neue Pauly.* Metzler Verlag.
CNBC. (2016, March 16). *Hot robot at SXSW says she wants to destroy humans.* Retrieved March 13, 2016, from https://www.youtube.com/watch?v=W0_DPi0PmF0.
Collins, T. (2018). *Boston dynamics: "Black Mirror" SpotMini robo-dog.* Retrieved January 31, 2019, from http://www.dailymail.co.uk/sciencetech/article-5416823/Boston-Dynamics-taunts-creepy-SpotMini-robo-dog.html.
Constantine VII Porphyrogenitus, Emperor of the East, & Reiske, J. J. (1829). *De cerimoniis aulae Byzantinae libri duo.* Bonnae : Ed. Weberi. Retrieved from http://archive.org/details/bub_gb_I9QFAAAAQAAJ.
Devecka, M. (2013). Did the Greeks believe in their robots? *The Cambridge Classical Journal, 59,* 52–69. https://doi.org/10.1017/S1750270513000079.
Dinter, A. (1979). *Der Pygmalion-Stoff in der europäischen Literatur: Rezeptionsgeschichte einer Ovid-Fabel.* C. Winter.
Diodorus Siculus. (1935). *The Library of History* (Vol. 2).
Elliott, L. (2018, February 1). Robots will take our jobs. We'd better plan now, before it's too late!Larry Elliott. *The Guardian.* Retrieved from https://www.theguardian.com/commentisfree/2018/feb/01/robots-take-our-jobs-amazon-go-seattle.
Euripides, J. B. (1882). *The Hecuba of Euripides.* Retrieved from http://archive.org/details/hecubaeuripides00unkngoog.
Frieß, P., & Steiner, R. (2003). Frömmigkeits-Maschinen in der Frühen Neuzeit. In *Automaten in Kunst und Literatur des Mittelalters und der Frühen Neuzeit* (pp. 223–246). Wiesbaden. Retrieved from https://historische-bibliographie.degruyter.com/hbo.php?F=titel&T=HB&ID=23011507.
Friedrich, U. (2003). Contra naturam. Mittelalterliche Automatisierung im Spannungsfeld politischer, theologischer und technologischer Naturkonzepte. In *Automaten in Kunst und Literatur des Mittelalters und der Frühen Neuzeit* (pp. 91–114). Wiesbaden. Retrieved from https://historische-bibliographie.degruyter.com/hbo.php?F=titel&T=HB&ID=23011507.
Heron Alexandrius. (1899). *Pneumatica et automata.* (W. Schmidt, Trans.) (Vol. 1). Leipzig.

Hesiod. (n.d.). *Works and days*. Retrieved from http://www.sacred-texts.com/cla/hesiod/works.htm.

Hill, P. (1974). *The book of knowledge of ingenious mechanical devices: (Kitāb fī ma 'rifat al-ḥiyal al-handasiyya)*. Springer Netherlands. Retrieved from https://www.springer.com/de/book/9789401025751.

Homer. (1990). *Ilias*. (J. H. Voß, Trans.). Frankfurt am Main: Insel Verlag.

Luidprand. (2007). *The complete works of Liudprand of Cremona* (P. Squatriti, Trans.). The Catholic University of America Press.

Mayor, A. (2018). *Gods and robots—Myths, machines, and ancient dreams of technology*. Retrieved from https://press.princeton.edu/titles/14162.html.

McGlynn, S., Snook, B., Kemple, S., Mitzner, T. L., & Rogers, W. A. (2014). Therapeutic robots for older adults: Investigating the potential of paro. In *Proceedings of the 2014 ACM/IEEE International Conference on Human-robot Interaction* (pp. 246–247). New York, NY, USA: ACM. https://doi.org/10.1145/2559636.2559846.

Mozur, P. (2018, August 7). Google's A.I. program rattles chinese go master as it wins match. *The New York Times*. Retrieved from https://www.nytimes.com/2017/05/25/business/google-alphago-defeats-go-ke-jie-again.html.

Ovid. (n.d.). Metamorphoses 1, 1 - 150. Retrieved from http://12koerbe.de/pan/ovid-met.htm.

Papadopoulos, J. (1994). Talos. In *Lexicon Iconographicum Mythologiae Classicae* (Vol. 2, pp. 834–837). Germany: Artemis & Winkler Verlag.

Pettit, H. (2018, October 29). *Sophia is granted the world's first robot VISA*. Retrieved January 31, 2019, from https://www.dailymail.co.uk/sciencetech/article-6328455/Creepy-AI-humanoid-Sophia-granted-worlds-robot-VISA-embarks-world-tour.html.

Poe, E. A. (1836). *Maelzel's chess-player*. Retrieved January 31, 2019, from https://www.eapoe.org/works/essays/maelzel.htm.

Rolfe, J. C. (Trans.). (1914). *Suetonius: The lives of the twelve Caesars* (Loeb Classical Library 38, Vol. 18). Cambridge, MA: Harvard University Press.

Sawhney, M. (2018). *As robots threaten more jobs, human skills will save us*. Retrieved January 31, 2019, from https://www.forbes.com/sites/mohanbirsawhney/2018/03/10/as-robots-threaten-more-jobs-human-skills-will-save-us/.

Scharfenberg, A., & Wolf W. (1952). *Der jüngere Titurel*. Retrieved from https://www.zvab.com/buch-suchen/titel/der-j%FCngere-titurel/.

Sigalos, J. U., & MacKenzie. (2018, June 5). *Sophia the robot's complicated truth*. Retrieved January 31, 2019, from https://www.cnbc.com/2018/06/05/hanson-robotics-sophia-the-robot-pr-stunt-artificial-intelligence.html.

Standage, T. (2003). *The Turk: The life and times of the famous eighteenth-century chess-playing machine*. New York: Berkley Trade.

Strocka, V. M. (2002). Der Apollon des Kanachos in Didyma und der beginn des Strengen Stils. *Jahrbuch Des Deutschen Archäologischen Instituts, 117*, 81–125.

Stroh, W. (2016). Citius altius fortius? *Gymnasium, 123*(2), 115–144.

Tegmark, M. (2017). *Life 3.0: Being human in the age of artificial intelligence* (01st ed.). London: Allen Lane.

Truitt, E. R. (2016). *Medieval robots: Mechanism, magic, nature, and art* (Reprint edition). University of Pennsylvania Press.

Veyne, P. (2002). Lisibilité des images, propagande et apparat monarchique dans l'Empire romain. *Revue historique, 621*(1), 3–30.

Wolfmayr, S. (2009). Die Schiffe vom Lago di Nemi. *Diplom*, 37–39.

Chapter 2
Designing Social Robots at Scales Beyond the Humanoid

Tarek H. Mokhtar

Abstract Social robots are expected to attract users and provide a high level of affinity with the aim of achieving the "healthy person's" level of interaction, i.e., overcoming the "uncanny valley." While embodiment in the form of humanoids has taken many steps to escape the "uncanny valley," certainly social robots are far from being "everyware," i.e., ubiquitous computing environments. This chapter represents the design and characteristics of social robots for what can be called "Non-Humanoid Social Robots," which will shape the future of our cities and societies. The social interactions that occur in these fascinating responsive, haptic, and multisensorial environments promise to define the evolution of social robotic systems. The chapter includes the *different types of social robots at scales beyond the humanoid*, i.e., from the mesoscale to macroscale robots; understanding their architecture; classifying the systems and representing the paradigms and styles for designing them. These different styles of non-humanoid social robots will provide the essential guidelines needed for designers and artists to start their creative explorations in this field of study.

Keywords Non-humanoid social robots (NH-SR) · Ubiquitous computing · "Everyware" · Characteristics of non-humanoid social robots · NH-SRs' paradigms and trends

2.1 Introduction

The last 20 years have witnessed a paradigm shift from the embodiment of social robotic interactions in the form of the humanoid toward the *"everyware,"* i.e., a new paradigm where computation disappear within everyday objects and environments (Greenfield, 2006). *The hybridization of bits and atoms that interact with humans* has evolved from humanoid robots to complex social and physical environmental

T. H. Mokhtar (✉)
College of Engineering, Alfaisal University, Riyadh, Kingdom of Saudi Arabia
e-mail: tmokhtar@alfaisal.edu

© Springer Nature Switzerland AG 2019
O. Korn (ed.), *Social Robots: Technological, Societal and Ethical Aspects of Human-Robot Interaction*, Human–Computer Interaction Series,
https://doi.org/10.1007/978-3-030-17107-0_2

systems which promise to act as assistants and partners to humans in their daily lives (Dourish, 2004).

Today, however, with our powerful electronic devices and gadgets, the use of social robots at scales beyond the humanoid, i.e., furniture and architectural spaces, as a physical interface for our social interactions, may seem quaint. The literature on the use of interactive interfaces/environments (Fox & Kemp, 2009; Green, 2016; Mokhtar, Green, & Walker, 2013; McCullough, 2005; Norskov, 2016; Weiser & Brown, 1997) is growing; nevertheless, there is a gap in the literature on designing and classifying the different systems for social robotic environments beyond the scale of the humanoid (Norskov, 2016).

Interactive social environments promise to behave like living organisms by being able to adapt to different humans' conditions and needs (McCullough, 2005; Green, 2016). In the past two decades or even less, roboticists and designers have taken this exercise of designing social robots to scales beyond the humanoid and have developed vibrant social environments that not only respond to humans' gestures and cues, but also create unique experiences for human–robotic interactions, that can be called "non-humanoid social robots" ("NH-SRs"). The basic approach in designing these environments is still not based on well-defined paradigms, i.e., styles, but on individual interpretations and techniques borrowed from design, art, or architecture (McCullough, 2005).

In the past 20 years, the interactive physical environments which create new types of social interactions (human–space interactions) have been exemplified in the many research activities in the fields of cyber-physical environments (CPS), human–robotic interactions (HRI), architectural robotics (AR), adaptive environments (AD), intelligent environments (IE), interactive environments (IE), or as TU-Delft's Hyperbody group termed them, the hyper-(space, environment, body, etc.) which interacts with humans to solve social or environmental needs. In this chapter, the term "non-humanoid social robots ("NH-SRs")" is used as the domain field in which humans interact with ubiquitous computing environments. Accordingly, these environments, i.e., NH-SRs, are responding, retuning and adapting to humans' inputs.

Concerning the growing interest in non-humanoid social robotic interactions (Fox and Kemp 2009; Green, 2016; McCullough, 2005), as described later in this chapter, there are two reasons to explain this growing phenomenon, the *Uncanny Valley* and *Multisensorial Interfaces*. Firstly, we investigated the "eeriness" sensation effect that Mori highlighted in his *Uncanny Valley* hypothesis for different types of humanoid robots (Mori, MacDorman & Kageki, 2012) concluding with two contradictory results about the readiness of social humanoid robots to interact with humans (Becker, 2017; Cafaro et al., 2014; Enz et al., 2011; Norman, 2010; Tschöpe, Reiser, & Oehl, 2017; Von der Pütten & Krämer, 2012; Strait et al., 2015). Secondly, the *interface of interaction* with the single humanlike body (i.e., humanoid) is crucial; as humans, we experience interactions as part of the built context in its multisensorial dimensions, not only in the two-dimensional scale (i.e., two humans, or a human and humanlike robot). Moreover, in architectural, physiological, and psychological terms, the spatial qualities of the built environment including the physical and the phenomenological dimensions have influences on shaping the behaviors of users, and the place for interactions/events (Rykwert, Leach, & Tavernor, 1988; Tschumi, 2012). "Physical spaces" play a critical social role as multisensorial contexts for human interactions, whether interactive or static, as was also found by phenomenologists, e.g., Heidegger and Schutz, supporting the importance and effect of the physical settings on humans' perceptions and behaviors. These reasons can help explain the growing interest of users and researchers in the different social robotic mediums, i.e., spatial robotic interactions at scales beyond the humanoid that interact with humans.

The question is not whether we consider interactive physical environments, i.e., ubiquitous computing, as social robots or not, nor the degree of intelligence they inherit from robotics and computation. In the past decade, as described in detail in the following sections, NH-SRs have been becoming popular in built projects worldwide, and in research venues such as human–robotic interaction, interactive environments, and human–computer interaction, among others. The questions that have not been yet answered, and that are explored here, are: *What are the robotic and artistic characteristics that define non-humanoid social robots?* and *What are the different styles (paradigms) we use to design them?* This chapter is intended to answer these questions, and to provide guidelines for developing non-humanoid social robots; specifically, to help educators, researchers, artists and designers in the different transdisciplinary fields to design social robots in their different scales and scopes.

2.2 Toward Ubiquitous Social Robots

In its broader definition, social robots are robotic systems comprised of sensors, actuators for moving masses and artificial intelligence (AI) to process this information and configure the actions. The actions expected are to engage humans in conversation with robots, human–robotic interactions (HRI). While the art of designing social robots has been focused on the anthropomorphic or the zoomorphic forms, "[but] this does not have to be the case," according to Cynthia Breazeal et al., (2008).

In his insightful study on the future of social robots, Duffy argues for the need to understand intelligence and be more concerned with the type of activities the social robot can perform. He claims that we can build on the anthropomorphic physical and social characteristics of humans, but the robots should look as if they are "built in their own image" (Duffy, 2003). In "Sharing a Life with Harvey," the long-term study of peoples' likeliness and perceptions of social robots showed contradictory affinities to anthropomorphism in social robotic interactions (De Graaf et al., 2015; De Graaf et al., 2015a; Sakamoto & Ono, 2006). Today, the limitation of the anthropomorphic/zoomorphic figure is evident, and the need for exploring the different bodies and mediums that robots can inhabit when interacting with humans is growing (Ames, 2015; Norman, 2010; Reeves & Nass, 2003; Tinwell & Grimshaw, 2009) (Fig. 2.1).

Social robots at scales beyond the humanoid are not silent environments, but vibrant and interactive. They interact with us in subtle ways, i.e., calmly, creating

Fig. 2.1 Philip Beesley's shivering and vibrating structures in the 18th Biennale of Sydney. On the left, a child interacting with the "Hylozoic" structure; and, on the right, people interacting with "Hylozoic" as a "living structure for living in!" (Courtesy of Philip Beesley Architect Inc.)

beautiful habitats for human–space interactions (HSI) (McCullough, 2005; Norman, 2007).

In the Uncanny Valley, Masahiro Mori left a great distance between the humanoids and what he described as the "healthy person." This distance can be bridged by developing cybernetic systems in which robots interact with humans (Fig. 2.2), without imposing their presence, to build better affinity and communication; i.e., they behave more like living organisms (McCullough, 2005; Pask, 1969).

In his dream of ubiquitous environments, Mark Weiser envisioned information technology that will move away from being flattened virtual systems to play more social roles by interacting with us in real-life settings, i.e., social environments (Weiser, 1991). Recently, social robots have been developing toward Weiser's Dream of Ubiquitous Environments. We witnessed social robotic systems in the field of human–computer interaction, human–Robotic interaction, the Internet of things, interactive architecture (IA), architectural robotics, and the bordering field of cyber-physical systems (CPS).

These social robots are direct expressions of Weiser's ubiquitous environments, only if they dissolve and disappear into our everyday things and places. Adam Greenfield called them the "everyware" systems (Greenfield, 2006). The dream for ubiquitous computing environments is to have computers that dissolve and are everywhere and in everything. Social robots in any form beyond the humanoid that achieve these goals and dreams will be defined here as non-humanoid social robots ("NH-SR").

2.3   What Are Non-humanoid Social Robots "NH-SRs"?

Non-Humanoid Social Robots "NH-SR"s have *spatial, physical, and robotic features* to interact with or respond to users, either virtually or physically or using digital-physical hybrid systems. The human–robotic interactions in this type of robotic

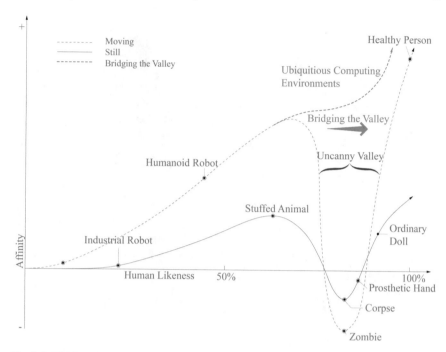

Fig. 2.2 NH-SRs/ubiquitous computing environments are one way of bridging the Uncanny Valley—Modified Version of Mori's Uncanny Valley

environment create social dialogs in the form of lights, sounds, and movements, i.e., from the physical architectural level. Users, too, start to engage using their senses by talking, touching, moving or by using different types of body language, creating gestures and cues to activate and interact with these ubiquitous computing environments (Fig. 2.3). *Movement* is critical in non-humanoid social robots to change the environment's state, from being a static system to being able to morph and transform dynamically. The movement can be in the form of mimicking biological systems, as described later in the biomimicry paradigm, or by creating unique artistic choreographic representations in space, e.g., the "hylozoic" movement (Fig. 2.1). NH-SRs have different flavors of AI, e.g., predetermined, self-aware, sentient, autonomous. The NH-SRs can embody the space on two levels, the mesoscale and the macroscale, but not including the microscale level. The microscale social robots have their character and paradigms which are beyond the scope of the spatial ubiquitous computing environments.

In the following sections, we will discuss the properties of NH-SRs, as in Fig. 2.3, the different types of social robots at scales beyond the humanoid from the mesoscale to the macroscale robots, understanding of their architecture, and the paradigms and styles used for designing them.

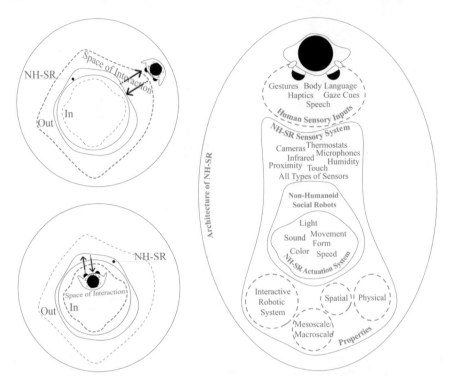

Fig. 2.3 Human interaction with NH-SRs from outside the robot's environment (top left) or inside the robot's environment (bottom left); and (right) the architecture of NH-SRs

2.4 The Characteristics of Non-humanoid Social Robots ("NH-SR")

The literature review used in this study on non-humanoid social robots and ubiquitous computing environments has been focused on reviewing articles and projects within the past two decades, i.e., 2000–2018, in the Institute of Scientific Information (ISI)-indexed references and real built projects in physical sites; see Fig. 2.4. The top-tier ISI indexed references used for this study are published in ACM, IEEE, and Elsevier journals, online conference proceedings and magazines. The real built ubiquitous computing environments in the past two decades were also reviewed and included in this study, but they did not include visionary drawings and perspectives or work-in-progress projects by artists, roboticists, and architects. The 59 projects that have been used to define the domain field of non-humanoid social robots (NH-SRs) were analyzed, and the following seven classifications and their sub-categories have been identified to ensure there is pairwise disjointness (non-overlapping) between categories and that the totality represents the entire domain space.

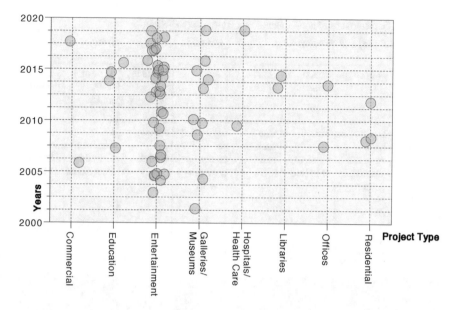

Fig. 2.4 Non-humanoid social robot's typologies per year between 2000 and 2018

Accordingly, the NH-SRs have been classified based on three main categories: the formal/structural, the functional, and the aesthetics of the artifacts. The *formal/structural characteristics* have been defined based on the artifacts' "scale," and "geometry/form." The *functional characteristics* have been designed based on the "scope/type," "robotic features," and the "number of users." The *aesthetics of NH-SRs* have been defined based on the "design patterns" and "type of user–space interactions." These characteristics constitute the form, scope, and character of the domain field of "NH-SRs" (Table 2.1).

Non-humanoid social robots "NH-SRs" interact as dynamic structures, more like living organisms, but not as humans (Fig. 2.5). They resemble features from biological systems and living organisms, e.g., as in the movement of plants in response to sunlight. They use many types of sensing and actuating technology to interact with users in social contexts. NH-SRs are embedded with AI technology for orchestrating the responses of the environment. The characteristics of NH-SRs are described in detail in the following sections.

Table 2.1 Characteristics of non-humanoid social robots (NH-SRs)

Basic components	Functional characteristics			Formal characteristics		Aesthetical characteristics	
	Scope/type	Robotic features	Number of users	Scale	Geometry	Type of user interaction	Design patterns
Sensors, actuators (mostly to move masses) and AI processing power	Education, health care, residential, commercial, office, museum, and libraries	Virtual, or physical, or hybrid	One users or multi-users	Mesoscale or macroscale environments	Simple or complex	One-dimensional or multi-dimensional	Function-driven or design-driven

Fig. 2.5 Human interactions with NH-SRs at the "Datagrove," by Future Cities Lab, Nataly Gattegno and Jason Kelly Johnson (Courtesy of Future Cities Lab)

2.4.1 Functional Characteristics

2.4.1.1 Scopes/Types of Non-humanoid Social Robots (NH-SR)

As architects have been using "contexts" to shape the behavior of users, non-humanoid social robots "NH-SRs" have been similarly designed in many types and for many contexts, i.e., typologies, to create social robotic interactions between users and their environments (McCullough, 2005). The interest in having these environments is based on either users' needs or design-driven technological explorations and contexts for this new type of social interaction, i.e., human–robotic interaction.

By embedding interactive and robotic technologies into the very fabric of our built environments, our physical and social spaces start to behave as if they are living bodies. These new types of intelligent physical environments respond to our growing interest in using robotics and information technology in our daily lives. In studying the various kinds of non-humanoid social robots, we define seven dominant types and contexts which are shaping this field. The categories defined here do not limit the explorations in a broader range of environments, but they are the contexts and types being used widely for NH-SRs (Fig. 2.5).

a. *Educational NH-SRs* include all types of educational environments such as kindergartens, schools, universities, classrooms, laboratories, auditoriums. The use of NH-SRs in education is shaping social interactions for users of certain ages, and with particular interests and behaviors. The "Open Columns" by the Situated Technologies Research Group and "ColorFolds" by Sabin Design Lab exemplify this type of human–robotic interaction in education.

b. *Healthcare NH-SRs* include all types of healthcare facilities and equipment. These environments promise to enhance patient–space interactions and provide assistance for elderly or persons in need in local houses or healthcare facilities. An example of NH-SR in healthcare is the "ART: Assistive Robotic Table" by the architectural robotics Lab.

c. *Residential/Housing NH-SRs* include environmental systems in housing and residential architecture that can be at the scale of furniture or walls or more massive structures, i.e., the whole space. "The Phalanstery Module" by Bureau Spectacular; the "Reconfigurable House" by Usman Haque and Adam Somlai-Fischer; and "Aether" by ElectroLand are among many others which exemplify this type of environment.

d. *Entertainment NH-SRs* represent the largest typology in respect to the number of constructed projects among all other typologies, including all spatial environments that create happiness, and entertain users. *Entertainment NH-SRs* include, but are not limited to, hotels, restaurants, cafes, parks and gardens, theaters, playgrounds, etc. Examples of *Entertainment NH-SRs* are the "Sky Ear" by Haque Design+Research; the "D-Tower" by NOX; "Reef" by Rob Ley Studio; "WIND 3.0" by Studio Roosegaarde; "Bubbles" by FoxLin; "Light Span" by ElectroLand; "Adaptive Room" by the Justin Goodyer_Bartlett School of Architecture; "Reflectego" by the Hyberbody Group, among many others.

e. *Commercial NH-SRs* include all types of NH-SRs in retail shops, grocery stores, shopping malls, and all sorts of commercial facilities. The "Target Interactive Breezeway" by ElectroLand and the "MESOLITE" by Behnaz Farahi are examples for this typology of social robots.

f. *Office NH-SRs* include systems in office spaces, such as desks, chairs, cabinets, or any spatial elements within that space. The "AWE: Animated Work Environment" by the Architectural Robotics Lab, and "BALLS" by Ruairi Glynn and Alma-nac Collaborative Architects are examples of this typology.

g. *Galleries and Museums NH-SRs* represent one of the most significant typologies, the second after the "entertainment NH-SRs." It includes environments that exhibit or embody all sorts of artworks, and many of these are artwork representations that have been widely used to interact with visitors. They include the "NSA Muscle Project" by ONL-Kas Oosterhuis; the "Party Wall" by nArchitects; the "Plinthos Pavilion" by MAB Architects; the "Lightswarm" by Future Cities Lab; and, the "Edge of Chaos" by Interactive Architecture Lab.
h. *Libraries NH-SRs* as in the "The LIT Room" and "CyberPLAYce" by the Architectural Robotics Lab.

2.4.1.2 The Robotic Features of Non-humanoid Social Robots (NH-SR)

Considering the three robotic components of the robot, the sensing, processing/controlling, and actuating components, all of the following features have the processing power to physically actuate the robot in linear or nonlinear movements to move masses by employing soft or hard actuators. The robotic features of NH-SRs have been categorized based on the medium by which users interact with the robot (human–robotic interaction), as in virtual, physical or hybrid users' interactions with the robots.

Virtual Interactions in NH-SRs include interactions with the robot using flat displays or gestural 2D interfaces, e.g., smartphone apps or wearables, as in the "D-Tower" by NOX, the "MURMUR" and "BITSTREAM" by the Future Cities Lab, and "Calgary Tram Stops" by ElectroLand.

Physical Interactions in NH-SRs include interactions with the robot using local or networked sensors embedded within the system, for example by using physical pushbuttons, microphones, potentiometers, etc. There are many projects in this category, such as the "Sky Ear" and "Burble" by Haque Design+Research, the "Party Wall" by nArchitects, "The Phalanstery Module" by Bureau Spectacular, "Alloplastic Architecture" and the "The Living Breathing Wall" by Behnaz Farahi, and "Waterlict" by Studio Roosegaarde.

Hybrid Interactions comprise both platforms for interaction, e.g., using graphical interfaces and pushbuttons on site. The "LIT Room" and "CyberPLAYce" by the Architectural Robotics Lab exemplify this type of robotic interaction.

2.4.1.3 Number of Users in Non-humanoid Social Robots (NH-SRs)

Social robots have been designed to interact with either one user as in the "Locus" by the Interactive Architecture Lab and "MESOLITE" by Behnaz Farahi, or multiple users at a time, as in the "NSA Muscle Project" by ONL-Kas Oosterhuis, the "Lightspan" by ElectroLand, and the "LOTUS 7.0" by Studio Roosegaarde.

2.4.2   Formal Characteristics

2.4.2.1   Scales of Non-humanoid Social Robots (NH-SR)

The scales of NH-SRs differ from mesoscale to macroscale robots. The microscale social robots do not have the same physical and phenomenological characteristics as the scaled robots; accordingly, they are not considered non-humanoid social robots.

Mesoscale NH-SRs are medium-scaled interactive architectural elements, such as furniture (tables, chairs, desks, sofas, couches, etc.); and all sorts of interior design accessories, e.g., lights and fixtures, curtains, shelving for art displays, and cabinets. Many NH-SRs are designed at this scale such as the "Interactive Façade" by Foxlin, "Locus" by the Interactive Architecture Lab, the "AWE: Animated Work Environment" by the Architectural Robotics Lab; "MESOLITE" by Behnaz Farahi; "Calgary Tram Stops" by Electroland, among many others.

Macroscale NH-SRs are large structures in the form of interactive walls and partitions, ceilings, floors, facades, or combinations of these, or take the form of a room or enclosed space that can be used by one person or more (Fig. 2.6). The "Sky Ear" by Haque Design+Research, the "D-Tower" by NOX, Lars Spuybroek, "4D-Pixel" by Studio Roosegaarde, the "Open Columns" by Situated Technologies Research Group, "Aerial Well Study" by Philip Beesley Architect, among many others are examples of macroscale NH-SRs.

2.4.2.2   The Geometry of Non-humanoid Social Robots (NH-SR)

The physical form of NH-SRs has two distinctive patterns, i.e., simple or complex geometries. The *simple form* resembles the primitive shapes as in cubical, cylindrical, pyramidical, conical, or spherical structures. The "Ada intelligent Room" by Paul Verschure, the "4D Pixel" by Studio Roosegaarde, the "Muscle Towers II" by the Hyberbody Group, and "AWE: Animated Work Environment" by the Architectural Robotics Lab represent simple geometrical environments of NH-SR.

The *complex form* of NH-SRs is interactive organic robots imitating or inspired by natural and biological systems (but, not all biomimetic designs are complex) or are multi-geometric compositions. Examples are the "Reef" by Rob Ley Studio, the "Sentient Chamber" by Philip Beesley Architect, the "Edge of Chaos" by the Interactive Architecture Lab, "ColorFolds" by Sabin Design Lab, and the "Hyposurface" by deCOi, "Lightswarm" by Future Cities Lab.

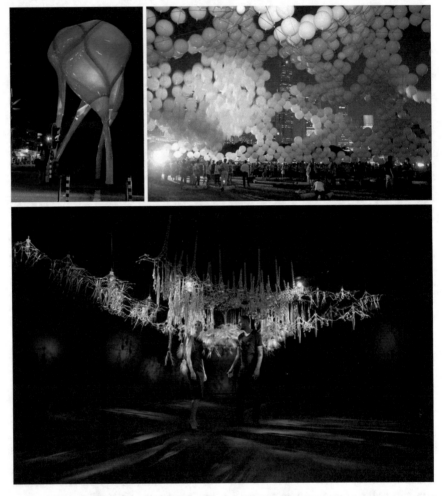

Fig. 2.6 Macroscale NH-SRs as in the D-Tower by NOX, Lars Spuybroek; "Burble" by Haque Design+Research, and the "Aerial Well Study" installation view by Philip Beesely (Courtesy of Lars Spuybroek; Haque design+research; and, Philip Beesley Architect Inc., respectively)

2.4.3 Aesthetical Characteristics

2.4.3.1 Type of User Interactions in Non-humanoid Social Robots (NH-SRs)

The user interactions with NH-SRs can be either *unidimensional* by employing the use of one sensor for interacting with the robot, e.g., gestural, or haptics, or visual, or speech, or can be *multi-dimensional* by using more than one sensory input as in the hybrid interactions described above.

Unidimensional NH-SRs have been widely used in many robots as in the "Party Wall" by nArchitects, "WIND 3.0" by Studio Roosegaarde, "Alloplastic Architecture" by Behnaz Farahi, and the "Aether" by ElectroLand.

Multi-dimensional NH-SRs have been used in the "Reconfigurable House" by Usman Haque, Adam Somlai-Fischer, and the "BALLS" Alma-nac with Ruairi Glynn.

2.4.3.2 Design Patterns for Non-humanoid Social Robots (NH-SR)

Non-humanoid social robots offer a unique platform for being active participants in our daily lives. The design patterns for NH-SRs can be classified into two main trends:

Function-Driven Non-Humanoid Social Robots (*FD_NH-SR*) address real-life problems for which there is an urgent need for solutions. These can be anything from a patient in need of help in a hospital, an elderly person needing to get her medicine, chatbots for collecting surveys from customers, airport systems for carrying luggage, and many more functions that need designers' interventions at scales beyond the humanoid.

Design-Driven Non-Humanoid Social Robots (DD_NH-SR) create experiences that promote innovations to enhance and support us in our daily lives. For example, the use of a robotic table which brings the patient's needed glasses or medicine to her bed. DD_NH-SR may also be exemplified in the use of the robotic office desk that supports classical, gaming and casual meetings. The DD_NH-SR can be seen in public spaces for communicating with the visitors by listening and responding to their interactions.

2.5 Social Robots at Scales Beyond the Humanoid

The above highlights the design principles for the NH-SRs, yet the sociological, psychological, environmental, and cultural values which NH-SR environments embody are still critical to the design and the development of social robotic environments; i.e., the context is important. The creative processes used by designers and architects to design non-humanoid social robotic environments follow nonlinear iterative procedures which can either follow a specific style, i.e., intentionally, or can become design exploration without a particular style. Thus, by learning about the different paradigms in designing these environments, designers will have a palette of styles to choose from, using which they may revolutionize the current design trends.

In recent years, as described before, the fields of architectural robotics, cyber-physical systems, and ubiquitous computing research have been growing, and we have literature that can be used to draw a map of design trends and paradigms of non-humanoid social robots. The following sections will describe, in the short space of this chapter, the essential characteristics of each trend.

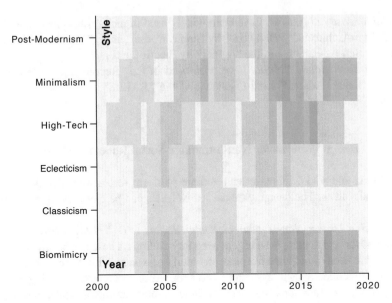

Fig. 2.7 Non-humanoid social robots classified based on the number of projects for each trend per year (blue intensity represents higher number of projects in specific year, and yellow represents no projects)

A critical difference between designing architectural spaces and designing socially interactive robots is the way we interact with them, i.e., interaction and adaptation. NH-SRs provide unique platforms for communication based on dynamic and interactive structures that retune and morph as users engage in a spatial dialog with them. The classification of NH-SRs' paradigms follows this study's comprehensive research on the different socially interactive robots and the typical design characteristics they share.

The design characteristics that were taken into consideration are fundamentally "formal," but include many phenomenological dimensions, as in the change of colors, light, shadows, movements, i.e., "external space" as defined by Merleau-Ponty. Accordingly, the analyses of fifty nine NH-SRs led to six design styles/paradigms which have been the primary drivers for the design of social robots at scales beyond the humanoid (Fig. 2.7). Sections 2.5.1 to 2.5-6 provide details about the NH-SR's six design styles/paradigms.

2.5.1 *"Minimalism" in Non-humanoid Social Robots*

In art and design as well as in architecture, minimalism was inspired to a great extent by Miesian's architecture, referring to the modern architect and pioneer Mies Van Der Rohe, and his *"less is more"* philosophy. The users' needs had highly concerned the Japanese minimalists in fulfilling the spatial and the building's functional require-ments and structures in their most literal direct representations—i.e., "uncluttered space" (Macarthur, 2002; Obendorf, 2009a).

"Less is more" summarizes the tendency of minimal design to overcome com-plexity and cluttered designs. The aim is to free the design from additional elements by having simple and clean designs. The contemporary architectural critic Charles Jencks refers to it as the art of "freeing the design from postmodern and classical references including culture and traditions."

Historically, minimalism in architecture has been recognized as a paradigm shift from postmodernism in the 1980s, but has been classified under the postmodern era. This contradiction was due to the postmodernist architects who insisted on providing an answer to those who had criticized the complexity of postmodern works, by introducing designs where there was a global clarity of meaning that could help people to live calmer lives, away from the repetitive designs of the mass production era or the pre-modern symbolic historical references (Stevanovic, 2013).

Minimalism, reflecting its unique character in interface design is an essential driver for the usability of IT and robotic systems, including social robots, since it helps with the aim of *achieving satisfying levels of aesthetics in both the machine's function and the interface's usability*, and in embracing effective human–robotic interactions. Minimalism has been discussed in depth in Hartmut Obendorf's book on minimalism, "Designing Simplicity," where it was described as a "good" design style for HCI (Obendorf, 2009b).

From a psychological perspective, Don Norman argues for the need to understand the complexity of systems and to distinguish the complexity of IT systems from the simplicity of their user interfaces. Hence, the need for simplicity and clarity of inter-actions with interfaces are fundamental to good designs, with the aim of achieving usefulness, accessibility, usability, and beautiful interactions. In "About Face 3" by Alan Cooper, R. Reimann, and D. Cronin, the authors claim that a simple interface design whether a virtual or physical interface is an essential tool for satisfying both the user and the business needs (Cooper, Reimann, & Cronin, 2007).

Because minimalism entails simplicity, aesthetics, and usability, minimalism in NH-SRs is both functional and user-friendly. Non-humanoid social robots that are minimalist, such as the "NSA Muscle Project" by ONL-Kas Oosterhuis, the "Bub-bles" by FoxLin, the "BITSTREAM" by Future Cities Lab, the "reEarth: Hortum machina, B" by Interactive Architecture Lab, among many others, have the following characteristics:

1. *A Transparent Design* of the system which can be either fully or partially exposed to its users. The transparency of the system is represented in the exposed sensing and actuating technologies used in the environment including, but not limited to, pulleys/strings, gears, linear actuators, servo motors, lights, and movements, and sensors of all types, including haptic, visual, and auditory technologies.
2. *Simplicity, Abstraction, and Reduction* of form by removing any clutter in the design, i.e., not functional or structural. The structures of minimalistic NH-SRs have simple geometry, such as rectangular, circular, triangular, or basic shapes,

but can also have simple organic forms and as clean surfaces as possible. The form is inspired by the reductionist approach in art, architecture, and music. Abstraction has also been the primary driver for the design of these systems by representing reality in its abstract shapes and by avoiding figurative compositions.

3. *Usability and Ease of Understanding* are essential for a minimal and straightforward user interface. The physical, digital, or hybrid interfaces achieve simplicity and minimal design through usability studies. The aim is to be able to interact with the interfaces effectively and immediately.

2.5.2 *"High-Tech" in Non-humanoid Social Robots*

The high-tech movement has been a by-product of industrialization and globalization with the aim of placing technology at the forefront of human achievements. In this movement, the designers of NH-SRs have sought the highest performance and stability of their systems and the use of new materials to achieve unique properties and interactions (Khan, 2008).

In art and architecture, the high-tech movement was influenced by late-modernism and structuralism. The structuralist philosophy of Ferdinand de Saussure and Claude Lévi-Strauss has changed our understandings of society and culture by focusing on the inner structure of all systems. According to structuralists, objects should have meanings which are based on their internal structures; then, these arrangements will be put together to form meaningful compositions. This approach to design has helped in the emergence of new sciences and systems, and in changing the way, we understand how we perceive built environments (Jencks, 1997).

The high-tech movement started in the 1970s, but its influence on contemporary architects and designers is still evident. As non-humanoid social robots embody spatial and physical characteristics inherited from the built environment, there have been similarities in the approaches and styles that architects and designers use to develop them. As NH-SRs are focused on innovation and advancements in robotics and information technologies, high-tech qualities have been used by designers and technologists to create state-of-the-art human–robotic interactions in these environments. Examples of "high-tech" NH-SRs can be seen in the works of Daan Roosegaarde in the "4D Pixel," Paul Verschure in the "Ada Intelligent Room," ElectroLand in the "Lightspan" and "Aether," and Situated Technologies Research in the "Light Walk" projects.

High-tech non-humanoid social robots have the following properties:

1. Use of *Digital Interfaces/Screens* as a platform to present the structure of the system and also to provide an innovative platform for interaction that the tech-savvy generation is using for many types of social interactions.
2. They reveal *the Architecture of the Robotic and IT Components* by exposing the circuitry, electronics, wiring, cables, LEDs, and pixels of the system.

3. They reveal *the Robot's Inner Structure* without altering the materials' original texture and color; but, in some cases the designers use polishing techniques to give new textures to the surfaces to make them either shiny, glossy, bright, or smooth.

4. The *Use of Innovative Materials* has been an important driver of this trend. Notably, this includes experimenting and prototyping by using new and innovative materials, and applying new techniques to change the materials' properties and reactions to contexts. Examples of new materials that have been recently used in the built environments are fiberglass, titanium, Corian, shape-memory alloys, and plastics in the form of transparent/translucent materials.

5. *Modular Innovative Claddings/Coverings* in the form of classical or evolutionary patterns represent a crucial feature in "high-tech" NH-SRs by employing parametric/algorithmic design. These complex patterns and materials are manufactured by using computer numerical control (CNC) machines, 3D printers, and digital fabrication technologies.

2.5.3 *"Biomimicry" in Non-humanoid Social Robots*

The main driving force for biomimicry or biomimetic design is the growing interest in understanding systems that have proven their abilities to sustain and evolve. Biomimicry started in the intersection between sciences, especially those concerned with the understanding of biological, agricultural, medical, and material systems. In this field of study, scientists explore how systems work; the underlying structures and behaviors; and, the intrinsic values that govern interactions within natural environments (Baumeister, 2014; Benyus, 2002; Lakhtakia & Martín-Palma, 2013).

Similarly, the interdisciplinary approach in design has inspired engineers and designers of non-humanoid social robots to learn from and be inspired by nature and biological systems in the intersections between biology, architecture, and robotics. In envisioning robots mimicking the behavior, or form, or structure and characteristics of natural biological systems, we have developed fascinating interactive social environments that behave more like living organisms than a "machine for living in," following a biomimetic design process.

Biomimicry has been the main paradigm/trend for non-humanoid social robots, as in the "Reef" by Rob Ley Studio, the "AWE: Animated Work Environment" by the Architectural Robotics Lab, the "HypoSurface" by deCOi, and the "Aerial Well Study" by Philip Beesley Architect Inc. (Fig. 2.6).

These biomimicry environments encompass the following characteristics:

1. *Biologically inspired environments* in their underlying structures, systems, behaviors, and the intrinsic values that govern human–space interactions and systems' responses to users.

2. *Mimicking biological systems' form* by replicating the underlying structure (scaled or at the same scale) or mechanisms of stability and movement. Then,

NH-SR's form is manufactured by employing the use of new materials and technologies.
3. *Abstraction* as an approach and design principle for learning from biological
   systems how they work and accordingly how we can use this to develop our
   systems (Cohen & Reich 2016).

2.5.4 *"Postmodernism" in Non-humanoid Social Robots*

Historically, postmodernism was a radical artistic movement to its predecessor "Modernism." Postmodernism continued the mannerist phase of modernism leading to
complexity, ambiguity, and contradiction of design (Venturi et al. (1), 1977; Jencks,
2011). Postmodernism in architecture was aimed to move the design of spaces from
mere "machines" to fulfill the social and popular needs of people in their complexity
and contradictions. Postmodern art can be understood by Jencks's "double code"
concept which embody duality, complexity and contradiction of meanings. It is an
art form that embodies both the old and the new. This can be seen in the inclusive
understanding of form as in Venturi's "both-and," e.g., an art that is both contemporary and use historical references to create culture-centered design, instead of being
exclusive as in the "either-or" character of modernism, which is to be either contemporary (modern) or classic (old). For postmodernists, users or people, in general,
have nostalgic connections with the beauty of the past and its main characters, but
they appreciate the silver shiny technological advancements of our contemporary
life, i.e., both the old and the new.

In "Learning from Las Vegas," the famous architect Robert Venturi and Denis
Scott Brown argue for the need of architecture and art, in general, that is more
popular and reflects the identities of its inhabitants (Venturi et al. (2), 1977). Thus, if
the twentieth-century modern designs had failed to appreciate the diversity of people
and accordingly incapable to satisfy their various needs, it was due to the fact, as
described by the famous Italian architect, historian, and theoretician Manfredo Tafuri,
that modern art and architecture "controlled artistic production or violently inserted
the irrational."

For designers of the twenty-first century, postmodernism can be seen as a retreat
toward the past or even the existing designs. To explain that in our context, the
complexities and contradictions in postmodern designs have added layers of technological advancements to the elements of our built environments, leaving their main
structure and design mostly unaltered; yet in a few cases, the designers have chosen to
change or re-proportion its elements. This trend can be seen in the "The Phalanstery
Module" by the Bureau Spectacular, the "Plinthos Pavilion" by MAB Architects, the
"Reflectego" by the Hyperbody Group, and the "LOTUS 7.0" by Daan Roosegaarde
(Fig. 2.8).

Accordingly, "postmodern" non-humanoid social robots have the following common characteristics:

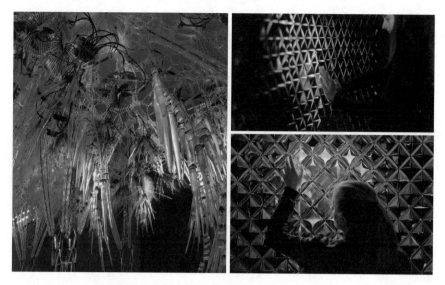

Fig. 2.8 "Sentient Chamber" by Philip Beesely and "Lotus 7.0" by Daan Roosegaarde (Courtesy of Philip Beesley Architect Inc. and Studio Roosegaarde)

1. *Double Code* by embodying opposites to communicate with the public; mimicking or replicating past experiences, objects, or artifacts but making changes to either their form, texture, color, materials, or proportions, or as described above, adding a technological touch to the existing objects.
2. *Complexity* by taking into consideration the different needs of all users by using "usability evaluation" techniques to be able to design the NH-SRs' interface. Complexity is represented in both the NH-SR's inner architecture and interface design. The interface is rich with different ways of interaction; i.e., it is a hybrid digital-physical system.
3. *Ambiguity* is fundamental in postmodernism due to NH-SR's complexity and double-coded features, which make the system open to different interpretations.

2.5.5 *"Classicism" in Non-humanoid Social Robots*

Classicism refers to the "traditional" style in art and architecture, where the objects have classical references that go back to Ancient Greek and Roman Architecture and continue to modern architecture. Classicism employs the use of traditional and classical formal compositions that are guided by the principles of symmetry, harmony, balance, elegance, and dignity. In our context of non-humanoid social robots, we have seen less interest in the use of that style due to its lack of a contemporary and technological look. The main driver for using the classical style is the function of the robot. The "ART: Assistive Robotic Table" by the Architectural Robotics Lab exemplifies the classical trend in non-humanoid social robots at the mesoscale, where the table has been designed for elderly and for post-stroke adults in hospital.

2.5.6 "Eclecticism" in Non-humanoid Social Robots

In architecture, eclecticism is a style that combines different styles in one artifact, i.e., a space or building. It can be seen as lacking clarity of character and identity, but it can also be seen as a creative platform for artists and designers to combine different styles and forms to create new products. Eclecticism has been widely criticized in art and architecture (Jencks, 1987). In our context, eclectic non-humanoid social robots are popular due to their ability in creating creative experiences and new typologies for interactive spatial interactions. The "Alloplastic Architecture" by Behnaz Farahi and the "ECO-29 Wedding Hall" by FoxLin exemplify this trend.

2.6 Coming Age of Ubiquitous Social Robots

The abundance of information technologies in everything and everywhere is shaping the future of our environments. "Coming age of calm technologies" have already started, where architects and designers are taking essential roles in determining how robotic technologies will shape our lives, i.e., to have calm environments! As humans are getting distracted by the overflow of information in their environments, there is a growing need for calm environments where robotics and IT become invisible.

As represented in this chapter, the design of non-humanoid social robots promises ubiquitous human interactions beyond the humanoid and in everything in our environments (Fig. 2.8). The dusk of petrified static architectural environments, i.e., unintelligent physical architecture, has already opened the door for many explorations in many types of spaces, i.e., intelligent physical environments as in the NH-SRs, and the expectations of more research are evident. Even as we speak, there are ongoing advancements in IT, sensing, processing, and actuating robotic technologies, and artificial intelligence will help us to create environments which can sense and feel our emotions, listen to us, be organic cybernetic systems, and respond to us as delicately and elegantly as living organisms.

The six styles of NH-SRs presented in this chapter, i.e., minimalism, high-tech, biomimicry, oostmodernism, classicism, and eclecticism, describe the different formal and architectural features used in designing NH-SRs. These styles can be used as design guidelines for designers, artists, and educators in the design and development of NH-SRs of all scopes and types, i.e., educational, health care, residential, entertainment, commercial, offices, galleries and museums, and libraries. The domain field of NH-SRs has the advantage of being at the intersection of many sciences including architecture, design, robotics, sociology, psychology, and computation. The advancements in any of these sciences will affect the whole domain field and create revolutionary/evolutionary paradigms of social robots at scales beyond the humanoid.

Notably, we still need more research to measure the effectiveness of social interactions within these NH-SR environments that coexist with humans and humanoid social robots. Social and ethnographic studies are needed to understand humans' preferences concerning NH-SRs and to discover what designers can do to help in creating new design trends for the age of calm environments and technologies.

References

Ames, M. G. (2015). Charismatic technology. *Aarhus Series on Human Centered Computing, 1*(1), 12. https://doi.org/10.7146/aahcc.v1i1.21199.

Baumeister, D. (2014). *Biomimicry resource handbook: A seed bank of best practices*. CreateSpace Independent Publishing Platform.

Becker, J. (2017). The uncanny valley: why robots make us feel uneasy. Anthropologue, Laboratoire d'anthropologie sociale, Collège de France, This article is published in collaboration with The Conversation. In *World Economic Forum*. https://www.weforum.org/agenda/2017/04/why-we-dont-trust-robots/.

Benyus, J. (2002). *Biomimicry: Innovation inspired by nature* (2nd ed.). Perennial.

Breazeal, C., Takanishi, A., & Kobayashi, T. (2008). Social robots that interact with people. In *Springer Handbook of Robotics* (pp. 1349–1369). Berlin: Springer.

Cafaro, F., Lyons, L., Roberts, J., & Radinsky, J. (2014). The uncanny valley of embodied interaction design. In *Proceedings of the 2014 Conference on Designing Interactive Systems—DIS '14* (pp. 1075–1078). Vancouver, BC, Canada: ACM Press. https://doi.org/10.1145/2598510.2598593.

Cohen, Y. H., & Reich, Y. (2016). *Biomimetic design method for innovation and sustainability*. Springer International Publishing.

Cooper, A., Reimann, R., & Cronin, D. (2007). *About face 3: The essentials of interaction design* (3rd ed.). Wiley.

De Graaf, M. M. A., Allouch, S. B., & Klamer, T. (2015a). Sharing a life with Harvey: Exploring the acceptance of and relationship-building with a social robot. *Computers in Human Behavior, 43*, 1–14. https://doi.org/10.1016/j.chb.2014.10.030.

De Graaf, M. M. A., Allouch, S. B., & van Dijk, J. A. G. M. (2015). What makes robots social? A user's perspective on characteristics for social human-robot interaction. In *Social robotics* (pp. 184–193). Cham: Springer. https://doi.org/10.1007/978-3-319-25554-5_19.

Dourish, P. (2004). *Where the action is: The foundations of embodied interaction*. MIT Press.

Duffy, B. R. (2003). Anthropomorphism and the social robot. *Robotics and Autonomous Systems, 42*(3), 177–190. https://doi.org/10.1016/S0921-8890(02)00374-3.

Enz, S., Diruf, M., Spielhagen, C., Zoll, C., & Vargas, P. A. (2011). The social role of robots in the future—Explorative measurement of hopes and fears. *International Journal of Social Robotics, 3*(3), 263. https://doi.org/10.1007/s12369-011-0094-y.

Fox, M., & Kemp, M. (2009). *Interactive architecture*. Princeton Press.

Green, K. (2016). *Architectural robotics: Ecosystems of bits*. Bytes and Biology: The MIT Press.

Greenfield, A. (2006). *Everyware: The dawning age of ubiquitous computing* (1st ed.). Berkeley: New Riders Publishing.

Jencks, C. (1987). Postmodern and late modern: The essential definitions. *Chicago Review, 35*(4), 31. https://doi.org/10.2307/25305377.

Jencks, C. (1997). *New Science = New Architecture?* (1st ed.). Wiley.

Jencks, C. (n.d.). 2000 July: Jencks' theory of evolution, an overview of 20th century architecture. In *Architectural review*.

Khan, O. (2008). *Reflexive architecture machines*. Matthew Hume.

Lakhtakia, A., & Martín-Palma, R. J. (Eds.). (2013). *Engineered biomimicry* (1st ed.). Elsevier.

Macarthur, J. (2002). The look of the object: Minimalism in art and architecture, then and now. *Architectural Theory Review, 7*(1), 137–148. https://doi.org/10.1080/13264820209478450.

McCullough, M. (2005). *Digital ground: Architecture, pervasive computing and environmental knowing*. MIT press.

Mokhtar, T., Green, K., & Walker, I. (2013). *Giving form to the voices of lay-citizens: Monumental-IT, an intelligent, robotic, civic monument*. In *The Proceedings of the International Conference of Human-Computer Interactions 2013*, Las Vegas, USA.

Mori, M., MacDorman, K. F., & Kageki, N. (2012). The uncanny valley [from the field]. *IEEE Robotics Automation Magazine, 19*(2), 98–100. https://doi.org/10.1109/MRA.2012.2192811.

Norman, D. (2007). *Emotional design: Why we love (or hate) everyday things* (1st ed.). Basic Books.

Norman, D. A. (2010). THE WAY I SEE IT: Looking back, looking forward. *Interactions, 17*(6), 61. https://doi.org/10.1145/1865245.1865259.

Norskov, M. (2016). *Social robots: Boundaries, potential, challenges, emerging technologies, ethics and international affairs*. Routledge.

Obendorf, H. (2009a). In search of "Minimalism"—Roving in art, music and elsewhere. *Minimalism: Designing Simplicity*, 21–64. https://doi.org/10.1007/978-1-84882-371-6_2.

Obendorf, H. (2009b). *Minimalism: Designing simplicity*. London: Springer.

Pask, G. (1969, Sep). The architectural relevance of cybernetics. *Architectural Design*, (pp. 496–496).

Reeves, B., & Nass, C. (2003). *The media equation: How people treat computers, television, and new media like real people and places* (New edition). Center for the Study of Language and Information, Stanford, California, CSLI Publications.

Rykwert, J., Leach, N., & Tavernor, R. (1988). *On the art of building. In ten books by Leon Battista Alberti*. Cambridge, MA: The MIT Press.

Sakamoto, D., & Ono, T. (2006). Sociality of robots: Do robots construct or collapse human relations? In *Proceeding of the 1st ACM SIGCHI/SIGART Conference on Human-Robot Interaction—HRI '06*. Salt Lake City, Utah, USA: ACM Press. https://doi.org/10.1145/1121241.1121313.

Stevanovic, V. (2013). A reading of interpretative models of minimalism in architecture. *METU Journal of the Faculty of Architecture and Visual Arts, Faculty of Architecture, University of Belgrade, SERBIA, 30*(2). https://doi.org/10.4305/METU.JFA.2013.2.10.

Strait, M., Vujovic, L., Floerke, V., Scheutz, M., & Urry, H. (2015). Too much humanness for human-robot interaction: Exposure to highly humanlike robots elicits aversive responding in observers. In *Proceedings of the 33rd Annual ACM Conference on Human Factors in Computing Systems—CHI '15* (pp. 3593–3602). Seoul, Republic of Korea: ACM Press. https://doi.org/10.1145/2702123.2702415.

Tinwell, A., & Grimshaw, M. (2009). Bridging the uncanny: An impossible traverse? In *Proceedings of the 13th International MindTrek Conference: Everyday Life in the Ubiquitous Era on—MindTrek '09*. Tampere, Finland: ACM Press. https://doi.org/10.1145/1621841.1621855.

Tschöpe, N., Reiser, J. E., & Oehl, M. (2017). Exploring the uncanny valley effect in social robotics. In *Proceedings of the Companion of the 2017 ACM/IEEE International Conference on Human-Robot Interaction—HRI '17*. Vienna, Austria: ACM Press. https://doi.org/10.1145/3029798.3038319.

Tschumi, B. (2012). *Architecture concepts: Red is not a color*. USA: Rizzoli.

Venturi, R., Izenour, S., & Brown, D. S. *Learning from Las Vegas - Revised Edition: The forgotten symbolism of architectural form*. The MIT Press.

Venturi, R., Scully, V., & Drexler, A. (1977). *Complexity and contradiction in architecture* (2nd ed.). New York: The Museum of Modern Art.

Von der Pütten, A. M., & Krämer, N. C. (2012). A survey on robot appearances. In *Proceedings of the seventh annual ACM/IEEE International Conference on Human-Robot Interaction—HRI '12* (p. 267). Boston, MA, USA: ACM Press. https://doi.org/10.1145/2157689.2157787.
Weiser, M. (1991). The computer for the 21st century. *Scientific American (94–104).* https://www.lri.fr/~mbl/Stanford/CS477/papers/Weiser-SciAm.pdf.
Weiser, M., & Brown, J. S. (1997). The coming age of calm technology. In P. J. Denning & R. M. Metcalfe (Eds.), *Beyond calculation: The next fifty years.* New York, NY: Copernicus.

Chapter 3
A Study on Robot-Initiated Interaction: Toward Virtual Social Behavior

H. P. Chapa Sirithunge, H. M. Ravindu T. Bandara,
A. G. Buddhika P. Jayasekara, D. P. Chandima
and H. M. Harsha S. Abeykoon

Abstract Human-like decision-making skills are sought after during the design of social robots. On the one hand, such features enable a robot to be easily handled by its non-expert user. On the other hand, the robot will have the capability of dealing with humans in such a way that the human will not be disturbed by the behavior of the robot. In an effort to introduce proxemics-based etiquettes to a social robot, we have used a teleoperated robot to find human interest in the intelligent proxemic and conversational behavior of a robot. Engagement of humans with a situation-cautious behavior of the robot upon a static approach behavior was examined during the study. During this approach, physiological cues displayed by the humans were used by the robot to perceive an encounter with humans. As the robot approached the subject after analyzing the physiological behavior of the subject, spatial constraints occur due to the movement of the robot could be demolished. Furthermore, utterances generated by the robot to initiate an interaction with the user were decided by predicting the intentions of that individual based on these displayed human cues. Results of the experiment confirm the fact that a more socially acceptable spatial and verbal behavior could be observed from the robot through situation-awareness than a static behavior.

H. P. Chapa Sirithunge (✉) · H. M. R. T. Bandara · A. G. B. P. Jayasekara · D. P. Chandima ·
H. M. H. S. Abeykoon
Intelligent Service Robotics Group, Department of Electrical Engineering, University of
Moratuwa, Katubedda 10400, Sri Lanka
e-mail: ra-chapa@uom.lk

H. M. R. T. Bandara
e-mail: ra-ravindu@uom.lk

A. G. B. P. Jayasekara
e-mail: buddhikaj@uom.lk

D. P. Chandima
e-mail: chandimadp@uom.lk

H. M. H. S. Abeykoon
e-mail: harsha@uom.lk

© Springer Nature Switzerland AG 2019
O. Korn (ed.), *Social Robots: Technological, Societal and Ethical Aspects
of Human-Robot Interaction*, Human–Computer Interaction Series,
https://doi.org/10.1007/978-3-030-17107-0_3

Keywords Situation-awareness · Social robots · Emotional intelligence · Human–robot interaction · Intelligent systems

3.1 Introduction

Social robots not only deal with humans in collaborative workspaces, but they also accompany humans in personal settings (Rantanen, Lehto, Vuorinen, & Coco, 2018; Rossi, Dautenhahn, Koay, & Saunders, 2017). Therefore, simulating events and appropriate behavior in social settings, as humans do when entering a certain environment, is an emerging requirement for robots. Simply, robots are required to have an emotional intelligence to match their actions with a particular encounter, especially with humans involved.

Robots deployed in social environments have to abide by society's unspoken social rules as humans do (Korn, Bieber, & Fron, 2018; Rossi et al., 2017). Respecting human personal space is one rule in this regard and people's reactions upon personal space change in various situations (Mead & Mataric, 2017). Familiarity between humans reduces the personal space and this remains the same for human–robot interaction (Walters, 2008). Therefore, robots need cognitive skills to decide favorable means of interaction to deal with humans. This involves understanding human behavior and predicting their intentions and expectations. During a robot-initiated interaction, approaching the user, and maintaining proxemics and conversational cues are identified based on this perception. When developing such robots, there are two aspects to consider; the first is how you are going to construct such a socially competent robot, and the second is how people respond to such behavior. Our work focuses on the second aspect, by studying the human responses toward the socially intelligent behavior of robots.

Displaying appropriate approaching behavior has several aspects; following societal norms, establishing psychophysical distancing with people, and using gestures and expressions, verbal utterances, etc. (Mumm & Mutlu, 2011). Furthermore, human approach behavior is shaped by one individual's psychophysical closeness to the other. These two aspects have to be perceived in order to facilitate effective human–robot interaction (HRI). Personality attributes are the most important when attention-seeking features and behavior displayed by the robots are considered (Walters, Syrdal, Dautenhahn, Te Boekhorst, & Koay, 2008). Furthermore, humans prefer robots with human-like personality attributes in human–robot interactions. Much of the previous work has explored the factors which affect human–robot engagement (Mead & Mataric, 2012; Wiltshire, Lobato, Velez, Jentsch, & Fiore, 2014). However, there is a setback in exploring the behavioral trends during the initiation and maintenance of a smooth flow of robot-initiated interaction. One under-explored factor within this domain is perceiving the nonverbal human behavior prior to an interaction. Hence, the robot may evaluate the possibility and appropriate approaches to initiate an interaction based on the user situation. In contrast, the lack of predictability and transparency in many modern robotic systems has negatively affected the trust and reliance of humans on robots.

In order to respect the space acquired by a human (Stark, Mota, & Sharlin, 2018) and to predict the intentions of that individual, his/her behavior has to be observed. In this work, we conducted a study to explore the tendencies in users when approving the behavior of his/her robotic companion under different conditions. Under various circumstances, the activities humans engage in may vary and activities entail various bodily movements. During this study, users were allowed to engage in a set of activities, and the movements of the users were tracked for a certain period. The properties of these movements such as speed, frequency, and fanning were used to determine the engagement of a particular user. As the engagement in an activity reduces the interaction demanded by the user from outsiders, the degree of affect in a highly engaged activity will be low and vice versa. This fact was evaluated during the study using various activities in daily chorus that humans usually engage in. Depending on the degree of affect, various factors considered before the interaction was evaluated. The type of conversation, proxemics between the two conversant and the approaching path were the factors considered in this study. Finally, this approach was compared with an ordinary interaction scenario by initiating a direct interaction with the human without considering the behavioral aspects of the situation. These two approaches were compared by obtaining user feedback during each scenario.

Accordingly, when the robot decides to engage in an assisting task for a human, the robot should decide the type of approach used to accompany the human being toward an interaction such as a conversation. This approach may depend on several factors such as attention level, type of activity, accessibility, and obstacles in the environment. In this work, we propose a method to identify the suitable approach for an assistive robot based on the modeling of factors in conceptual space related to a virtual cognitive frame. Human beings are capable of making decisions naturally based on the experience, and they are capable of virtually simulating the scenario in a conceptual space. Furthermore, this ability makes the human unpredictable and more interactive in an interesting manner. In order to gain this ability, the assistive robot should possess the same skill at least at a certain level. We conducted a series of experiments to recognize the conceptual factors that can affect the approaching behavior of a robot, and evaluation criteria for the assistive robot to model conceptual parameters in human behavior more intellectually. This work demonstrates a design space for proxemics development and for choosing conversational preferences in social domains.

3.2 Related Literature

Research has been conducted to discover how social cues displayed by robots are interpreted by humans. Of these, cues associated with proxemic behavior were found to significantly affect human's perception of the social presence of the robot and the emotional state in that encounter (Papadopoulos, Küster, Corrigan, Kappas, &

Castellano, 2017). On the other hand, proxemics is an aspect of improving a robot's perception of the environment where the robot manipulates the interpretation of distance (Obaid et al., 2016; Walters et al., 2009). According to the previous studies, there are many emotional and psychophysiological aspects to approach behavior, such as gender, social norms, and personality. Therefore, such aspects have to be taken into consideration before long-term interaction occurs. The social presence of the robot was more appealing to humans when its behavior was determined by gaze and proxemics aspects rather than when it just followed user commands (Wiltshire et al., 2013).

Proxemic behavior falls under an interdisciplinary taxonomy of social cues and signals in the service of engineered social intelligence in robots (Wiltshire et al., 2014). Robotic social cues such as approach behavior have an effect on interpersonal attributions during HRI (Wiltshire et al., 2015). Therefore, from the perspective of humans' consideration of proxemics by anyone in a social environment is important. A set of feature representations for analyzing human spatial behavior motivated by metrics used in the social sciences are given in (Mead, Atrash, & Mataric, 2013). Such methods do not cover smaller movements in joints such as hand tips or elbows, although many tasks involve such joints. In addition, probabilistic frameworks for proxemic control have been used (Mead & Mataric, 2012) for mobile human–robot interaction. In such approaches, sensory features experienced by an agent (robot or human) were evaluated to determine proxemic behavior (Satake et al., 2009). However, such approaches lack the capability of perceiving random human behavior.

Proxemics between two persons depends on the current behavior of the two as well as the context of interaction (Takayama & Pantofaru, 2009). Therefore, service robots should be capable of perceiving the behavior of their users and decide proxemics that are appropriate for the current context. Scaling functions have been introduced to alter robots' physical movements based on the proximity to the human subject (Henkel, Bethel, Murphy, & Srinivasan, 2014). Service robots that are capable of approaching customers in shopping malls have been developed (Kanda, Shiomi, Miyashita, Ishiguro, & Hagita, 2009; Satake et al., 2009). These are fixed approaching methods. However, these are not acceptable for a situation where the activities carried out by the users are significantly different from each other by means of movements. Mechanisms to understand the distancing behavior of people with robots based on speech and gaze have already been proposed (Mumm & Mutlu, 2011). A great number of studies have been carried out based on static models of users and contexts of interaction to decide the appropriate proxemics (Dragone, Saunders, & Dautenhahn, 2015; Marquardt & Greenberg, 2012). Many studies have been conducted to identify the user activities and behavior as well (Gaglio, Re, & Morana, 2015; Kanda et al., 2009; Sirithunge, Jayasekara, & Pathirana, 2017; Vitiello, Acampora, Staffa, Siciliano, & Rossi, 2017; Wu, Pan, Xiong, & Xu, 2014). Most of the present systems are capable of deciding the best distance between the user and the robot, although these do not evaluate human behavior to decide the type of conversation to be maintained at any particular moment. Spatial relationships used in human–robot interaction were investigated in (Huttenrauch, Eklundh, Green, & Topp, 2006). Robot's perception upon the physical formation of itself and the human was developed during this approach.

Perceived connection between human and the robot was evaluated in (Rich, Ponsler, Holroyd, & Sidner, 2010; Sanghvi et al., 2011) through comprehension of changes in interaction strategy. A number of physiological factors related to user behavior were evaluated in such systems. Engagement in a certain situation was measured by means of posture and movements, but this evaluation was used for special purposes, not for an HRI scenario in general. Therefore, occasions in which human behavior is monitored before a robot makes a general decision regarding the encounter are rare at present.

This study focuses on analyzing human physiological behavior. Finally, the evaluation of user behavior is used to make decisions regarding approaching and initiating an appropriate conversation with the user. Combining both utterances and proxemics to approach behavior will allow the interaction scenario to follow socially accepted etiquette.

3.3 System Overview

An overview of the presented approach is given in Fig. 3.1. The robot tracks observable human cues through sensory inputs and then processes this data using vision techniques. Required information is extracted, calculated, and analyzed by the Perception Model proposed in the chapter. A decision-making unit makes interaction decisions according to the model. Interaction decisions include the approach path or the direction, mutual distancing, and the type of conversation or the conversational preference which are most appropriate to the scenario. The scenario is analyzed by means of body-based movements adopted by the user.

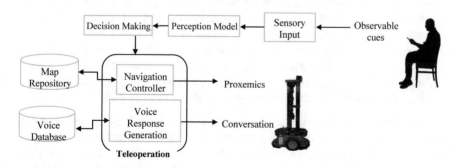

Fig. 3.1 Overview of the system

3.4   Rationale Behind the Approach

The key points that stimulated the study of user responses toward an emotionally intelligent robot and the requirement for situation-awareness in such a system can be summarized as follows.

a. Utilization of service robots in social environments has increased over the past few decades. Therefore, the intelligence of such robots has to be developed to cater to human needs in close encounters. These systems should not cause any disturbance to humans through their behavior.
b. In most robotic systems, there are more limitations in emotional intelligence than in physical capabilities and efficiency. However, emotional intelligence plays a major role in a social environment. Monitoring human behavior is required before the robot initiates an interaction with a human unless the individual demands a specific service.
c. Humans have to restrict their behavior when a robot lacks the capability to perceive his/her situation adequately. For example, if a robot invades the personal space of a human while approaching that individual, he has to limit his movements so that he will not hit the robot. Such inaccuracies could be avoided if the robot perceived its surroundings in a human-friendly manner.
d. In real-world applications, there are occasions on which humans seek association. Such situations which are favorable for a robot to initiate an interaction can be predicted through the situation awareness of a robot. This will further enhance the relationship between humans and robots. In addition, robots will be accepted by their users for a longer duration.
e. The reliability of the robot in terms of behavior can be improved by gaining emotional intelligence. Hence, the occasions in which humans get disturbed by verbal or proxemic behavior of a robot can be reduced. In addition, perception of human behavior has many applications other than social robotics. Caretaker robots, healthcare robots, rescue robots, and robots deployed in extreme conditions such as disaster sites can make use of this capability to track and identify humans and predict their intentions, to complete the robots specified task accordingly. Furthermore, it is important to identify human behavior in order to generate the most appropriate and timely responses during emergencies.

Accordingly, the objectives of this work focus on identifying the effect of factors which affect the perception of nonverbal human behavior and human responses toward a situation-aware robot with the above-mentioned capability. The authors achieved these objectives by conducting a user study by means of a wizard-of-oz experiment.

3.4.1   Theoretical Approach

Our psychological state is what guides our behavior. This psychological state is displayed to the outside through both verbal and nonverbal behavior. Body-based behavior is the result of cognitive processes developed in the human brain. Behavior can be analyzed as an interplay of mental states and actions. Put simply, thoughts and emotions provoke actions. In addition, cognitive elements such as facial expressions and verbal phrases fall under "behavior" that includes both verbal and nonverbal aspects. Furthermore, brain activities such as internal states of mind, cognition, and emotions are responsible for one's actions. Proper interaction between brain activity and actions, not only makes a person perceive the world around him, but also enables the others around to perceive him. Behavioral responses can be either voluntary or sometimes involuntary. Furthermore, many involuntary behaviors are nonverbal. There are many psychological theories behind both voluntary and involuntary human behavior. Out of them, the theory of planned behavior and the theory of reasoned action provide a reasonable and a justifiable basis for reasoning out human behavior under various environmental conditions and circumstances.

3.4.2   Theory of Planned Behavior

According to the theory of planned behavior, one's beliefs and behavior are linked. This causes reasoned actions based on a controlled or restricted behavior. An individual's intention or expectation of a certain behavior at a specific time or place, etc., are based on the rules that humans follow. These regulations take three forms: behavioral, normative, and control. Furthermore, the theory of planned behavior comprises six constructs which collectively present the actual control of a person over the behavior (Ajzen, 1985). These can be listed as attitudes, behavioral intention, subjective norms, social norms, perceived power, and perceived behavioral control.

A human–robot scenario can be explained using the same concept, as follows. A human's intention within the environment changes depending upon the factors that prevail in the surroundings. This perception will be based on the individual's beliefs as well. Hence, the user's reaction to an interaction initiated by the robot will take different forms in various scenarios. During the study, user responses were recorded and analyzed. Responses that were most likely to be displayed from a human during a human–robot interaction scenario could be used to evaluate human behavior during the study (Breazeal, 2004). Such observable responses are listed below.

- Gaze—maintaining or returning gaze, e.g., looking at the robot or looking away
- Gestures—using hand gestures, e.g., waving the hand, calling
- Postures–posture changes, e.g., changing from sitting to standing posture
- Utterances—verbal responses, e.g., Hello, How are you doing?
- Movements—random or intentional movements associated with tasks
- Expressions—basic facial expressions, e.g., smile, frown.

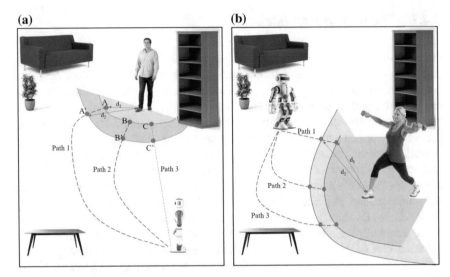

Fig. 3.2 Two different encounters in a social environment are shown. **a** Robot encounters with a relaxed user. The robot has a number of options as regards path and proxemic distances to reach the user. The robot can take paths 1, 2, or 3 to approach from the right, front, or left of the user. Positions that the robot can take after approaching are shown by A, A', B, B', C, and C'. Depending on how close the robot should be, it can acquire A or A', keeping a distance of d_1 or d_2, given that it takes path 1. **b** Robot encounters with a user engaged in an activity. The user has wide movements due to her activity. Hence, the robot has to keep more distance than in the previous situation, when approaching the user

These responses contribute to perceiving and evaluating attitudes, attention, expectations, subjective norms, and perceived behavioral control mainly out of the six constructs of theory of planned behavior. It is assumed that the type of response conveyed by the user changes when the task and other environmental factors change. This fact was evaluated by the human study conducted in the form of a WoZ experiment, and the findings were used for further decision evaluation regarding an interaction.

The basic idea of this study is to evaluate the requirement of providing the robot with the ability to understand situations, or "cognition." Hence, the cognition in situations will relate to the connection between a human's state and his/her behavior. This will facilitate a dynamic interplay of flexibility and adaptation in the robot. This approach is illustrated in Fig. 3.2. In the two scenarios shown, as the user adopts a static behavior in (a), the robot can move closer to the user. In contrast, in (b), the user adopts more dynamic movements. Therefore, the robot is expected to keep a distance so that it will not disturb the user's task during its approach.

3.4.3 Theory of Reasoned Action

The theory explains the relationship between one's attitudes and actions (Sheppard, Hartwick, & Warshaw, 1988). This theory can be used to predict the behavior of a particular human based on the pre-existed attitudes and behavioral intentions of that individual. The theory serves to understand an individual's voluntary behavior. Therefore, intentions and motivation for a certain behavior drive the individual. The individual's decision to engage in a certain behavior depends on the intentions of that individual upon the outcomes of that behavior. Hence, this fact is deployed in exploring the tendencies in human behavior in the presence of the robot used in the wizard-of-oz study.

3.5 Human Behavior

3.5.1 Measuring Behavioral Responses

In order to observe and interpret human behavior, there are various techniques which use an adequate collection of data indicative of movements, cognitive states, and random behavior. These techniques deploy qualitative, quantitative procedures, or both. Experimental setups are based on the assumption that there is a relationship between one's cognitive state and his/her actions. A qualitative study will be based on measurable human features such as facial expressions and staring. As a robot cannot perceive such features without analyzing body geometry, facial contours, etc., we focused on quantitatively studying the motion of the human body for a period of time, before choosing decisions regarding his/her behavior. Hence, we took an effort to translate behavioral observations that the robot made into discrete values and statistical outputs to perceive their meaning. Even so, our experiment includes both qualitative and quantitative approaches since it uses a quantitative measure to extract behavioral information and qualitative measure to evaluate robot's behavior through user feedback. Evaluation of verbal responses will only be possible after starting a conversation. In this scenario, the robot requires to evaluate human behavior before an interaction. Therefore, only nonverbal cues were selected to evaluate the situation for the appropriateness of an interaction.

3.5.2 Observable Nonverbal Human Behavior

Out of the observable human cues, the following cues were used to analyze a situation. Body-based movements were used as the factors which drive robot's decisions regarding interaction. The most fit variables that can define motions of a human were selected for the study. The robot perceives its environment through qualitative information derived from quantitative information. In the same way, three major features included in bodily movements were used as cues to perceive the situation during this study. These features are mentioned below.

1. Speeds of selected body joints
2. Positioning of body joints
3. Maximum occupied areas in space around the individual.

These features were measured quantitatively to determine emotional cues associated with the movement. Here, the parameters which involve emotional state included the priority given to the task, usage of activity space, and the user's engagement. All the above factors are determined mostly by the task of the individual. Even so, humans adopt both rational and irrational behaviors in the same environment. When ordinary human behavior is considered rational behavior is the influence of emotions or thoughts, while irrational behavior is not. Therefore, there can be illogical movements associated with irrational human behavior. Random movements in the hands and legs are an example of such movements. But these movements last a very short time. Misinterpretation of such behavior is avoided by analyzing the usage of *maximum occupied areas* around the individual over the period of observation.

Body joints considered for monitoring human behavior in this way are shown in Fig. 3.3. These joints include the head, the spine base and elbow, and the wrist and ankle of the right and left sides of the body. Joints which are mostly utilized during a task were tracked during this study. Variables associated with each motion are shown in Fig. 3.3b. These variables are the distance from the reference and joint speeds. Here, the vertical going through the spine base was chosen as the reference to measure distance, as the spine base is least subjected to movements due to the inertia of the human body.

These variables observed throughout the period of observation and are plotted against time in Fig. 3.4. This shows the *speed, fanning,* and *maximum occupied areas* of the joint with the highest speed. The reason for this is that this joint is the one most actively used during the task. Hence, the robot considers adjusting the interaction scenario using these properties associated with the user. As the robot determines proxemic behavior in addition to conversation, analysis of the spatial behavior of the user will play an important role in being situation-cautious.

3.5.3 Decision-Making Criteria

Two decisions, concerning proxemic behavior and conversational preference to suit the occasion, were taken after an analysis of the observed cues. The proxemic behavior includes the approach direction or the path to be followed and the mutual distance between the human and the robot. Conversational preference includes the type of conversation to have with the user. Conversational preferences are categorized according to the length of the conversation.

(a) **(b)**

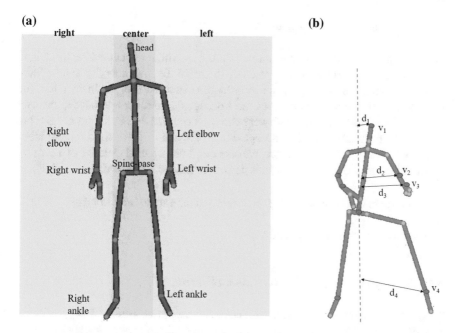

Fig. 3.3 Skeletals extracted from a standing and a seated person are shown. The joints used to track movements are marked and the right, center, and left regions are shown in (**a**). **b** Shows the distances measured from the line going through the spine-base of the person. Distances to head joint, left elbow, left wrist, and left ankle are marked as d_1, d_2, d_3, and d_4. The corresponding joint speeds are marked as v_1, v_2, v_3, and v_4. Similarly, the distances and speeds of joints on the right side are also considered for analysis

(a) **(b)** **(c)**

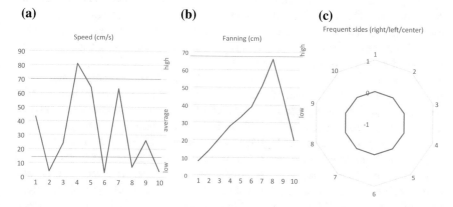

Fig. 3.4 Nonverbal behavioral responses observed during the experiment **a** joint speed (here, the speed of the right wrist is shown) **b** the distance from the vertical drawn through the spine-base joint to the considered joint is marked against the time of observation **c** a radial graph showing the highest occupied area. Areas were denoted as right, center, and the left of the user. Right, center, and left were denoted as 1, 0, and −1, respectively. The responses observed by the robot over 10 s are shown

3.5.3.1 Determining the Proxemics

The approach direction was chosen so that the maximum occupied region (right, center, or left) and was least obstructed by the robot. For example, if the maximum occupied region was the right, the robot approached the user on his/her left and vice versa. If the center was the area most occupied, the robot approached the user on the right. If both right and left were equally occupied during the task, the robot approached from the front. The accepted interpersonal distance for a mutual inter- action is within 1–1.5 m. Therefore, we chose an interpersonal distance of 1.2 m for a motionless user. Otherwise, the interpersonal distance is calculated as in (3.1).

$$\text{Interpersonal distance} = 1.2 + \text{maximum fanning to the front (m)} \qquad (3.1)$$

The units are indicated in meters.

3.5.3.2 Determining the Conversational Preference

Types of conversations robot chose to have with each user were as follows. For the ease of future reference, these types of conversational preferences are shortened.

- No interaction—NI
- Greeting—GRT
- Asking to deliver a service—SER
- Small talk—STLK

The conversational preference was based on the speed of the joints and fanning. This was due to the fact that the maximum occupied area will have no impact on the conversation that the user intends to have with the robot. Joint speeds and fanning were chosen as the demonstrators of engagement in a specific task. These decisions are shown in the decision grid in Fig. 3.5.

The values of these variables were categorized as "high," "average," and "low" for ease of analysis. The exact figures for these boundaries were chosen by trial and error, repeating the experiment a number of times. For example, while reading a newspaper, "low" speed and "high" fanning were observed. Low speed was a result of the still nature due to high user engagement in work. On such occasions, the conversational preference was NI as such behavior was recorded in tasks which required higher user engagement. A similar interaction was preferred in tasks such as doing an exercise or arranging some items on a shelf. These tasks involved "high" fanning as well as "high" joint speeds. For tasks such as desk activities, "low" speed and "low" fanning were recorded. Hence, GRT was selected as the conversational preference as the least disturbance was expected while maintaining friendliness in the encounter. In tasks such as working on a laptop or listening to a song, SER was selected as the conversational preference. This is because users preferred robots to offer a service while they were engaged in an important or highly prioritized activity. In such situations "high" speeds and "low" spanning could be observed. STLK was

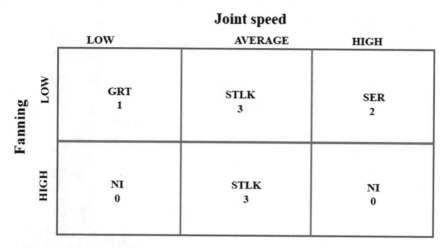

Fig. 3.5 Decision grid: Conversational preference chosen at each occasion are shown

preferred in more relaxing tasks such as resting on a chair and dining, where an "average" speed was observed. Conversational preferences were decided using the criteria given in Fig. 3.5.

3.6 MiRob: The Social Robot

The experiment was conducted in a simulated social environment within the laboratory. The participants in this experiment were students and some outsiders in the age range 23–28 (Mean—25.64, SD—1.68). There were 17 participants, and all were in good mental and physical health and able to make decisions. The effects of gender and the age of the user were not considered within the scope of this study. The participants were given instructions regarding the process and the behavior of the robots in the laboratory upon arrival. Users were allowed to engage in a selected set of tasks as part of the experiment.

The experiment was conducted using a service robot called MIRob (Muthugala & Jayasekara, 2016). The robot is a visually and verbally capable robotic platform which can approach a user, make conversation and manipulate small objects. It is a Pioneer 3DX MobileRobots platform equipped with a Cyton Gamma 300 manipulator and a Kinect camera. The required maps to navigate around the environment were created with Mapper3 Basic software. This platform is equipped with a microphone and a speaker to listen to and respond to users. The robotic platform is shown in Fig. 3.6.

Fig. 3.6 Service robot
platform used in the
experiment: MIRob

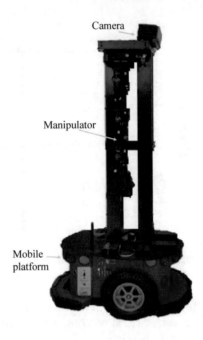

Camera

Manipulator

Mobile
platform

3.6.1 Monitoring Human Behavior

The experiment was conducted to monitor the behavioral responses of users to
robot-initiated interactions. The robot initiated the interaction after considering the
movement-based behavior of its user. Movements of the head joint, spine joint, right
and left wrist, elbow and foot joints were monitored for a period of time. Decisions
of the robot included appropriate proxemic and conversational behavior. Since the
user responses depend on the priority given by the user to his/her current activity,
we selected a set of tasks to evaluate this fact. These set of tasks are as shown below.
Tasks encountered in typical domestic and laboratory environments were selected
for the study. These tasks were as follows.

- Engaged in a desk activity
- Resting on a chair
- Doing an exercise
- Cleaning the floor
- Standing, relaxed
- Having breakfast
- Engaged in laboratory work
- Listening to a song
- Making a phone call.

3.6.1.1   Experimental Setup

The robot was remotely navigated toward a particular participant while he/she was engaged in the task. Before approaching, the robot observed the behavior of the individual for a duration of 10 s. The robot initiated a conversation after approaching that particular user. Each user was allowed to perform at least three tasks listed above. Path planning and navigation tasks of the robot were autonomous while response generation and tracking the user were carried out through teleoperation by the experimenter. Hence, the teleoperator instructed the robot where to approach and what to say. A single user participated in the experiment at a time, and the interaction process was repeated for each participant separately.

As MIRob was monitored by a human operator, its responses were generated with respect to the responses from the user. During the experiment, the responses of the robot to its user included proxemic and verbal cues only. If the situation was not favorable for interaction in the form of displayed nonverbal behavior of the user, the robot was instructed to leave the situation without interaction. In such a situation, it is assumed that the user does not prefer to interact.

The independent variables used in the study were the movements made by the user. The response of the humans toward the interaction initiated by the robot was the dependent variable during the analysis stage.

Joint coordinates of a particular user were measured by the Kinect SDK. All the behavioral changes were identified according to the explained criteria, through the algorithm. The map of the environment was predefined in the simulation. Therefore, the robot navigated to the target positions and its orientation was defined by the operator.

As each participant was asked to perform three out of the nine tasks mentioned above, 51 different scenarios in total were encountered during the experiment. MIRob was allowed to observe each individual once the activity was started. After the evaluation of user behavior according to the model, the approach behavior of the robot was rated by the participant. The user rated the robot's behavior based on the convenience or the discomfort he/she felt as the robot approached. Here, a period of observation of 10 s was determined so that an adequate amount of data was obtained for the analysis and so that the user did not feel the discomfort that might result from a longer observation. Visual information was extracted at a rate of two sets of information per second. Proxemic decisions were taken so that highly engaged areas were least obstructed.

The study included two experiments. One to evaluate user responses for a direct approach behavior with a fixed conversational and proxemic behavior. The second as a situation-based approach behavior adopted by the robot which is referred to as adaptive approach behavior.

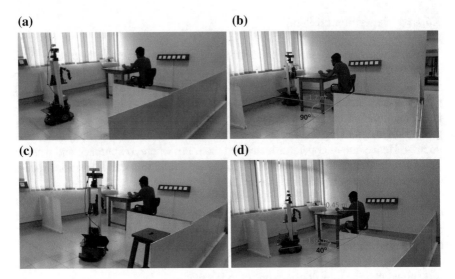

Fig. 3.7 Two occasions from the experiment **a**, **b** Show the direct approach behavior. **c**, **d** Show the adaptive approach behavior after an analysis of the movements made by the user

3.6.2   Experiment 1

The direct approach method was used in the first experiment. Hence, the approach behavior of the robot was not adaptive in this experiment. The robot always reached the user from his/her front, and the conversation was STLK. A mutual distance of 1.2 m was kept between the robot and the user during the interaction process. After this behavior, the user was asked to rate the behavior of the robot in the form of a feedback score (out of 10). This approach scenario is shown in Fig. 3.7a. In Fig. 3.7a, b, the robot approached the user from the front, with a mutual distance of 1.2 m between them. Figure 3.7c, d shows the approach behavior of the robot while the user was dining. As the user's right hand does most of the work when eating, the maximum occupied area was to the user's right. Hence, the robot approached from the left. As the user keeps his hands in the front region, the hands are far from the body. Therefore, the robot measured that distance before maintaining the mutual distance. Hence, the mutual distance was greater than that in Fig. 3.7b.

3.6.3   Experiment 2

This experiment was conducted on two occasions as follows.

Occasion 1: The robot approached the user after an analysis of his/her behavior. Hence, the approach direction, mutual distance, and conversational preference were adapted according to the conclusions derived after the observation. The user was

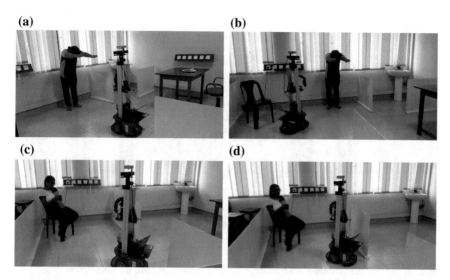

Fig. 3.8 Two occasions in which adaptive approach behavior was implemented. **a** The robot observes the user for a time. The user was exercising. **b** After the observation, the robot approached the user from the right and initiated the conversation. **c** The robot observed the user who was listening to a song. **d** The robot approached the user from the front and initiated interaction

asked to rate the behavior of the robot as in experiment 1. Meanwhile, the participants expected behavior from the robot was also recorded.

Occasion 2: After participating in occasion 1, the user was allowed to take part in experiment 1 again and give a feedback score.

The adaptive behavior of the robot during the two tasks can be observed in Fig. 3.8. Figure 3.8b shows the adaptive approach from the user's right, as the front region is obstructed by his hands. In Fig. 3.8c, d, the robot approaches from the front, as right and left are equally occupied by the user. This decision system is as described in Sect. 4.3.

The three feedback scores received for experiment 1 and occasions 1 and 2 in experiment 2 were compared to find human trends toward each type of interaction.

3.7   User Responses Toward the Robot

A comparison of feedback scores for experiment 1 and occasion 1 in experiment 2 is shown in Fig. 3.9. When the feedback scores are considered, occasion 1 in experiment 2 received a higher feedback score during most of the tasks. The users were impressed with the fact that the robot respected their activity space while approaching them. Another reason for the higher feedback score was the selection of appropriate verbal interaction. In tasks such as desk activity and resting on a chair, where fewer bodily

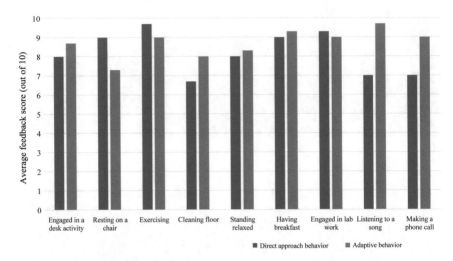

Fig. 3.9 A comparison between the average feedback scores received for the behavior of the robot during the direct approach only versus adaptive behavior after comparing both the situations; direct approach and adaptive behavior

motions could be observed, users preferred shorter conversations. This preference remained the same for tasks such as exercising and cleaning the floor, where speedy and wide motions could be observed. Motions were at the two ends (very fast or very slow) when the users were highly engaged in their task. Motions of average speed and width could be observed for ordinary tasks where user engagement is not very high. In such scenarios, the robot used conversations such as SER or STLK. In the tasks: "resting on a chair," "exercising," and "engaged in laboratory work," the second experiment received a lower feedback score and the difference was considerable in "resting on a chair." The reason behind this trend was that users preferred to talk to the robot while "resting." However, the conversational preferences during experiment 1 and occasion 2 in experiment 2, were STLK and GRT, respectively. Hence, it can be observed that there is a requirement to comprehend the complete task by the robot, in addition to user behavior.

A comparison of feedback scores for occasion 1 and 2 in experiment 2 is shown in Fig. 3.10. After users were allowed to encounter the proxemic aware approach behavior of the robot in experiment 2, the feedback scores for the experiment decreased considerably. It could be observed that the users were impressed with the situation awareness of the robot. This trend had an anomaly in "resting on a chair" as in the previous comparison. Again, the reason for this is that users preferred a longer conversation, but movement-based analysis suggested an interaction limited to just a few words.

In experiment 2, a confusion matrix was created in order to evaluate user satisfaction regarding the evaluation of the robot. This is shown in Table 3.1. The experiment was designed to meet the assumptions mentioned below. In the confusion matrix,

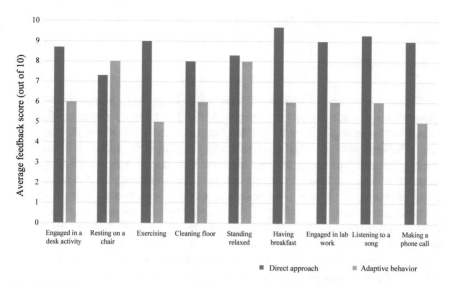

Fig. 3.10 Two occasions from experiments **a**, **b** show the direct approach behavior. **c**, **d** Show the approach behavior after an analysis of the movements made by the user

Table 3.1 An analysis of the confusion matrix generated from the results of experiment 2

Observed Cappa	Standard error	0.95 confidence interval	
0.722		Lower limit	Upper limit
Method 1	0.0841	0.5572	0.8868
Method 2	0.082	0.5613	0.8827
0.9382	Maximum possible unweighted kappa, given the observed marginal frequencies		
0.7696	Observed as proportion of maximum possible		

Cohens kappa value was calculated with linear weighting. Used weights were equal in the confusion matrix. Since the kappa values for the systems were over 0.6, it can be proved that the systems were working properly and were capable of making good judgments that substantially agree with those of the user. Hence, the adaptive approach scenario can effectively replace the direct approach scenarios utilized by most of the robotic platforms.

3.8   Conclusions and Implications

3.8.1   Conclusions

Intelligent decision-making prior to interaction and perceiving human-involved situations have become demanding aspects of HRI. In this study, we report findings related to human responses to a robot-initiated interaction with situation awareness. These findings were used to decide parameters regarding the approach behavior of a robot, namely proxemics, interpersonal distance, and utterances. This approach behavior of the robot facilitates interaction after evaluating the physiological behavior of the human for a time. User observation by a robot was used to rearrange the conversation, proxemics, and the path followed to approach the user. This study evaluates observable human cues such as body-based movements as demonstrators of the degree of engagement in a human–robot encounter. For comparison purposes, responses from humans during an interaction initiated without any pre-concerns of the situation were analyzed to support the major experiment. The experiment was intended to explore the requirement of robot's situation-awareness in social domains. Interesting facts regarding user expectations on proxemics and conversational preferences were revealed during the study.

The results confirm that the capability of the robot to perceive human behavior makes an impact on sustaining interaction between a robot and its user. Moreover, there are still other factors influencing humans which may affect their degree of engagement in a certain situation, such as their internal state of mind, traditions, beliefs, and norms followed by that human. However, as these aspects are non-observable, there is still a huge space for novel methods to evaluate similar psychophysiological behavior in humans. This proved to be helpful in determining the appropriate approach behavior for the robot when a human was encountered. As a robot evaluates human behavior in all physical, social, and emotional aspects according to this experiment, the robot was able to behave with situation awareness. Hence, user dissatisfaction that occurs due to inappropriate behavior of the robot, such as disturbing the user without invitation or invading the personal space, has been eliminated. This fact is validated using the results of the experiment and suggestions to improve future robotic models for initiating interaction with humans are stated. The experimental analyses have shown the possibility of embedding nonverbal-behavior-based situation awareness into a social robot. The empirical results of this study can be used to explore the human preferences for a robot with situational awareness. Hence, the guidelines below are proposed based on the findings of the study.

3.8.2   Implications for Theory

The findings of the study were mostly based on the assumption that people prefer similar proxemic rules when interacting with robots as they do while having interactions with other humans. Several factors which influence personal space, such as gender, previous experience, and familiarity with the robot, were not considered within the context of this experiment. As all the participants were known to the authors, the personalities of individuals (introversion/extroversion) might have an effect upon the approach behavior. Such psychological parameters could not be measured by this model. As a matter of fact, this experiment evaluates a selected number of human psychophysiological cues of a human. Therefore, this proposed system could not replicate all parts of the HHI into the HRI scenario. However, we were able to identify several relationships between preferred robot behavior and demonstrated cues which hold true for the HRI scenario as well. Furthermore, human cues are sometimes deceiving. Humans use ironic responses in certain situations, and perception of such behaviors is still difficult for a robotic platform. For example, differentiating between a smile intended for sarcasm and a smile-showing affection is still difficult for a robot.

Human tendencies during the two experiments show that humans were delighted with the proxemic-cautious behavior of robots. Understanding human behavior and act accordingly can create a lively personality within the robot. Hence, the perception of the robot as a machine will be transformed into an idea of a robot as a creature.

In the experiment, we simulated a prototype of a social environment. We believe that there will be other factors which could be influential on human preferences. Such factors related to the environment, the robot, and the user should also be considered in the development of cognitive robots. In addition, patterns of speech, physical appearance and personality traits of the robot might influence the acceptance of the robot. Therefore, the evaluation of such factors is also important. The conceptual design of a robot's cognition must consider these factors before implementing them in general applications.

The findings of the study were based on the assumption that reaction of humans toward to HRI is similar to that toward HHI. However, there are communities that have a different, most of the time, negative opinions upon robots and such incidents may create complications in perceiving such encounters. It is important to overcome these limitations to define a rather completed physical, behavioral, and emotional state of a human–robot encounter.

3.8.3   Implications for Design

Based on these findings, we can state that this evaluation offers better means of determining an appropriate approach behavior for a robot to initiate interaction with a user based on nonverbal cues displayed. As users prefer not getting to be disturbed or disrupted by the behavior of robots, the first design guideline proposed by the study is improving the ability to respect the personal space of an individual based on both physical and emotional aspects. It is vital to design social robots that do not violate user expectations. This will gently allow the user to establish the idea of accompanying a social companion when using a machine in his work. The feature of maintaining appropriate approach behavior can be used as a form of etiquette for social robots deployed in human environments. Therefore, this can be introduced as the second design guideline for robots.

The third guideline proposed by the study is to consider human cues as much as possible. It is important to get a clear presentation of the situation and perceiving that situation. In addition, considering more factors related to the surroundings as well as the user, can be a plus for better situation awareness. Therefore, it is necessary to equip the robot with hardware having adequate capabilities to do this. Reaction to the presence of the robot may differ from user to user. Therefore, the perception of different forms of psychophysiological human behavior will help make a robot companion situation-aware. This should be considered the fourth guideline derived from the study.

Often, movements of humans are generated as a result of their tasks. Recognition of the task is an important aspect of determining proxemics behavior. It is important in order to predict the fanning of body parts and objects involved in the task, as well as to determine the appropriateness of a conversation in the particular situation. This will be the fifth design guideline derived from the study.

Acknowledgements The authors would like to acknowledge the participants who volunteered in the study. This work was supported by the University of Moratuwa Senate Research Grant SRC/LT/2018/20.

References

Ajzen, I. (1985). From intentions to actions: A theory of planned behavior. In *Action control* (pp. 11–39). Berlin: Springer.

Breazeal, C. (2004). Social interactions in HRI: The robot view. *IEEE Transactions on Systems, Man, and Cybernetics, Part C (Applications and Reviews), 34*(2), 181–186.

Dragone, M., Saunders, J., & Dautenhahn, K. (2015). On the integration of adaptive and interactive robotic smart spaces. *Paladyn, Journal of Behavioral Robotics, 6*(1).

Gaglio, S., Re, G. L., & Morana, M. (2015). Human activity recognition process using 3-D posture data. *IEEE Transactions on Human-Machine Systems, 45*(5), 586–597.

Henkel, Z., Bethel, C. L., Murphy, R. R., & Srinivasan, V. (2014). Evaluation of proxemic scaling functions for social robotics. *IEEE Transactions on Human-Machine Systems, 44*(3), 374–385.

Huttenrauch, H., Eklundh, K. S., Green, A., & Topp, E. A. (2006). Investigating spatial relationships in human-robot interaction. In *2006 IEEE/RSJ International Conference on Intelligent Robots and Systems* (pp. 5052–5059). New York: IEEE.

Kanda, T., Shiomi, M., Miyashita, Z., Ishiguro, H., & Hagita, N. (2009). An affective guide robot in a shopping mall. In *Proceedings of the 4th ACM/IEEE International Conference on Human Robot Interaction* (pp. 173–180). New York: ACM.

Korn, O., Bieber, G., & Fron, C. (2018). Perspectives on social robots: From the historic background to an experts' view on future developments. In *Proceedings of the 11th PErvasive Technologies Related to Assistive Environments Conference* (pp. 186–193). New York: ACM.

Marquardt, N., & Greenberg, S. (2012). Informing the design of proxemic interactions. *IEEE Pervasive Computing, 11*(2), 14–23.

Mead, R., Atrash, A., & Mataric, M. J. (2013). Automated proxemic feature extraction and behavior recognition: Applications in human-robot interaction. *International Journal of Social Robotics, 5*(3), 367–378.

Mead, R. & Mataric, M. J. (2012). A probabilistic framework for autonomous proxemic control in situated and mobile human-robot interaction. In *Proceedings of the Seventh Annual ACM/IEEE International Conference on Human-Robot Interaction* (pp. 193–194). New York: ACM.

Mead, R., & Mataric, M. J. (2017). Autonomous human–robot proxemics: Socially aware navigation based on interaction potential. *Autonomous Robots, 41*(5), 1189–1201.

Mumm, J. & Mutlu, B. (2011). Human-robot proxemics: Physical and psychological distancing in human-robot interaction. In *Proceedings of the 6th International Conference on Human-Robot Interaction* (pp. 331–338). New York: ACM.

Muthugala, M. A. V. J. & Jayasekara, A. G. B. P. (2016). MIRob: An intelligent service robot that Learns from interactive discussions while handling uncertain information in user instructions. In *Moratuwa Engineering Research Conference (MERCon), 2016* (pp. 397–402). New York: IEEE.

Obaid, M., Sandoval, E. B., Zotowski, J., Moltchanova, E., Basedow, C. A., & Bartneck, C. (2016). Stop! That is close enough. How body postures influence human-robot proximity. In *2016 25th IEEE International Symposium on Robot and Human Interactive Communication (RO-MAN)* (pp. 354–361). New York: IEEE.

Papadopoulos, F., Küster, D., Corrigan, L. J., Kappas, A., & Castellano, G. (2017). Do relative positions and proxemics affect the engagement in a human-robot collaborative scenario? *Interaction Studies, 17*(3), 321–347.

Rantanen, T., Lehto, P., Vuorinen, P., & Coco, K. (2018). The adoption of care robots in home care—A survey on the attitudes of Finnish home care personnel. *Journal of Clinical Nursing, 27*(9–10), 1846–1859.

Rich, C., Ponsler, B., Holroyd, A., & Sidner, C. L. (2010). Recognizing engagement in human-robot interaction. In *2010 5th ACM/IEEE International Conference on Human-Robot Interaction (HRI)* (pp. 375–382). New York: IEEE.

Rossi, A., Dautenhahn, K., Koay, K. L., & Saunders, J. (2017). Investigating human perceptions of trust in robots for safe HRI in home environments. In *Proceedings of the Companion of the 2017 ACM/IEEE International Conference on Human-Robot Interaction* (pp. 375–376). New York: ACM.

Sanghvi, J., Castellano, G., Leite, I., Pereira, A., McOwan, P. W., & Paiva, A. (2011). Automatic analysis of affective postures and body motion to detect engagement with a game companion. In *Proceedings of the 6th International Conference on Human-Robot Interaction* (pp. 305–312). New York: ACM.

Satake, S., Kanda, T., Glas, D. F., Imai, M., Ishiguro, H., & Hagita, N. (2009). How to approach humans? Strategies for social robots to initiate interaction. In *Proceedings of the 4th ACM/IEEE International Conference on Human Robot Interaction* (pp. 109–116). New York: ACM.

Sheppard, B. H., Hartwick, J., & Warshaw, P. R. (1988). The theory of reasoned action: A meta-analysis of past research with recommendations for modifications and future research. *Journal of Consumer Research, 15*(3), 325–343.

Sirithunge, C., Jayasekara, B., & Pathirana, C. (2017). Effect of activity space on detection of human activities by domestic service robots. In *Region 10 Conference, TENCON 2017–2017 IEEE* (pp. 344–349). New York: IEEE.

Stark, J., Mota, R. R., & Sharlin, E. (2018). Personal space intrusion in human-robot collaboration. In *Companion of the 2018 ACM/IEEE International Conference on Human-Robot Interaction* (pp. 245–246). New York: ACM.

Takayama, L. & Pantofaru, C. (2009). Influences on proxemic behaviors in human-robot interaction. In *IEEE/RSJ International Conference on Intelligent Robots and Systems, 2009. IROS 2009* (pp. 5495–5502). New York: IEEE.

Vitiello, A., Acampora, G., Staffa, M., Siciliano, B., & Rossi, S. (2017). A neuro-fuzzy-Bayesian approach for the adaptive control of robot proxemics behavior. In *2017 IEEE International Conference on Fuzzy Systems (FUZZ-IEEE)* (pp. 1–6). New York: IEEE.

Walters, M. L. (2008). *The design space for robot appearance and behaviour for social robot companions*. Ph.D. thesis.

Walters, M. L., Dautenhahn, K., Te Boekhorst, R., Koay, K. L., Syrdal, D. S., & Nehaniv, C. L. (2009). An empirical framework for human-robot proxemics. In *Proceedings of New Frontiers in Human-Robot Interaction*.

Walters, M. L., Syrdal, D. S., Dautenhahn, K., Te Boekhorst, R., & Koay, K. L. (2008). Avoiding the uncanny valley: Robot appearance, personality and consistency of behavior in an attention-seeking home scenario for a robot companion. *Autonomous Robots, 24*(2), 159–178.

Wiltshire, T. J., Lobato, E. J., Garcia, D. R., Fiore, S. M., Jentsch, F. G., Huang, W. H., & Axelrod, B. (2015). Effects of robotic social cues on interpersonal attributions and assessments of robot interaction behaviors. In *Proceedings of the Human Factors and Ergonomics Society Annual Meeting* (Vol. 59, pp. 801–805). Los Angeles, CA: Sage.

Wiltshire, T. J., Lobato, E. J., Velez, J., Jentsch, F., & Fiore, S. M. (2014). An interdisciplinary taxonomy of social cues and signals in the service of engineering robotic social intelligence. In *Unmanned Systems Technology XVI* (Vol. 9084, p. 90840F). International Society for Optics and Photonics.

Wiltshire, T. J., Lobato, E. J., Wedell, A. V., Huang, W., Axelrod, B., & Fiore, S. M. (2013). Effects of robot gaze and proxemic behavior on perceived social presence during a hallway navigation scenario. In *Proceedings of the Human Factors and Ergonomics Society Annual Meeting* (Vol. 57, pp. 1273–1277). Los Angeles, CA: Sage.

Wu, H., Pan, W., Xiong, X., & Xu, S. (2014). Human activity recognition based on the combined SVM & HMM. In *2014 IEEE International Conference on Information and Automation (ICIA)* (pp. 219–224). New York: IEEE.

Chapter 4
The Role of Gesture in Social Telepresence Robots—A Scenario of Distant Collaborative Problem-Solving

Dimitra Anastasiou, Christoph Stahl and Thibaud Latour

Abstract Human–robot interaction is a well-studied research field today; robots vary from tele-operators and avatars to robots with social characteristics. In this review paper, first we present related work on tele-operation, mobile robotic telepresence, and social robots. Then, we focus on the role of gestures and body language in robotics, and more precisely their importance for communication in collaborative settings. In our collaborative setting scenario, we have a group of multiple human users working on collaborative problem-solving around a tangible user interface (TUI). A TUI employs physical artifacts both as "representations" and "controls" for computational media. We have the same situation in a separate spatial location. We extend this specific scenario by having an avatar robot in each one of the two locations which represents remote team members and mirrors their actions, gaze, and gestures. Our goal in this paper is to give an overview of current solutions that provide a sense of being in a different place and to describe our future scenario of having an avatar robot solving a problem on a TUI collaboratively with human users. We present a discussion about technical and social questions related to the acceptance of avatar robots at work considering which properties they should have, to what extent the current state of the art in social robotics is applicable, and which additional technical components need to be developed.

Keywords Service robots · Social robots · Tele-operation and avatar robots · Telepresence robots · Mobile robotic telepresence · Tangible user interfaces (TUI)

D. Anastasiou (✉) · C. Stahl · T. Latour
Luxembourg Institute of Science and Technology, 5, avenue des Hauts Fourneaux, Esch/Alzette 4362, Luxembourg
e-mail: dimitra.anastasiou@list.lu

C. Stahl
e-mail: christoph.stahl@list.lu

T. Latour
e-mail: thibaud.latour@list.lu

© Springer Nature Switzerland AG 2019
O. Korn (ed.), *Social Robots: Technological, Societal and Ethical Aspects of Human-Robot Interaction*, Human–Computer Interaction Series,
https://doi.org/10.1007/978-3-030-17107-0_4

4.1 Introduction

As a consequence of globalization, working situations are becoming more systemic and complex than ever. The problems people have to tackle often cannot be solved individually, but require strong collaboration between various experts from different fields. In the meantime, many cities are suffering from the ecological and economic consequences of high traffic volume caused by daily commuting workers and are actively considering the promotion of teleworking and suburban coworking centers. Both situations share a common need of effective support for good collaboration between spatially separated teams that, by definition, must be mediated by adequate technologies. In this paper, we discuss the emerging role of telepresence robots to assist with communication and collaboration between employees over a distance.

Videoconferencing systems are widely used today, where participants use their individual personal device, such as a laptop computer or smartphone. The quality of the experience depends on how the teams are distributed. In situations where all participants are distributed in different locations, all have to overcome the same limitations to express themselves by means of camera and display. On the one hand, a situation where most participants are collocated and one or a few are distant yields a significant asymmetry in the collaboration experience between participants. In this case, it is easier for the collocated participants to follow the conversation and be concentrated, since they can turn their heads toward the speaker and can signal their willingness to make a statement, for example, by raising their hand. For the isolated distant participant, the same gesture on a screen might not be recognized as easily as in a face-to-face meeting. The same situation occurs at home in private communications: if a remote caller, e.g., the grandmother, uses videoconferencing software on her mobile device to communicate with the family. She must ask the local family members to hand each other the phone or tablet to establish eye contact with them or produce some sound that might not be noticed easily by the group party. From the isolated participant, another drawback appears, especially when the group party is large. While wide-angle lenses are available, their use requires very large screens and ultra-high definition resolutions to show multiple faces at acceptable quality. Another issue with current videoconferencing systems is the interaction with content for collaboration. While most systems support the sharing and editing of documents, it is difficult to find a suitable arrangement of content and video streams on the display that allows productive work comparable to face-to-face settings, e.g., using pen and paper or a whiteboard. In addition, combining both visibility of people and content simultaneously and symmetrically requires sophisticated technologies, especially when interacting with non-digital content.

In this paper, we summarize arguments from the literature regarding the importance of formal and informal gesture and body language for communication and collaboration. We argue that humanoid telepresence robots with a capability to express gestures could play an important role in teleworking and collaboration over a distance, which will gain importance for reducing the need for mobility and traffic. Based on current work in our research group on collaborative problem-solving on

TUIs, we sketch a scenario where a social avatar is replacing a human in two spatially separated spaces, where two groups solve a collaborative problem. We argue that social robots go beyond a traditional interface by enriching both interhuman communication and (non-digital) content interaction through the distance, playing the role of social mediators with rich gestures, postures, and manipulations.

This paper is laid out as follows: We begin with related work on robotics, and specifically on service robots (Sect. 4.2.1), social robots (Sect. 4.2.2), tele-operation and avatar robots (Sect. 4.2.3), and mobile robotic telepresence (Sect. 4.2.4). Section 4.3 highlights the role of gesture in human–robot interaction, and in Sect. 4.4, we present a scenario where a robot is replacing a human in solving a collaborative problem on a TUI, including various aspects, such as collaborative problem-solving, gesture analysis, and design and technical challenges. We conclude the paper with a discussion and future work in Sect. 4.5.

4.2 Related Work

In the next subsections, we refer to related work concerning service robots (Sect. 4.2.1), social robots (Sect. 4.2.2), tele-operation and avatar robots (Sect. 4.2.3), as well as mobile robotic telepresence (Sect. 4.2.4).

4.2.1 Service Robots

The International Organization for Standardization defines a "service robot" as a robot "that performs useful tasks for humans or equipment excluding industrial automation applications" (ISO 8373). According to ISO 8373, robots require "a degree of autonomy," which is the "ability to perform intended tasks based on current state and sensing, without human intervention."

According to the executive summary of World Robotics 2018 about service robots, the total number of professional service robots sold in 2017 rose considerably by 85% to 109.543 units up from 59.269 in 2016. The main applications of professional service robots are logistic systems, defense applications, and public relation robots. The total number of service robots for personal and domestic use increased by 25% to about 8.5 million units in 2017. The value raised by 27% to reach US$ 2.1bn. Those robots are mainly in the domestic (household) domain, which includes vacuum and floor cleaning, lawn-mowing robots, and entertainment and leisure robots, including toy robots, hobby systems, education, and research. The executive summary of World Robotics 2018 forecasts a rise in the annual supply of industrial robots from 376,000 units in 2017 to 621,000 units in 2021. Research on humanoid service robots is still in its infancy, and in this paper, we try to explore what robots already exist and how they can be used in the future in a working environment to connect users working collaboratively.

4.2.2 Social Robots

The main goal of social robots is to allow humans to socially interact with them in a similar manner to the social interaction between humans. A definition of social robots by Dautenhahn and Billard (1999) follows:

> Social robots are embodied agents that are part of a heterogeneous group: a society of robots or humans. They are able to recognize each other and engage in social interactions, they possess histories (perceive and interpret the world in terms of their own experience), and they explicitly communicate with and learn from each other.

The history and many examples of social robots are presented in Fong, Nourbakhsh, and Dautenhahn (2003). Fong et al. made a survey of the so-called socially interactive robots and presented a taxonomy of design methods and system components used to build such socially interactive robots. Fong et al. (2003) described robots that exhibit the following "human social" characteristics:

- express and/or perceive emotions;
- communicate with high-level dialog;
- learn/recognize models of other agents;
- establish/maintain social relationships;
- use natural cues (gaze, gestures, etc.);
- exhibit distinctive personality and character;
- may learn/develop social competencies.

In the context of social learning in robotics research, Breazeal and Scassellati (2002) introduced the social and task-oriented aspects of robot imitation. Presented the role of emotion and expressive behavior in regulating social interaction between humans and expressive anthropomorphic robots. Breazeal (2000) designed *Kismet*, which is capable of learning from people from social interactions and learns to be more socially sophisticated in the process.

Regarding design frameworks for social robots, Bartneck and Forlizzi (2004) proposed a design-centered framework viewing social robots as products that facilitate co-experience and social interaction. They proposed the following properties: form, modality, social norms, autonomy, and interactivity. They stated that social robots should recognize, respond to, and employ where possible all modalities that humans naturally use to communicate. These include verbal cues, such as speech, intonation, and tone of voice, and nonverbal cues, such as gesture, posture, and stance, among others. They provided the following definition of the social robot:

> [A] social robot is an autonomous or semi-autonomous robot that interacts and communicates with humans by following the behavioral norms expected by the people with whom the robot is intended to interact.

Dautenhahn (2007), speaking about "socially intelligent" robots, set evaluation criteria that the application domain and the nature and frequency of contact with humans have to take into account in order to decide which social skills are required for a robot.

Fig. 4.1 Spectrum of evaluation criteria to identify requirements on social skills for robots

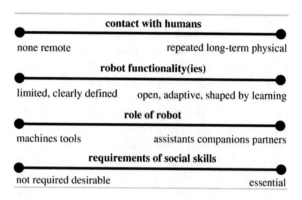

contact with humans

none remote   repeated long-term physical

robot functionality(ies)

limited, clearly defined   open, adaptive, shaped by learning

role of robot

machines tools   assistants companions partners

requirements of social skills

not required desirable   essential

As Fig. 4.1 presents, contact with humans ranges from no remote contact (e.g., for robots operating in deep-sea environments) to long-term, repeated contact potentially involving physical contact, as is the case with assistive robotics. The functionality of robots ranges from limited, clearly defined functionalities (e.g., vacuum cleaning robots) to open, adaptive functions that might require robot learning skills (e.g., applications such as robot partners, companions, or assistants). Depending on the application, domain requirements for social skills vary from not required (e.g., robots designed to operate in areas spatially or temporally separated from humans, such as on Mars, or patrolling warehouses at night) to possibly desirable (even vacuum cleaning robots need interfaces for human operation) to essential for performance/acceptance (service or assistive robotics applications).

As far as the evaluation or analysis of humanoid robots is concerned, Eyssel and Kuchenbrandt (2012) investigated effects of social category membership on the evaluation of humanoid robots. They used anthropomorphism to rate and document that people even apply social categorization processes and subsequent differential social evaluations to robots. Meltzoff, Brooks, Shon, and Rao (2010) examined social robots as psychological agents for 12- to 18-month-old infants. They used eye gaze and particularly gaze following to test whether infants would treat the robot as a psychological agent that could see. Indeed, infants want to look at what the robot is seeing and thus shift their visual attention to the external target.

4.2.3   Tele-operation and Avatar Robots

Tele-operators are remote-controlled manipulators, e.g., robotic hands, that were conceived in the 1950s to be used by operators to work in remote places that are either too dangerous (high radiation levels in nuclear plants) or too distant (planet Mars) to allow for the presence of humans.

The *real haptics avatar robot* with a general-purpose arm (GPA) developed by Takahiro Nozaki at Keio University (Fukushima et al., 2017) can recognize the shape and composition of materials (soft or hard), position objects in 3D space, and manipulate them according to real-time instructions from a remotely located user, where the arm acts as a real-time avatar. Instead of telepresence or tele-operation, Keio University uses the term **avatar**, which in Hinduism refers to the material appearance or incarnation of a deity on earth.

Another recent avatar robot is the *third-generation humanoid robot T-HR3* developed by Toyota Motor Company (Toyota, 2017), which is controlled from a *Master Maneuvering System* that allows the entire body of the robot to be operated instinctively with wearable controls that map hand, arm, and foot movements to the robot, and a head-mounted display that allows the user to see from the robot's perspective. The robot's 29 body parts mirror all movements of the human operator, getting very close to the idea of an avatar body as described in the 2009 motion picture "Avatar" directed by James Cameron. Partially, such technology could also be used for prosthetic hands, as described in Fukushima et al. (2017).

4.2.4   Mobile Robotic Telepresence

While social robots aim to interact with humans in a natural way, the rationale behind robotic telepresence is to use robots as avatars to improve communication between humans over a distance. Minsky (1980) introduced the term **telepresence** as sense of "being there" and expected the biggest challenge to lie in the coupling of artificial devices with the sensory mechanisms of human organisms. Today, almost 40 years later, telepresence is still a subject of research. Nowak and Biocca (2003) defined three dimensions of presence: (i) telepresence, (ii) copresence, and (iii) social presence. The concept of mediated presence, or **telepresence**, is usually defined succinctly as the sensation of "being there" in the virtual or mediated environment (Heeter, 1992). **Copresence** "renders persons uniquely accessible, available, and subject to one another" (Goffman, 1966). Short, Williams, and Christie (1976) provided the theoretical background to the concept of social presence. They defined **social presence** as "the degree of salience of the other person in the interaction and the consequent salience of the interpersonal relationships."

In the following paragraphs, we focus specifically on mobile robotic telepresence. Telepresence robots have been used in different domains, such as the workplace, the medical and domestic domains, and education. We begin with the definition of **mobile robotic telepresence** (MRP) given by Kristoffersson, Coradeschi, and Loutfi (2013) and then provide specific examples:

> Mobile robotic telepresence (MRP) systems are characterized by a video conferencing system mounted on a mobile robotic base. The system allows a pilot user to move around in the robot's environment. The primary aim of MRP systems is to provide social interaction between humans. The system consists of both the physical robot (sensors and actuators) and the interface used to pilot the robot.

- A **pilot user** is a person who remotely connects to the robot via a computer interface. The pilot who is embodied in the MRP system can move around in the environment where the robot is located and interact with other persons.
- A **local user** is the user that is situated in the same physical location as the robot. Local users are free to move around while interacting with the pilot user who is visiting them via the robot.

Kristoffersson et al. (2013) also gave an overview of various MRP systems and their experience in different application domains, such as research, healthcare, school, and the office. All systems use a display and speaker to show the face of the pilot user, and a camera and microphone to transmit the response of a local user. Most critical for the usability of the robots is their navigation interface. In the simplest case, the robot is piloted using the keyboard, which requires the full attention of the user and can distract from the communication. Most systems provide a secondary, down-facing camera to help to avoid obstacles on the ground. More advanced robots use sensors, such as laser-range scanners, to implement driving assistance and autonomous driving to destinations via pointing on a touch screen. Some systems also provide a remote-controllable laser pointer for deictic gestures, such as the telepresence robots *MantaroBot* (MantaroBot, 2016) and the *GestureMan* by Kuzuoka et al. (2000).

Specifically with regard to MRP use in the workplace, Lee and Takayama (2011) run a study where the *Texai Alpha* (Willow Garage, 2012) prototype connected remote coworkers who live approximately 1800 miles apart. The coworkers used the MRP for 2–18 months. Lee and Takayama (2011) showed how remotely controlled mobility enabled remote workers to live and work with local coworkers almost as if they were physically there. Figure 4.2 presents the impact of MRP usage on workplace activities based on the coworkers' experience.

Apart from workplace, robots have been also applied in formal education scenarios. Most recently, Cha, Chen, and Mataric (2017) identified several research and design themes that must be addressed to make telepresence a usable technology in K-12 education: communication quality, inclusion, embodiment, interaction, and interface. As far as the challenges related to interaction are concerned, they highlighted that the most telepresence robots cannot cope with rich, dynamic interactions

Fig. 4.2 Impact of MRP usage on workplace activities (Lee & Takayama, 2011)

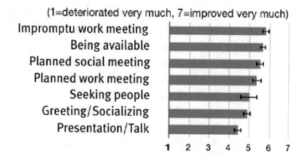

that happen in face-to-face communication. They pointed out that users are unable to utilize many typical nonverbal signals, such as gesture, as this functionality is missing. While facial expression and linguistic cues provide a wide range of information, they are dependent on audio and video quality and the robot's positioning.

4.2.5 Telepresence Robots with Physical Embodiment

In our context of telework and collaboration in office environments, the focus is on sharing and working on digital documents and artifacts. As already mentioned in the introduction, several problems can occur when distributed teams use videoconferencing systems to discuss and edit documents. If a remote user communicates with a local, collocated group through a video connection, the static arrangement of camera and display makes it difficult to follow and participate in the conversation. This effect is known as **presence disparity.** Tang, Boyle, and Greenberg (2005) describe presence disparity as follows: A participant's perception of the presence of a fellow collaborator differs depending on whether that collaborator is physically co-located or remote. This disparity disrupts group collaborative and communication dynamics. They suggest that one of its causes is that consequential communication (i.e., visibility of another's body) between remote participants is inadequate.

According to a recent report in the Steelcase magazine (Steelcase, 2016), presence disparity is more than just a nuisance. It can undermine the benefits of having a diverse, distributed team and hurt their productivity. The overall collaboration experience can easily become unpleasant and taxing, with participants feeling strained physically, cognitively, and emotionally. Meanwhile, as the pace of work has intensified, people often find themselves in a "mixed presence" work mode; i.e., they are physically present in one conversation, while being virtually present in one or more synchronous conversations using an array of technologies to text, chat, email, etc., having as consequence distractions, misunderstandings, misinterpretations, and conflicts. In Sect. 4.3, we will give further details that explain the effect by highlighting the importance of presence and embodiment on conversational awareness.

One current approach to overcoming presence disparity and to achieving a sense of telepresence is the use of dedicated conference rooms that provide multiple displays, supported by microphone arrays, and featuring a symmetrical architecture and furnishing between the local and remote room. Environments, such as the *Polycom RealPresence OTX* solution (Polycom, 2018), achieve an effect as if both groups were sitting together around a table in the same room. Of course, this requires that both parties of a videoconference have access to such a room, which is usually not the case. Moreover, such a family of solutions is hardly usable by nomadic collaborators traveling outside corporate premises.

Seeking more portable and versatile solutions, we concentrate on the recent developments of **telepresence robots**, which have been specifically designed to extend videoconferencing with a physical embodiment. Figure 4.3 presents two examples of full-size MRP robots with physical embodiment. *Double 2* (Double, 2018) from

Fig. 4.3 Examples of MRP (Double, Ava—top) and social robots (QT, Care-O-Bot—bottom)

Double Robotics has a pan and tilt unit and can also adjust the height to match the local user. Another telepresence robot developed recently is *Ava* (Ava Robotics). Ava is an autonomous robot designed to coexist with humans in workplaces and other large spaces.

Apart from these full-size robots, some smaller mobile platforms exist, such as the *PadBot T1* (PadBot), which is similar to a docking station on wheels that is capable of carrying a smartphone over a desktop. *Keecker* (Keecker) is a mobile robot that combines a projector with a camera. It can be used to project a video conversation anywhere at home on a wall.

Other platforms are stationary, yet allow remote users to pan and tilt a mobile device in order to participate in conversations with multiple local users, overcoming presence disparity. *Kubi* (Kubi) is such a robotic stand for the desktop that can pan and tilt a tablet computer. The *Swivl* (Swivl) platform's original purpose was to track persons during video recordings, but the makers also suggest that it can be used by students to attend school from home. *Jibo* (Jibo) embodies an intelligent agent in a non-mobile robotic platform that allows the head to be freely spun and tilted its head, and which shows an eye animation or any other information. It could also probably serve as telepresence robot through apps.

It is notable that the vast majority of the aforementioned mobile robotic telepresence robots have limbs, such as arms and legs. Presumably, their creators see no need for manipulators on a robot that is designed and built to support videoconferencing rather than tele-operation. On the other hand, most autonomous humanoid social robots, such as *Nao* or *Pepper* from Aldebaran Robotics, do not have a display as face, to show the face of a pilot user. Yet, *Pepper* is a humanoid robot capable of recognizing the principal human emotions and adapting its behavior to the mood of its interlocutor. It supports speech recognition and dialog in 15 languages. *Pepper* could potentially support deictic gestures. Apart from *Pepper*, two other promising exceptions for telepresence robots which could potentially support gesturing are the robots *QT* and *Care-O-Bot*. *QT* (Fig. 4.3) is a humanoid social robot from LuxAI, designed as a robot tutor for teaching emotional abilities to children with autism spectrum disorder. *QT* has a TFT display as its face, presenting animated characters with emotions. The robot has 14 degrees of freedom to present upper-body gestures with its arms. The *Care-O-Bot 4* (Mojin Robotics) (Fig. 4.3) features two robotic arms and a head with a touch display. Mojin Robotics mention that they address telepresence and telemanipulation systems through a modular hardware that supports interactivity through body gestures and pointing gestures by a laser pointer. Last but not least, *MeBot* (Adalgeirsson & Breazeal, 2010) is a physically embodied and expressive telerobot. It was designed through an iterative design process (head and neck control, arm and shoulder control, navigation, eye contact). Adalgeirsson and Breazeal (2010) indicated that telepresence technologies, particularly in business collaborative meetings, could benefit from enabling their users to express their nonverbal behavior in addition to simply transmitting audio and video data.

4.3   The Role of Gesture in Human–Robot Interaction

In this section, we focus on gestures and body language, and more precisely their importance for communication in collaborative settings, first generally and then in human–robot interaction (HRI).

According to Mehrabian (1980), communication related to emotional topics comprises the following elements:

$$\text{Total feeling} = 7\% \text{ verbal feeling (spoken words)} + 38\% \text{ vocal feeling (voice, tone)} + 55\% \text{ facial feeling (incl. body language).}$$

Mehrabian (1980) said that, generally speaking, a person's nonverbal behavior has more bearing than his words on communicating feelings or attitudes to others.

When any nonverbal behavior contradicts words or speech, people rely more heavily on actions (distance, eye contact, postures, gestures) to infer another's feelings.

Being able to sense feelings and emotions is, of course, highly important for negotiations, but it is also relevant in meetings with collaborators as concerns are often left unspoken. Mehrabian (1980) also referred to the immediacy principle: People are drawn toward persons and things they like, evaluate highly, and prefer; they avoid or move away from things they dislike. While it is often not possible to physically move, for example, in a meeting, most people still use abbreviated forms of approach; they lean forward or lean back according to their level of interest in a speaker.

The use of videoconferencing systems can be critical in this respect, since users may behave differently in front of a laptop than in a physical meeting, due to restrictions of the camera angle, microphone, etc. Mobile telepresence robots can help to overcome the limitations to a certain extent. They allow the head unit to pan and tilt, so they enable a pilot user to establish eye contact with local people. Furthermore, they can be moved by the pilot, adding a spatial dimension that could hypothetically express emotions through immediacy and proximity. However, what is missing in social robots nowadays, however, is a capability to express facial expressions, hand gestures, and body posture, in general.

According to Mehrabian's 7/38/55 rule where gesture outweighs words, lacking such capability seriously limits the quality of communication and collaboration. Baker, Greenberg, and Gutwin (2001) gave heuristics for the purpose of discovering problems in shared visual work surfaces for distance-separated groups. Among other issues, they highlighted the importance of gesture and body language in collaborative work settings and stated the following requirements:

i. A system must provide the means for **intentional verbal communication**: This is usually implemented through voice calls, instant messaging, emails, and video channels.
ii. A system must provide the means for **intentional and appropriate gestural communication.** People use the following types of intentional gestures to support the conversation:

- Illustration emphasizes speech.
- Emblems replace words with actions.
- Deictic references are often used in combination with speech to identify objects by pointing at them.

For a gesture taxonomy based on philological foundations, see McNeill (1992). It is important that a groupware system maintains the relationship between objects and the voice communication. Besides video images, gestures are often embodied as telepointers that make interactions with the workspace visible to all participants, for example, by multiple cursors in a document. In virtual reality environments, avatar characters are used to embody users, including their location, orientation, and actions.

iii. A system must provide the means for **unintentional (consequential) communication of an individual's embodiment**. Position, posture, and movements of the head, arms, hands, and eyes unintentionally "give off" information, which is picked up by others. Unintentional body language can be divided into two categories:

- Actions coupled with the workspace (e.g., gaze awareness);
- Actions coupled to conversation (e.g., head nods, eye contact, or gestures emphasizing talk, intonation, pauses).

These visual and verbal cues provide conversational awareness that helps people maintaining a sense of what is happening in a conversation.

Gesturing also facilitates group communication, since pointing or motioning toward a shared object during a discussion provides a clear spatial relationship to the object for the gesturer and the group members; see Bekker, Olson, and Olson (1995). Björnfot and Kaptelinin (2017) recently conducted an empirical study to explore design solutions for providing telepresence robots with deictic gesturing capabilities to improve the communication between the pilot and local users. Anastasiou and Stahl (2012) ran empirical studies to examine the modality of gesture in communication between a human user and an autonomous robotic wheelchair, and in controlling devices in an Ambient Assisted Living laboratory. In these studies, Anastasiou and Stahl (2012) asked the users sitting in the wheelchair to act as in a real-life everyday scenario of controlling devices available at home, such as closing and opening an automatic sliding door, turning a ceiling light on and off, raising and lowering the position of the bed's head- and footrest. They found that the swipe gesture type was the most prevalent gesture type, as it was used both for opening/closing the door and for turning on/off the ceiling light. It was followed by raising and lowering the arm to raise and lower the bed's head- and footrest accordingly. In fact, in the tasks concerning the door and the bed, there were "standard" gesture types that were performed by the majority of the participants, while big gesture variation was shown in turning on/off the light.

4.4 Scenario of a Telepresence Robot Around Tangible User Interfaces

This section is about sketching an application scenario based on our work concerning collaborative problem-solving on TUIs in multitouch tabletop displays.

This scenario requires research from different disciplines, such as social group detection, gesture recognition, and dialog management, but also disciplines related to conventional robotics challenges, such as navigation, mapping, and motion planning.

We have seen in the previous section that intentional and consequential communication can be even more important than verbal communication in certain situations.

Now, we will discuss how anthropomorphic telepresence robots could enhance collaboration in office work scenarios by adding embodied gesture and body language to the communication channel. Some applications of such robots include (i) collaborative meetings at workplace with remote colleagues, (ii) collaborative skill tests in formal education settings or in large-scale surveys such as PISA (Program for International Student Assessment), and (iii) smart offices with geographically dispersed colleagues.

4.4.1   Collaborative Problem-Solving on Tangible Interfaces

Typical workspaces are usually oriented either horizontally, such as desktops and tabletop displays, or vertically, such as smartboard, whiteboard, or projections on walls. The Human Dynamics in Cognitive Environments (HDCE) research unit at the Luxembourg Institute of Science and Technology (LIST) has been working for many years with tabletop displays that use tangible objects as the input modality, following the TUI paradigm that was established by Ullmer and Ishii (2000) as follows:

> [TUIs] give physical form to digital information, employing physical artifacts both as 'representations' and 'controls' for computational media.

TUIs further provide "tangible representations to digital information and controls, allowing users to quite literally grasp data with their hands" (see Shaer & Hornecker, 2010).

According to Hornecker and Buur (2006), TUIs are the optimum medium for collaborative problem-solving, as they enhance social and contextual interactions, such as collaboration. Hornecker and Buur (2006) designed a framework that contributes to understanding the social user experience of tangible interaction: tangible manipulation, spatial interaction, embodied facilitation, and expressive representation.

Our unit developed the *COPSE* (COllaborative Problem-Solving Environment) framework (Maquil, Tobias, Anastasiou, Mayer, & Latour, 2017) for the rapid implementation of TUI-based applications, mostly with an collaborative learning and testing objective. COPSE (see Fig. 4.4) is a novel and unique software framework for instantiating *microworlds* as collaborative problem-solving activities on tangible tabletop interfaces. A *microworld* is defined as the instantiation of an artificial environment that behaves according to a custom set of mathematical rules or scientific subdomains (Edwards, 1991). The aim of COPSE is to simplify the processes of creating, adjusting, and reusing custom microworld scenarios, where tangible objects represent the input parameters of optimization problems that influence each other. COPSE provides three types of building blocks: *widgets*, *equations*, and *scenes*, which can be specified in the form of structured text. The *scenes* have an *image* or an image sequence (filling the tabletop background), a *drawing priority* (which drawing is on top), as well as a *trigger condition* (when the scene is drawn). *Scenes* generally display the story and narrate the problem. *Widgets* are physical objects that the users manipulate, together with their associated zones on the tabletop. They materialize physically and digitally input or output variables of the problem. *Equations* model the possibly complex function that relates the independent (input) variables to the dependent (output) ones. Equations are hidden, and solving the problem consists in exploring the variable to understand the model and exploit such understanding to find out the configuration of dependent variables that best match a target state among the dependent variables. *Microworld* scenarios are run with an XML-based configuration file and are changed on the fly.

One example scenario developed by Maquil et al. (2017) supports collaborative urban planning processes. Policy makers, urban planners, citizens, and other stakeholders can gather around the tangible tabletop to view geographical data (such as heat maps, population density, or available areas for installing solar panels), discuss this data, and make decisions for urban planning projects.

4.4.1.1 Gesture Analysis on Tangible Interfaces

We have seen above that gestural interaction is an essential part of social interaction during collaborative tasks both in human–human and human–robot interaction. Anastasiou, Ras, and Fal (2019) in her current project *Gestures in Tangible User Interfaces* (GETUI), explored the gestural performance of users while interacting on a TUI in a collaborative problem-solving task (see Fig. 4.5).

An asset of this project is its holistic gesture data set including both *touch-based/manipulative* gestures related to the interface and 3D mid-air *freehand* gestures (Anastasiou et al., 2019). Some examples of manipulative gestures are tracing, rotating, holding, etc., while freehand gestures include mainly human–human gestures, such as pointing, iconic gestures, adaptors, and emblems. This taxonomy can be found in (Tables 4.1, 4.2).

As far as the recognition of these gestures is concerned, the touch-based gestures can be recognized using our developed application that can analyze the number of object manipulations with respect to the timing axis, the subject, and the handedness. The human–human gestures encompass many fine-grained gestures, often including multiple fingers, and are difficult to recognize automatically. Thus, a manual coding has been performed with the software ELAN (Wittenburg, Brugman, Russel, Klassmann, & Sloetjes, 2006), a professional tool for the creation of complex annotations on video and audio resources (Table 4.1).

The gestures presented in Table 4.2 are single-user gestures. A multi-user gesture including tangible objects can be giving/taking an object to/from another user.

4.4.2 An Avatar Robot for Collaborative Problem-Solving

As a next step, we want to investigate how distributed teams can collaboratively solve problems using tangible tabletops. In this scenario, we assume that two teams

Fig. 4.4 Initialization of COPSE by a configuration file showing intervening libraries (Maquil et al., 2017)

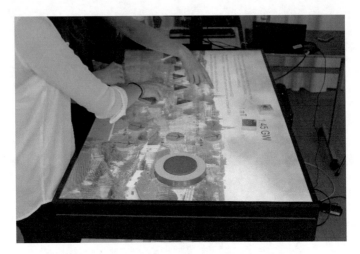

Fig. 4.5 Example of gesturing during collaborative problem-solving on a TUI

Table 4.1 Taxonomy of *freehand gestures* while solving a collaborative problem

Pointing	Object(s)
	TUI
	Other participant(s)
	Self-pointing participant
Iconic	Encircling with whole hand
	Encircling with index finger
	Moving an open hand forward/backward
	Moving an open hand downwards vertically
Adaptor	Head scratching
	Mouth scratching
	Nail biting
	Hair twirling
Emblems	Thumps up
	Victory sign
	Fist(s) pump

in different locations need to work together on digital content that is presented on an interactive display. Two identical tables are connected and synchronize their content in real time, and the users are either standing or sitting around their tabletop. The TUI concept would require that physical objects are duplicated on both tabletops. In this situation, the physical objects on each table must reflect the state changes that are made by the users at the other tabletop.

One solution that we are currently investigating is motorized objects than can actively change their position and orientation. This concept is described by Ishii

Table 4.2 Taxonomy of *touch-based gestures* while solving a collaborative problem

Touch-based/manipulative	Placing
	Removing
	Tracing
	Rotating
	Resizing
	Tapping
	Sweeping
	Flicking
	Holding

(2008) as **Tangible Telepresence**, meaning "the synchronization of distributed objects and the gestural simulation of presence artifacts, such as movement or vibration, allowing remote participants to convey their haptic manipulations of distributed physical objects."

We believe that the interaction would feel natural and intuitive if there were **avatar robots** at each tabletop, performing the same actions with the tangible objects as the remote users were. The situation is illustrated in Fig. 4.6, where two users collaborate to solve a problem on two synchronized instances of a tangible tabletop (TUI A and B) over a distance. In both environments, an avatar robot replaces the remote user and mirrors their actions on the tabletop. The robots act as mediators between both places and make the local users aware of the remote parties' intentions, i.e., when they look at an object and/or approach an object with their hands. This helps to coordinate actions and to intuitively avoid conflicts; e.g., both parties recognize when they try to manipulate the same object at the same time.

The biggest difference between our scenario and mobile robotic telepresence, as described in Sect. 4.2.3, is the symmetry between the environments; both users interact directly with their own instance of the shared workspace—there is no role of a "pilot" or "local" user—as if they were at the same table and in the same room.

4.4.2.1 Design and Technical Challenges

In order to realize such a scenario from a technical perspective, it is necessary to synchronize the objects on the tabletop and to mirror the behavior of the human collaborators by their avatar robots.

The advantage of the existing TUI tables is that the tangible objects are visually tracked with high precision by cameras inside the tabletop, so their position and orientation are always known to the system. The current project GETUI further tracks the pose of the users in 3D, using a Microsoft Kinect sensor, in order to figure out who is manipulating an object. It is an object (tangible) and gesture recognition application which can automatically recognize object manipulation in real time with regard to (i) *which object* has been manipulated, (ii) *when*, (iii) by *whom*, and (iv) using *which*

Fig. 4.6 Two users
collaborating over a distance
using two avatar robots that
mirror their actions and pose

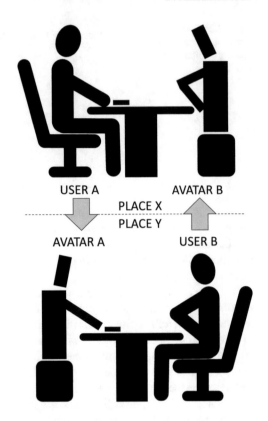

hand. The application is fully replicable for Kinect™ SDK users and TUI holders with a tracking software framework. The method we follow for object and gesture recognition is to merge the logging files from our TUI software framework, COPSE (object recognition) with the Kinect log files (gesture recognition) in one file. The application converts the TUI objects' screen coordinates to the Kinect coordinates system. All active objects are continuously controlled by the application in order to check whether the participants' hand coordinates clash with the objects' coordinates (Anastasiou et al., 2019).

Having this input, it should be possible to coordinate the pose and gesture of the robotic avatars with animated tangible objects that are able to autonomously move and rotate. In this case, the robot would not even need to physically touch and manipulate the objects. The arms and hands could be designed to be quite simple in contrast to expensive tele-operators. On the other hand, the interaction with the tabletop makes it also difficult or even impossible to capture a video image of the faces of the users, since their focus is expected to change between the tabletop and robot.

In Tables 4.3 and 4.4, we abstracted the gestures presented in Tables 4.1 and 4.2 for the robot's perspective.

Table 4.3 Freehand gestures that a robot can mirror and mimic compared to a human

Freehand gestures				
Pointing			Human	Robot
		Object(s)	X	X
		TUI	X	X
		Other participant(s)	X	X
Iconic	Self-pointing participant		X	X (self-pointing robot)
	Encircling with whole hand		X	X
	Encircling with index finger		X	X
	Moving an open hand forward/backward		X	X
Adaptor	Moving an open hand downwards vertically		X	X
	Head scratching		X	
	Mouth scratching		X	
	Nail biting		X	
Emblems	Hair twirling		X	
	Thumps up		X	
	Victory sign		X	
	Fist(s) pump		X	

Table 4.4 Manipulative gestures that a robot can mirror and mimic compared to a human

Touch-based/manipulative gestures			Human	Robot
		Placing	X	X
		Removing	X	X
		Tracing	X	X
		Rotating	X	
		Resizing	X	
		Tapping	X	
		Sweeping	X	
		Flicking	X	
		Holding	X	X

As can be seen in Table 4.3, many of the gestures are feasible to be performed, as soon as the robot has arms or in the best case, fingers. However, good finger manipulation is difficult for adaptors and emblems are more related to human parts.

Moreover, head movements can also be used by the robot mainly to increase the engagement of the operator. Furthermore, it is planned that the robot should share the eye gaze, body posture, and gesture of the remote user, which would not only support consequential communication, but also help to communicate feelings and emotions (sadness, joy, etc.) more reliably than a video channel, as suggested by Mehrabian's *7/38/55* rule.

4.5   Discussion and Future Prospects

We have introduced the problem of presence disparity in the context of teleworking and collaboration over a distance and have explained the concepts of telepresence and mobile robotic telepresence. We have further highlighted the importance of body language (immediacy principle) and gesture for intentional and non-intentional communication, including conversational awareness.

While humanoid avatar robots are being developed to remotely manipulate objects, they do not explicitly address communication and collaboration. We have presented a future scenario where avatar robots mirror the actions and gestures of humans to support collaboration on a tangible table over a distance.

Our aim in this chapter is to foster a discussion about technical and social questions related to the acceptance of avatar robots in the workplace. Current social robots are often designed to be small and cute to arouse mainly positive emotions. In collaborative scenarios, the robots would need to have the same size as an average human; otherwise, they would not be able to reach and manipulate objects in the workspace with their hands. For acceptance in a business scenario, it is probably also important that the design and style of the avatar robot represents positive attributes of the remote user, such as professionalism and competence. Last but not least, the display of the social status can be as important as the intrinsic, practical utility of a device. This could be achieved by sophisticated design, technology, and materials.

Regarding communication, we have discussed five elements of telepresence: audio, video, mobility, body posture, and gesture. The majority of current products combine audio and video with mobility to overcome presence disparity. It remains an open question how collaboration is sensed with a robot that does not provide a display as a video channel, but supports gestures instead. This combination would be a new experience and needs to be investigated in empirical studies.

References

Adalgeirsson, S. O., & Breazeal, C. (2010). MeBot: A robotic platform for socially embodied presence. In *Proceedings of the 5th ACM/IEEE International Conference on Human-Robot Interaction* (pp. 15–22).

Anastasiou, D., Ras, E. & Fal, M. (2019). Assessment of collaboration and feedback on gesture performance. In *Proceedings of the Technology-Enhanced Assessment* (TEA) *Conference*. Berlin: Springer (to be published).

Anastasiou, D. & Stahl, C. (2012). Gestures used by intelligent wheelchair users. In *Proceedings of the International Conference on Computers Helping People with Special Needs* (ICCHP-12) (pp. 392–398). Berlin: Springer.

Baker, K., Greenberg, S., & Gutwin, C. (2001). Heuristic evaluation of groupware based on the mechanics of collaboration. In *Engineering for human-computer interaction* (pp. 123–139). Berlin: Springer.

Bartneck, C., & Forlizzi, J. (2004). A design-centred framework for social human-robot interaction. In *IEEE International Symposium on Robot and Human Interactive Communication (RO-MAN 2004)* (pp. 591–594).

Bekker, M. M., Olson, J. S., & Olson, G. M. (1995). Analysis of gestures in face-to-face design teams provides guidance for how to use groupware in design. In *Proceedings of the 1st Conference on Designing Interactive Systems: Processes, Practices, Methods, & Techniques* (pp. 157–166).

Björnfot, P., & Kaptelinin, V. (2017). Probing the design space of a telepresence robot gesture arm with low fidelity prototypes. In *2017 12th ACM/IEEE International Conference on Human-Robot Interaction (HRI)* (pp. 352–360).

Breazeal, C. (2000). *Sociable machines: Expressive social exchange between humans and robots, Ph.D. Thesis*. Cambridge, MA: Department of Electrical Engineering and Computer Science, MIT.

Breazeal, C., & Scassellati, B. (2000). Robots that imitate humans. *Trends in Cognitive Sciences, 6*(11), 481–487.

Cha, E., Chen, S., & Mataric, M. J. (2017). Designing telepresence robots for K-12 education. In *IEEE International Symposium on Robot and Human Interactive Communication (RO-MAN 2017)* (pp. 683–688).

Dautenhahn, K. (2007). Socially intelligent robots: Dimensions of human–robot interaction. *Philosophical Transactions of the Royal Society of London B: Biological Sciences, 362*(1480), 679–704.

Dautenhahn, K., & Billard, A. (1999). Bringing up robots or-the psychology of socially intelligent robots: From theory to implementation. In *Proceedings of the Third Annual Conference on Autonomous Agents* (pp. 366–367).

Double Robotics. https://www.doublerobotics.com/. Retrieved February 14, 2018.

Edwards, L. D. (1991). The design and analysis of a mathematical microworld. *Journal of Educational Computing Research, 12*(1), 77–94.

Eyssel, F., & Kuchenbrandt, D. (2012). Social categorization of social robots: Anthropomorphism as a function of robot group membership. *British Journal of Social Psychology, 51*(4), 724–731.

Fong, T., Nourbakhsh, I., & Dautenhahn, K. (2003). A survey of socially interactive robots. *Robotics and Autonomous Systems, 42*(3–4), 143–166.

Fukushima, S., Sekiguchi, H., Saito, Y., Iida, W., Nozaki, T., & Ohnishi, K. (2017). Artificial replacement of human sensation using haptic transplant technology. *IEEE Transactions on Industrial Electronics, 65*(5), 3985–3994.

Goffman, E. (1966). *Behavior in public places: Notes on the social organization of gatherings*. New York: Free Press.

Heeter, C. (1992). Being there: The subjective experience of presence. *Presence: Teleoperators & Virtual Environments, 1*(2), 262–271.

Hornecker, E., & Buur, J. (2006). Getting a grip on tangible interaction: A framework on physical space and social interaction. In *Proceedings of the SIGCHI Conference on Human Factors in Computing Systems (CHI '06)* (pp. 437–446). New York: ACM.

Ishii, H. (2008). Tangible bits: Beyond pixels. In *Proceedings of the 2nd International Conference on Tangible and Embedded Interaction* (pp. xv–xxv).

ISO 8373:2012, Robots and robotic devices—Vocabulary. https://www.iso.org/standard/55890.html. Retrieved February 14, 2018.

Jibo. https://www.jibo.com/. Retrieved March 21, 2018.

Keecker. https://www.keecker.com/. Retrieved March 21, 2018.

Kristoffersson, A., Coradeschi, S., & Loutfi, A. (2013). A review of mobile robotic telepresence. *Advances in Human-Computer Interaction, 3*.

Kubi. https://www.revolverobotics.com. Retrieved March 21, 2018.

Kuzuoka, H., et al. (2000). GestureMan: A mobile robot that embodies a remote instructor's actions. In *Proceedings of the 2000 ACM Conference on Computer Supported Cooperative Work* (pp. 155–162). New York: ACM.

Lee, M. K., & Takayama, L. (2011). Now, I have a body: Uses and social norms for mobile remote presence in the workplace. In *Proceedings of the SIGCHI Conference on Human Factors in Computing Systems* (pp. 33–42).

LuxAI. 2018. QT Robot. Retrieved March 21, 2018 from http://www.luxai.com.

MantaroBot. 2016. Retrieved February 13, 2019 from http://www.mantarobot.com/.

McNeill, D. (1992). *Hand and mind: What gestures reveal about thought.* Chicago: University of Chicago Press.

Maquil, V., Tobias, E., Anastasiou, D., Mayer, H., & Latour, T. (2017). COPSE: Rapidly instantiating problem solving activities based on tangible tabletop interfaces. *Proceedings of the ACM on Human-Computer Interaction, 1*(1), 6.

Minsky, M. (1980, June). Telepresence (essay). In *Omni Magazine* (Vol. 2, No. 9) (pp. 45–52). Omni Publications International Ltd.

Mehrabian, A. (1980). *Silent messages: Implicit communication of emotions and attitudes.* Belmont: Wadsworth Publishing Co Inc.

Meltzoff, A. N., Brooks, R., Shon, A. P., & Rao, R. P. (2010). "Social" robots are psychological agents for infants: A test of gaze following. *Neural Networks, 23*(8–9), 966–972.

Mojin Robotics. (2018). Care-O-Bot 4. Retrieved March 21, 2018 from http://www.mojin-robotics.de.

Nowak, K. L., & Biocca, F. (2003). The effect of the agency and anthropomorphism on users' sense of telepresence, copresence, and social presence in virtual environments. *Presence: Teleoperators and Virtual Environments, 12*(5), 481–494.

PadBot. (2018). PadBot Telepresence Robot. Retrieved March 21, 2018 from http://www.padbot.com.

Polycom. (2018). RealPresence OTX Studio. Retrieved March 21, 2018 from http://www.polycom.com/products-services/hd-telepresence-video-conferencing/realpresence-immersive/realpresence-otx-studio.html.

Shaer, O., & Hornecker, E. (2010). Tangible user interfaces. Past, present and future directions. *Foundations and Trends in Human-Computer Interaction, 3*(1–2), 4–137.

Short, J., Williams, E., & Christie, B. (1976). *The social psychology of telecommunications.* London, New York: Wiley.

Steelcase. (2016). Making Distance Disappear-Unleashing the Power of Distributed Teams. Steelcase360 Magazine, Issue 10, 8–13.

Swivl. (2018). Retrieved March 21, 2018 from https://www.swivl.com.

Tang, A., Boyle, M., & Greenberg, S. (2005). Understanding and mitigating display and presence disparity in mixed presence groupware. *Journal of Research and Practice in Information Technology (JRPIT), 37*(2), 193–210.

Toyota Global Newsroom. (2017). Toyota Unveils Third Generation Humanoid Robot T-HR3. (21 November 2017). Retrieved March 21, 2018 from https://newsroom.toyota.co.jp/en/download/20110424.

Ullmer B., & Ishii, H. (2000, July). Emerging frameworks for tangible user interfaces. *IBM Systems Journal, 39*(3.4), 915–931.

Willow Garage. (2012). Texai Remote Presence System, from http://www.willowgarage.com/pages/texai/overview.

Wittenburg, P., Brugman, H., Russel, A., Klassmann, A., & Sloetjes, H. (2006). ELAN: A professional framework for multimodality research. In *5th International Conference on Language Resources and Evaluation (LREC 2006)* (pp. 1556–1559).

World Robotics, Executive Summary. (2018). Retrieved November 20, 2018, https://ifr.org/downloads/press2018/Executive_Summary_WR_2018_Industrial_Robots.pdf.

Chapter 5
Unobtrusive Vital Data Recognition by Robots to Enhance Natural Human–Robot Communication

Gerald Bieber, Marian Haescher, Niklas Antony, Florian Hoepfner
and Silvio Krause

Abstract The ongoing technical improvement of robotic assistants, such as robot vacuum cleaners, telepresence robots, or shopping assistance robots, requires a powerful but unobtrusive form of communication between humans and robots. The capabilities of robots are expanding, which entails a need to improve and increase the perception of all possible communication channels. Therefore, the modalities of text- or speech-based communication have to be extended by body language and direct feedback such as non-verbal communication. In order to identify the feelings or bodily reactions of their interlocutor, we suggest that robots should use unobtrusive vital data assessment to recognize the emotional state of the human. Therefore, we present the concept of vital data recognition through the robot touching and scanning body parts. Thereby, the robot measures tiny movements of the skin, muscles, or veins caused by the pulse and heartbeat. Furthermore, we introduce a camera-based, non-body contact optical heart rate recognition method that can be used in robots in order to identify humans' reactions during robot-human communication or interaction. For the purpose of heart rate and heart rate variability detection, we have used standard cameras (webcams) that are located inside the robot's eye. Although camera-based vital sign identification has been discussed in previous research, we noticed that certain limitations with regard to real-world applications still exist. We identified artificial light sources as one of the main influencing factors. Therefore,

G. Bieber (✉) · M. Haescher · N. Antony · F. Hoepfner
Fraunhofer-Institut für Graphische Datenverarbeitung, Rostock, Germany
e-mail: gerald.bieber@igd-r.fraunhofer.de

M. Haescher
e-mail: marian.haescher@igd-r.fraunhofer.de

N. Antony
e-mail: niklas.antony@igd-r.fraunhofer.de

F. Hoepfner
e-mail: florian.hoepfner@igd-r.fraunhofer.de

S. Krause
Institut für Informatik, Universität Rostock, Rostock, Germany
e-mail: silvio.krause@uni-rostock.de

© Springer Nature Switzerland AG 2019
O. Korn (ed.), *Social Robots: Technological, Societal and Ethical Aspects
of Human-Robot Interaction*, Human–Computer Interaction Series,
https://doi.org/10.1007/978-3-030-17107-0_5

we propose strategies that aim to improve natural communication between social robots and humans.

Keywords Vital data · Activity · Recognition · Autonomous computing · Optical · Camera · Webcam · Wearable computing · Assistive technology

5.1 Motivation

Nowadays, robots pervade many areas of technology and daily life. Examples of these, such as mechanical workers in the important field of assembly, manufacturing, and production show the possibilities that emerge by applying robots. Since developments in machine learning and artificial intelligence drive the capabilities of robots, new application fields arise, such as that of social assistance. Social robots will entertain, train, educate, or simply interact with users in the same way as humans do (Korn, 2018). During a conversation between a human and a social robot, much information has to be exchanged (even exceeding the modality of speech). This information includes optical, tactile, and acoustical modalities such as facial expressions, prosody of speech, and body language. Moreover, in a conversation, it is necessary to know if the counterpart is nervous or somehow affected by the discussion. To assist in this non-verbal communication, emotion recognition enabled by detecting human vital signs would be very beneficial. Usually, the recognition of vital signs is performed by applying wearable sensors attached to the human body. Depending on the parameters measured, theses sensors tend to be obtrusive (e.g., wearable electrocardiography (ECG) sensors, heart rate chest straps) and require constant usage in order to provide gapless recording of data. Therefore, a touchless vital data recognition system for social robots is very beneficial for health care or communication purposes. In addition, or for special purposes, a light touch of the social robot on the user's body enables the robot to feel vital data, and this provides additional capabilities to enhance the natural communication (Fig. 5.1).

Due to the possible mobility of social robots, new environment-based requirements arise that lead to certain considerations with regard to the analysis of the transmitted signal quality. This leads to the following research questions:

- How can communication between a social robot and a human become more natural?
- Which of the vital data modalities can be assessed touchlessly?
- Which parameters influence the quality of touchless vital data recognition (via cameras) in the application field of social robots?
- In particular, what is the most relevant confounding factor for camera-based heart rate recognition and how can we deal with it?

Fig. 5.1 Cameras in the eye of the robot detect pulse rate and heart rate variability in order to enable natural communication between the social robot and the human

5.2 Related Work

Since robots are performing tasks in the home environment, the user desires the robots to have natural language capabilities (Goodrich, 2007). Speech interfaces support a natural communication modality and therefore support the identification of a user by recognizing individual speech habits or voice and language characteristics (Zissman, 1996). The combination of communication modalities enhances the total understanding and reliability of information exchange (e.g., McGurk effect) (Nath, 2012). We propose the identification of vital data for emotion detection since physiological signals show a strong correlation to emotions. However, whether emotions can be recognized reliably from physiological signals is still a matter of research (Jerritta, 2011). The most common signals for emotion detection are ECG signals for heart rate and heart rate variability, skin conductivity, respiration rate, and skin temperature. These parameters provide good results in terms of classification of emotions (Haag, 2004). Simple emotion detection can be achieved even with a reduced feature set (e.g., by analyzing ECG and respiratory signals only) (He, 2017).

The assessment of vital data by social robots is possible by direct body contact and by a touchless sensing. Such vital data assessment should be as unobtrusive as possible. This supports the natural communication and an agreeable feeling. During a normal conversation between humans, it is normal to touch the hand or arm of the interlocutor. It is of interest, if a robot is also accepted to assess relevant data to judge the human's feeling, mood, or emotion.

5.2.1 Touchy Sensors

Social robots might look like humans, but they can also be pet- or Muppet-like, comic figures, or even androids. The physical contact between robot and human is mainly by soft touch and body contact (Figs. 5.2 and 5.3).

Fig. 5.2 Care Robot Paro, it is touch sensitive and interacts with users while being fondled *Source* CC-BY-SA-2.0 (Biggs 2005)

Fig. 5.3 Toy-like robots

When a human is stroking the fur of a pet or is holding the hand of a robot, the integrated sensors of the social robot are able to measure simple but also complex vital data. A short contact is sufficient to measure the temperature and galvanic skin response very easily. In addition, the assessment of more advanced information is also possible. Direct contact with the skin enables electro-technical-based assessment methods. For example, an electrocardiogram (ECG) helps to detect the heart rate (HR) or heart rate variability (HRV) and provides basic stress parameters. Capacitive sensing might identify the respiration rate, and electromyography (EMG) is used for muscle activity detection.

Furthermore, remarkable parameters are the frequency and amplitude of muscle vibrations. Each muscle of a mammal performs tiny movements, resulting in a light vibration. This low-amplitude muscle vibration was first reported in the early 1960s (Rohracher, 1964), and this phenomenon is correlated to some body conditions, e.g., level of stress, medication, temperature distribution or hints of diseases such as Parkinson's or other neural degeneration diseases. Muscle activity can also be measured by electromyography (EMG) (Clancy, 2002), but recent accelerometry is also sensitive enough to detect muscle vibration (Bieber, 2013). While electro-technical methods usually need multiple body contacts to detect potential differences, accelerometry needs only one.

The measurement of a single point acceleration of the skin even provides information about the heart (Matthies, Haescher, Bieber, Salomon, & Urban, 2016). The physical movement of the heart and the blood flow through the body also cause movements of the body and skin. These tiny movements have characteristic patterns and may describe heart anomalies. This technique of measuring forces on the heart is called ballistocardiography or seismocardiography (Inan, 2015).

5.2.2 Optical Sensors

A touchless technology for the identification of vital data is the usage of optical information. The first non-invasive blood oxygen saturation meter SpO2 was invented in 1935 (Matthes, 1935). With this, the skin of the ear was illuminated in order to measure the amount of light passing through the tissue. For the optical and volumetric measurement of the skin, only one frequency band (the color of the light spectrum) is needed (Fig. 5.4).

This technique is referred to as photoplethysmography (PPG) (Hertzman, 1937). It can be performed by analyzing the reflected light (reflective PPG) or light that shines through the tissue (transmissive PPG). Medical oxygen saturation meters (SpO2) attached to the finger mainly apply the transmissive approach, while fitness trackers, smart bands, or smart watches, located on the wrist, mainly use the reflective approach. All of the devices use light-emitting diodes (LEDs) as the source of light for an appropriate illumination. The applied colors vary between the light of the green or red light diodes.

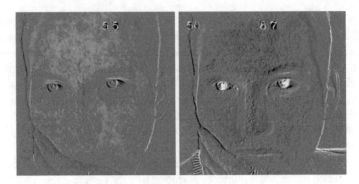

Fig. 5.4 False color image of a face with and without oxygen-enriched blood

The general concept for heart rate recognition with cameras is based on the identification of periodic change in skin color. Therefore, the camera detects skin-reflected light illuminated by sources within the surroundings of the user (Poh, 2011). An additional LED or comparable dedicated light source is not needed but provides better results. Blood with a higher oxygen saturation reflects light differently than blood with a lower oxygen saturation. With every heartbeat, the saturation changes and so does the light reflection (Kong, 2013). The facial skin shows a high degree of perfusion and therefore reflects light differently during the cardiac cycle. The effect of reflection characteristics is influenced by multiple factors, including varying tissue volume, tissue tension, and other side effects. Cameras detect the heart rate as a change in color, which is not visible to the human eye (Wu, 2012).

The optimal position for detecting changes in skin color is the forehead. This position is favorable because in a conversation with a robot, the head of the human is usually pointed toward the robot. Therefore, an integrated face detection algorithm in the social robot identifies the position of the eyes and the forehead region quite easily. The average of the green color channel values of the detected forehead region changes with every pulse cycle. The pulse rate can be determined by analyzing the resulting data stream. For the sake of data processing, we selected only part of the forehead image, the so-called region of interest (ROI). We recorded data with a camera (camera model IDS UI-306xCP-C) in a laboratory setting at constant lighting. The camera was mounted statically.

A social robot that is equipped with a camera for pulse rate detection should be able to move around in order to interact in different rooms or surroundings. Hence, the accuracy of pulse detection should be tolerant of user-specific effects (e.g., head movements) and environmental constraints. Therefore, we need to consider the main influencing effects of touchless vital data recognition via cameras.

The identification of heart rate by examination of the skin color depends on two general categories of parameter:

- Technical Parameters

- Environmental Parameters.

Both categories will be discussed in the following sections.

5.2.2.1   Technical Parameters

The quality of camera pictures depends on several factors. These include the image sensor, lens, processing hardware, and other factors.

Image sensor size: Digital cameras vary in design, size, energy consumption, and image quality. High-end cameras consist of an image sensor with a large physical size in comparison with compact cameras. When the image sensor is larger, more light can reach the individual pixel areas on the sensor. The Advanced Photo System type-C (APS-C) is an image sensor format approximately equivalent in size to the Advanced Photo System "classic" negatives of 25.1 × 16.7 mm. In contrast, cameras of compact devices such as the iPhone 5S have an image sensor with the size of 4.54 × 3.42 mm.

F-factor: Another parameter that determines how much light reaches the sensor is defined by the aperture size. The f-number of an optical system such as a camera lens is the ratio of the system's focal length to the diameter of the entrance pupil (Smith, 2007). It is a dimensionless number that is a quantitative measure of shutter speed and therefore an important concept in photography. It is also known as the focal ratio, f-ratio, or f-stop (Smith, 2005). The higher the f-ratio, the better the exposure. This applies to most applications. An iPhone 5S camera has an f/2.2 lens.

Photosensitivity: Analog film provides specific sensitivity to light. This sensitivity is measured and numbered as an ISO speed. The product of ISO and shutter speed controls the brightness of the photo. The base ISO describes the speed of the highest image quality, minimizing as much noise as possible. Digital sensors only have a single sensitivity, which is mainly defined by the signal-to-noise ratio (SNR). The SNR is measured in decibels (dB). The higher the SNR, the better. A good value is about 40 dB (Baer, 2000).

Speed: The digital image sensor needs time to take a photo or to sense the frame of a video. For the recognition of pulse or respiration rate, at least a double sampling rate is necessary in order to meet the Nyquist–Shannon sampling theorem. Almost every digital video sensor is capable of providing 24 frames per second as the sampling rate (Etoh et al., 2001). Therefore, vital data recognition is possible.

Resolution: Higher pixel density is often correlated to better video quality. Since each camera has its own parameter set (screen size, field-of-view, etc.), we have to focus on the resolution of the face itself and not on the entire screen. Since we are focusing on the change in color of the green channel, the resolution of the ROI is relevant but is not of substantial importance. The number of pixels within the ROI might be 100, 1000, or even higher but is not the defining quality parameter. Hence, the low resolution of a standard video graphics array (VGA) video is sufficient for pulse recognition (Mestha, 2014).

Automatic functions: For analyzing the change in color within the region of interest, stable recording is required. Some cameras perform automatic white calibration or brightness adjustment for enhanced imaging (Weng, 2005). Due to discontinuity or changes in color, the automatic functions affect the vital data recognition and lead to errors or additional noise.

5.2.2.2   Environment Parameters

Vital data recognition via reflective PPG approaches with cameras works well in laboratory settings (Irani, 2014). Therefore, it has to be considered that real-world scenarios with social robots might involve additional challenges. These can be classified as follows:

Motion artifacts: The image sensor of the robot is not mounted in a fixed frame. Therefore, the camera might experience vibration caused by a cooling fan, a power transformer, or by the motions of the robot itself. Furthermore, the communicating counterparts' movements while speaking or performing natural body language have to be considered.

Optical considerations: During a conversation, the spatial constellation between robot and dialog partner might vary due to the change of distance or the optical angle. Hair or makeup might cover the region of interest. Moreover, glasses worn by the user might disturb the face recognition algorithm.

Light source: The communication between a social robot and a human can take place in an indoor environment. In that case, the brightness of the light and the light source itself may lead to signal noise or disturbances. Artificial light sources particularly influence the signal noise.

Temperature: The sensing of changes in color depends on the perfusion of the skin. In addition to this, the temperature of the environment also influences the blood flow. Other effects include physical parameters of the user (e.g., skin flexibility, drugs, coffee, etc.).

A camera-based touchless vital data recognition system must be aware of the influencing parameters. Moreover, the recognition system has to have implemented algorithms for identifying the major disturbances in order to adapt.

5.3   Detection Algorithm

In order to measure the pulse signal, a video stream has to be captured by a camera first. Therefore, it is necessary to detect the human's face, then identify the forehead and a suitable part of it (ROI). Subsequently, we determine the average intensity of the green color channel within the red-green-blue (RGB) signal in the ROI. This signal is the basis for recognizing the pulse wave. Therefore, we first need to track the face within the video stream. To accomplish this, the face-tracking algorithms of the OpenCV library can be applied (Bradski & Kaehler, 2008). In order to reduce motion artifacts, one could perform face tracking on every frame. As this would result in reduced performance, the face-tracking frequency is reduced (once every 25 frames) and a larger main ROI is defined instead. This main ROI has the size of the whole forehead with another smaller region inside of it. This inner region moves with the head movement in each frame. This way, it follows the movement of the forehead without constant face tracking. This means a low requirement for computational power and ensures a stable sampling and frame rate.

Before evaluating the values of the ROI, it is necessary to remove motion artifacts with the help of filtering. Therefore, we check each pixel inside the ROI to determine if their green value was an outlier compared to the averaged value of the ROI from the preceding frame. Outliers are defined as pixel values that are beyond the $3 * \sigma$ (standard deviation) threshold within one frame.

Let now $f(t)$ be the recorded raw signal of the ROI at a time t and $m_{f(t)}$ the mean of $f(t)$. For the detection of outliers, we use low-pass and high-pass filters with sliding window. Butterworth filtering is also an option. For sliding window, we are not concentrating on just the current frame of the footage but on the mean of the last four frames.

Let ROI be a frame of $p * q$ pixels. Then the mean of one frame is calculated as:

$$m_{f(t)} = \frac{1}{p * q} * \sum_{i=0}^{p} \sum_{j=0}^{q} f(t)[i, j].$$

With these values, we can apply our filters:

- Low-pass filter:

$$m_{l(t)} = m_{f(t-1)} * (1 - \alpha) + m_{f(t)} * \alpha, \quad \text{with } \alpha = 0.05$$

- High-pass filter:

$$m_{h(t)} = m_{f(t)} - m_{l(t)}$$

Finally, we apply the sliding window to the filtered value to get our filtered mean value:

$$m = \frac{1}{4} * \sum_{i=0}^{3} m_{h(t-i)}.$$

If a Butterworth filter is applied, we recommend cutting off frequencies below and above normal heart rates (sampling rate = frames per second (FPS), lower cutoff frequency = 0.52, upper cutoff frequency = 5.02).

After removing all outliers, the average green value of all leftover pixels is determined.

The resulting pulse curve allows us to determine a reliable pulse signal (Fig. 5.5). Subsequently, a fast Fourier transformation and peak detection serve to identify the heart rate and heart rate variability.

5.4 Optimization Strategies

Social robots usually apply cost-efficient cameras. These customary web cameras provide moderate resolutions and frame rates. They are usually optimized for video conferencing and reduced data traffic. In contrast to this, heart rate or heart rate variability detection scenarios require a focus on image quality.

Motion within a video sequence leads to major artifacts in the heart rate signal. Therefore, face recognition and head tracking technologies support the readjustment of the ROI and the assessment of a change in color. Furthermore, reference regions allow motion compensation as well as general changes in the color or brightness. Therefore, reference regions in the face might compensate for automatic functions or may stabilize the lighting situation (Fig. 5.5).

The change in color within the ROI leads to a periodic signal that corresponds to the heart rate, as presented in Fig. 5.6.

The frame rate of most customary cameras is sufficient for vital data recognition since less than 10 Hz are needed for sampling the heart rate and heart rate variability, or for providing respiration rate recognition (RR).

Natural daylight provides almost white light that consists of a sufficient amount of green light for our study. Furthermore, daylight is a continuous light source and provides a setting for very good measurements. In contrast, artificial light highly influences the recorded data and produces signal noise.

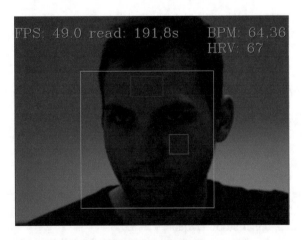

Fig. 5.5 Region of interest (blue) and reference region (red)

The analysis of artificial light in our measurements showed a tremendous change in brightness for higher frequencies. The normal power supply of the lights in our lab (located in Germany) is alternating current (AC) with a frequency of 50 Hz. This means that the voltage changes polarity 50 times per second. Thus, the light gets brighter with the maximum voltage and less bright in the zero-crossing zone. The zero-crossing happens 100 times per second so that the lights have a pulsation of 100 times per second. The intensity of the maximum brightness and the least brightness depends on the light technology. A neon light loses 50% of its intensity during the zero-crossing (Brundrett, 1974). Modern LED lights are affected even more by the pulsating current than neon lights or standard bulbs. The pulsating effect is also dependent on the ballast unit used. Summering up almost every artificial light pulsates (Fig. 5.7).

The changing brightness leads to an aliasing effect and influences the quality of data. In contrast to Figs. 5.4 and 5.5, Fig. 5.8 illustrates the high noise effect on the red and green zones caused by the pulsation of artificial light.

During our research with robots, we applied the robots' camera for heart rate and heart rate variability recognition. This research was performed mainly in indoor environments. We identified the surrounding light source and implemented a filter to minimize the aliasing effect. By using a digital filter, we could reduce the noise due to the very frequent changes in brightness. Furthermore, we modified the environmental lighting as soon as we noticed that the light conditions were insufficient for our measurements.

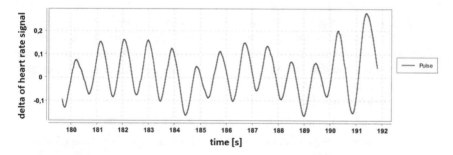

Fig. 5.6 Heart rate signal of the subject

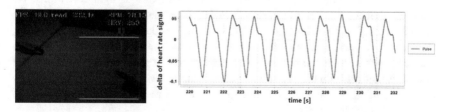

Fig. 5.7 Pulsating green LED light (left) and the received camera data (graph on the right)

Fig. 5.8 Aliasing effect of
the signal due to the
pulsation of artificial light

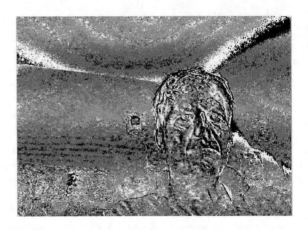

5.5 Study

In order to evaluate the performance of camera-based human–robot interaction, we
conducted a lab study with eight participants. The study included two different camera
systems: a Philips SPC 1300NC (webcam) and an IDS UI-306xCP-C (professional
camera). The webcam was integrated into our social robot (Fig. 5.1) as an eye. The
social robot was designed by us and originally used as a physical avatar (Sauer &
Gobel, 2003). Although it was possible to measure a reliable pulse wave with the
Philips camera and the social robot, we mostly applied the IDS since we could store
video streams for later processing and analysis at higher resolutions and frame rates.

The setting for the study was a normal office workplace; the IDS camera was
mounted on the monitor. The participants were advised to work at the computer. The
average measurement duration was 15 min. The participants had to behave normally,
as if they were not being recorded.

Our studies had two main purposes:

To identify which light intensity provides the highest heart rate accuracy
To estimate the percentage of time in which valid data is measurable using the heart
rate recognition algorithm provided.

In total, we recorded about 100,000 samples under various light conditions. We
found out that the heart rate accuracy is highly dependent on the brightness of the
surrounding light (Fig. 5.9). Lower light intensity results in more dominant noise,
which leads to varying light and color data. On the other hand, light which is too
intense results in total reflectance and therefore overexposure of the skin. Optical
saturation hinders a change in light intensity due to overexposure. The trend line
(Fig. 5.7) indicates that, in order to achieve optimal results, the most useful light
intensity is in the upper quartile.

In our study, we also investigated the amount of time required for valid identification of the pulse rate during camera surveillance. Therefore, we measured the total
amount of time with valid and invalid pulse rates for all subjects. The subjects had to

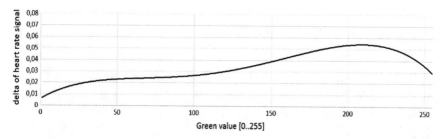

Fig. 5.9 Trend line of the delta of the heart rate signal (y) proportional to the measured green value of the pixels (x)

behave naturally while performing normal computer work. They were allowed to go to the restroom, talk to colleagues, or read printouts. In our tests, we found that for 50% of the time valid pulse detection was possible. 33% of the pulse detection was invalid because the OpenCV algorithm was not able to detect a face. This happened because the subject was unavailable, or the face could not be detected due to rotation, movement, or bad exposure. Unrealistic pulse rates were detected in the remaining 17% of the time and therefore excluded.

We also tested a scenario in which the subjects had to keep the head motionless while facing the camera the whole time. We identified that by using the aforementioned restriction under good light conditions, the pulse recognition was valid 95% of the time. Of course, keeping the head motionless is not a reasonable scenario for real-life applications since the many tasks one performs involve plenty of head motions. In addition to this, some of the subjects reported neck pain after several minutes.

5.6 Discussion

We were able to identify that face-to-face communication between a robot and a human enables a direct view of the subject's forehead. During our research with robots, we applied the robots' camera for heart rate and heart rate variability recognition. We found out that social robots could measure stress, strain, emotions, or medical parameters. This leads to the question of in which social situations this technology should be used. We think that social robots perfectly meet the care demands for elderly who are lonely or suffering from dementia. With the increasing potential of artificial intelligence, social robots will become very useful for entertainment but also as acquaintances or even friends. Their emotion detection leads to better understanding of the human by the robot, though of course, we hope that a robot will never be a better friend than a human is. The capability of vital data detection may also be very useful in hospital or care environments, so in future, rather than impersonal systems, nice robots will monitor patients.

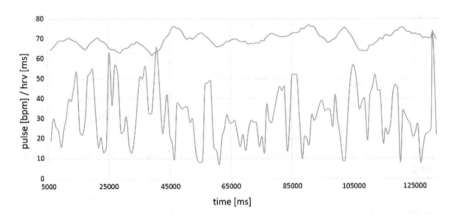

Fig. 5.10 Detected heart rate variability (HRV, red) and pulse rate (blue) via a camera-based reflective PPG approach

Unfortunately, this technology may have some social implications. Some people might feel uncomfortable with robots collecting their vital signals (either through or without touch). Furthermore, we can imagine that companies will use social robots to perform job interviews. The robot could ask specific questions and act like a polygraph, a lie detector. This scenario is also possible in a medical setting where doctors using social robots to obtain true answers from patients during the anamnesis. We should be aware that we are giving a further piece of capability to a robot that only a human had before. This might lead to the circumstance that a human can be assisted but can also be replaced.

Improving the face-tracking algorithm and lighting would greatly increase the amount of valid heart rate values. Our study shows that the recognition of heart rate and heart rate variability (Fig. 5.8) is possible. Therefore, camera-based vital data recognition allows touchless emotion recognition. In our study, we achieved assessment of a valid pulse rate for only half the time the measurement was performed. This is only a very rough estimation but indicates that the concept has a high potential and can be improved (Fig. 5.10).

We consider artificial light sources as well as movement artifacts and brightness change as the main noise in vital data recognition. A possible improvement might be achieved by transforming the RGB data into another color space, e.g., hue-saturation-value color space (HSV). This is currently under examination.

A simple recording of the environment in slow motion (e.g., iPhone 6s with 240 frames per second) demonstrates the varying light conditions. Social robots might also illuminate the person with whom they are communicating in the future. Furthermore, robots might use the invisible light spectra or infrared to extend their scanning possibilities.

The authors of "Emotion recognition using bio-sensors: First steps towards an automatic system." (Haag, 2004) states that wearing biosensors is less disturbing than being "watched" by a camera. We think that a friendly-looking social robot that is interacting with the interlocutor is not perceived as annoying or indiscrete.

5.7   Conclusion

In order to achieve natural communication between social robots and humans, important modalities have to be addressed. In the process of communication, social robots might apply a camera to identify the heart rate, heart rate variability or respiration rate of a user to enable the detection of emotional states. Since social robots will mostly be used indoors and in the homes of users, many sources of noise and disturbance might affect the camera-based vital data recognition. As one of the main noise factors, we identified the artificial light that surrounds the social robot. Aliasing filters can be used to reduce that noise in combination with an adapted frame rate to avoid side effects.

A sensitive conversational partner is capable of reacting to changing emotional states during a conversation. Our approach involves the integration of sensitivity in order to measure and understand the feelings of the interlocutor. In future applications, we envision social robots changing their facial expressions or skin color according to their emotional state to enable an exchange of emotional states with other social robots and humans. We are aware of the fact that social robots might receive more information about the emotional state than the interlocutors may want. This leads to interesting future scenarios that might involve social robots in job interviews, patient anamneses, and social or chaplain tasks, or even in polygraph (lie-detection) applications.

Our future work will focus on improving the vital sign recognition with cameras as well as the natural communication between social robots and humans.

References

Baer, R. L. (2000). CCD requirements for digital photography. In *IS and TS Pics Conference, Society for Imaging Science & Technology*, pp. 26–30.

Bieber, G. H. (2013). Sensor requirements for activity recognition on smart watches. In *Proceedings of the 6th International Conference on PErvasive Technologies Related to Assistive Environments (PETRAE'13)*, ACM.

Biggs, A. (2005). Paro therapeutic robot seal. Published on Flickr as CC-BY-SA-2.0, http://www.flickr.com/people/ehjayb/ehjayb.

Bradski, G., & Kaehler, A. (2008). *Learning OpenCV: Computer vision with the OpenCV library*. O'Reilly Media, Inc.

Brundrett, G. W. (1974). Human sensitivity to flicker. *Lighting Research & Technology, 6*(3), 127–143.

Clancy, E. A. (2002). Sampling, noise-reduction and amplitude estimation issues in surface electromyography. *Journal of Electromyography and Kinesiology, 12*(1), 1–16.

Etoh, T. G., Takehara, K., Okinaka, T., Takano, Y., Ruckelshausen, A., & Poggemann, D. (2001). Development of high-speed video cameras. In *SPIE 4183, 24th International Congress on High-Speed Photography and Photonics.*

Goodrich, M. A. (2007). Human-robot interaction: a survey. *Foundations and Trends in Human-Computer Interaction, 1*(3), 203–275.

Haag, A. G. (2004). Emotion recognition using bio-sensors: First steps towards an automatic system. In *Tutorial and Research Workshop on Affective Dialogue Systems* (pp. 36–48). Berlin, Heidelberg: Springer.

He, C. Y. (2017). An emotion recognition system based on physiological signals obtained by wearable sensors. *Wearable Sensors and Robots* (pp. 15–25). Singapore: Springer.

Hertzman, A. B. (1937). Photoelectric plethysmography of the fingers and toes in man. *Proceedings of the Society for Experimental Biology and Medicine, 37*(3), 529–534. https://doi.org/10.3181/00379727-37-9630.

Inan, O. T. (2015). Ballistocardiography and seismocardiography: A review of recent advances. *IEEE Journal Biomedical and Health Informatics, 19*(4), 1414–1427.

Irani, R. N. (2014). Improved pulse detection from head motions using DCT. In *2014 International Conference on Computer Vision Theory and Applications (VISAPP)* (pp. 118–124). IEEE.

Jerritta, S. M. (2011). 7th International Colloquium on physiological signals-based human emotion recognition: A review. In *Signal Processing and its Applications (CSPA)* (pp. 410–415), IEEE.

Kong, L. Z. (2013). Non-contact detection of oxygen saturation based on visible light imaging device using ambient light. *Optics Express, 21*(15), 17464–17471.

Korn, O. B. (2018). Perspectives on social robots: from the historic background to an experts' view on future developments. In *Proceedings of the 11th PErvasive Technologies Related to Assistive Environments Conference.*

Matthes, K. (1935). Untersuchungen über die Sauerstoffsättigung des menschlichen Arterienblutes. *Naunyn-Schmiedebergs Archiv für experimentelle Pathologie und Pharmakologie, 179*(6), 698–711.

Matthies, D. J. C., Haescher, M., Bieber, G., Salomon, R., & Urban, B. (2016). SeismoPen: Pulse recognition via a smart pen. In *International Conference on Pervasive Technologies Related to Assistive Environments, PetraE 2016*, ACM, Corfu Island, Greece. https://dx.doi.org/10.1145/2910674.2910708.

Mestha, L. K. (2014). Towards continuous monitoring of pulse rate in a neonatal intensive care unit with a webcam. In *2014 36th Annual International Conference of the IEEE Engineering in Medicine and Biology Society (EMBC)* (pp. 3817–3820).

Nath, A. R. (2012). A neural basis for interindividual differences in the McGurk effect, a multisensory speech illusion. *Neuroimage, 59*(1), 781–787.

Poh, M. Z. (2011). Advancements in noncontact, multiparameter physiological measurements using a webcam. *IEEE Transactions on Biomedical Engineering, 58*(1), 7–11.

Rohracher, H. (1964). Microvibration, permanent muscle-activity and constancy of body temperature. *Perceptual and Motor Skills, 19*(1), 198–198.

Sauer, S., Gobel, S. (2003). Focus your young visitors: kids innovation–fundamental changes in digital edutainment. In *Museums and the Web 2003: Selected Papers from an International Conference (7th, Charlotte, NC, March 19–22, 2003).*

Smith, W. (2005). *Modern lens design.* Publisher McGraw-Hill Professional Engineering.

Smith, W. (2007). *Modern optical engineering* (4th ed.). Publisher McGraw-Hill Professional Engineering.

Weng, C. C. (2005). A novel automatic white balance method for digital still cameras. In *IEEE International Symposium on Circuits and Systems, ISCAS 2005* (pp. 3801–3804).

Wu, H. Y. (2012). *Eulerian video magnification for revealing subtle changes in the world.* https://doi.org/10.1145/2185520.2185561.

Zissman, M. A. (1996). Comparison of four approaches to automatic language identification of telephone speech. *IEEE Transactions on Speech and Audio Processing, 4*(1), 31.

Chapter 6
Interacting with Collaborative Robots—A Study on Attitudes and Acceptance in Industrial Contexts

Sarah L. Müller-Abdelrazeq, Kathrin Schönefeld, Max Haberstroh and Frank Hees

Abstract Through combining robots' power, consistency, and accuracy with humans' creativity and flexibility, human–robot interaction offers new ways of manufacturing. For the successful introduction of human–robot interaction in manufacturing, it is important not only to consider the necessary change of qualification, but also to create a positive attitude toward this new technology as expectations trigger behavior and consequently influence the quality of work. The study reported in this chapter analyzes how attitudes toward collaborative robots are influenced through interaction with an industrial robot.

Keywords Human–robot interaction · Industry 4.0 · Attitude toward robots · Collaborative assembly

6.1 Introduction

As technology improves, human–robot interaction is spreading in industry. A first listing of certain actions that can be more effectively performed by humans or machines was made back in 1951 (Fitts, 1951). This listing is still relevant today (e.g., humans are more flexible and judging, whereas machines have more power and are better at computation). To be able to use the advantages of robots in power, consistency, and accuracy as well as those of humans in creativity and flexibility, the advantages can be combined. Accordingly, human–robot interaction can increase its efficiency

S. L. Müller-Abdelrazeq (✉) · K. Schönefeld · M. Haberstroh · F. Hees
RWTH Aachen University, Aachen, Germany
e-mail: sarah.abdelrazeq@ima-ifu.rwth-aachen.de

K. Schönefeld
e-mail: kathrin.schoenefeld@ima-ifu.rwth-aachen.de

M. Haberstroh
e-mail: max.haberstroh@ima-ifu.rwth-aachen.de'

F. Hees
e-mail: frank.hees@ima-ifu.rwth-aachen.de

through parallel and interactive task execution by man and robot, and this lowers costs compared to stationary, fully automated or manual solutions. By using the specific strengths of man and robot, ergonomics can be improved too. Furthermore, human–robot interaction increases flexibility and adaptability with regard to the placement of the handling technology, capacity, experience, and knowledge of the personnel as well as the type and scope of the task (Helms & Meyer, 2005).

Despite the advantages of human–robot interaction, the spread of collaborative robots has so far been low. In Germany, for example, the proportion of collaborative robots in the total market for industrial robots is less than 2%, which corresponds to less than one hundred collaborative robots in Germany (Buchenau, Höpner, & Wocher, 2016). In addition to technical reasons, there are also sociopsychological reasons for the low prevalence of collaborative robots, such as the lack of acceptance of robots among employees (Görke et al., 2017).

Acceptance has attitudinal and behavioral components (Arndt, 2011; Louho, Kallioja, & Oittinen, 2006). Accordingly, one can accept or reject (attitude), and use or not use a technology (behavior). Attitude and behavior do not necessarily coincide—a person can have a certain attitude toward a certain technology without ever really experiencing it. When using collaborative robots for the first time, people cannot rely upon previous experience. If experience is missing, expectations trigger behavior (Madhavan & Wiegmann, 2007) which is based on the general attitude. Especially when interacting with a robot, individual behavior might be influenced by the individual's general attitude and the overall societal opinion (Weiss, Bernhaupt, Tscheligi, & Yoshida, 2009). It can be assumed that negative expectations negatively influence the behavior toward the collaborative robot and lead to rejection. An aversion to new technology, for whatever reason, may lead to erroneous operation which in turn leads to a decreased quality of work (Buche, Davis, & Vician, 2012). However, expectations of new technology can be manipulated (Mayer, Fisk, & Rogers, 2009). Thus, aversion can be counteracted by positive experiences; negative experiences, on the other hand, can confirm negative expectations and reinforce them. Technology experience in general is positively related to the evaluation of technology (Melenhorst & Bouwhuis, 2004).

The present work deals with the influence of experience on attitude toward collaborative robots. Both general robot-related experiences (e.g., through the media) in recent years are considered and the current experience of direct interaction with a robot. For this purpose, an experiment with an alternative-treatments design was conducted. The gathered insights give hints on training and sensitization possibilities that increase the acceptance of employees and thus facilitate the successful introduction of human–robot interaction.

In the next section, human–robot interaction and attitude are defined, and the related work regarding the attitude toward robots and some influential factors are reported. In Sect. 6.3, the study's method is described in detail. In Sect. 6.4, the results are reported. In the last section, the results and limitations are discussed.

6.2 Related Work

6.2.1 Human–Robot Interaction

Current industry standards and international standards (e.g., ISO) often do not distinguish between different forms of human–robot interaction. In scientific publications, on the other hand, a distinction is usually made between coexistence, cooperation, and collaboration (see Fig. 6.1).

The coexistence of humans and robots is the weakest form of human–robot interaction (Schmidtler et al., 2015). Onnasch et al. (2016) understand coexistence as "an episodic meeting of human and robot, the interaction is very limited in time and space". This form of interaction does not imply a common goal. The purpose of the interaction is to avoid mutual damage and collisions. Therefore, both a temporal and a spatial variable are important for coexistence (ibid.).

The term "cooperation" is defined by Roschelle and Teasley (1995) as cooperative work accomplished by the division of labor, where both human and robot are responsible for a certain portion of the task. Accordingly, cooperation is based on a common goal. The overall task is subdivided into subtasks; because of the clear division of labor, robot and human do not depend on each other.

The term "collaboration" describes a process in which humans and robots work together on one part of the final result and are in direct contact with each other (Schmidtler et al., 2015). The difference with cooperation is that in collaboration both human and robot are involved in the production of all results of a project. In this respect, there are immediate coordination requirements. The distribution of subtasks takes place continuously and, if necessary, directly during the collaboration. Collaboration is also characterized by the creation and use of synergies (Onnasch et al., 2016).

Coexistence	Cooperation	Collaboration
work time	work time	work time
working area	working area	working area
	work objective	work objective
		contact

Fig. 6.1 Forms of human–robot interaction (see Onnasch, Maier, & Jürgensohn, 2016; Schmidtler et al., 2015)

The application areas of (collaborative) robots are divided into industry and service (DIN Deutsches Institut für Normung e. V., 2010). Service robots provide services useful to the well-being of humans. Industrial robots are used in industrial applications; they are usually equipped with grippers or tools. The present work focuses only on industrial collaborative robots. Industrial robots are automatically guided, multi-purpose manipulators equipped with at least three freely programmable motion axes. They are stationary or mobile and are used in industrial applications. Industrial robots can use different tools, depending on the given task (DIN Deutsches Institut für Normung e. V., 2010).

Collaborative robots or cobots are a special type of industrial robot. They can interact directly with humans and are therefore usually not enclosed (Cherubini, Passama, Crosnier, Lasnier, & Fraisse, 2016; Kadir, Brodberg, & Conceicao, 2018). They must meet special safety requirements. If these are met, use of cobots allows close interaction between humans and robots.

6.2.2 Attitude Toward Robots

The word attitude comes from the Latin *aptus* and means suitable or adapted. When comparing the numerous definitions of attitude, a common elementary component emerges: attitude refers to a mental state of preparation for action (Allport, 1935; Nomura, Kanda, Suzuki, & Kato, 2004). Certain persons, objects, institutions, situations, or themes are perceived in a certain, usually bipolar, formulated way as good or bad, harmful or useful, pleasant or unpleasant, sympathetic or unappealing. Pure perception of the object is a cognitive process, while the following assessment is affective (Bergman, 1998). This evaluation usually leads to a certain behavior or reaction.

The origin of attitudes is experiential and can be found in culture, family or personal experiences (Chaplin, 1985). Through positive or negative consequences of a certain behavior, i.e., through a socialization process, the individual develops attitudes as a relatively stable and permanent predisposition (Fishbein & Ajzen, 1980).

Explicit and implicit attitudes are often distinguished (Wilson, Lindsey, & Schooler, 2000). Explicit attitudes are conscious attitudes, while implicit attitudes are unconscious. Explicit and implicit attitudes can be contradictory.

Regarding attitudes toward robots, a big European survey aimed to determine the European public's attitude. The Eurobarometer is a regular public opinion survey commissioned by the European Commission and carried out in all countries of the EU. The 2012 Eurobarometer survey (Special Eurobarometer 382, 2012) reports that only 12% of EU citizens have experience with using robots. However, more than two-thirds of EU citizens have a positive opinion of robots. The survey finds that 88% of respondents agree that robots are necessary and helpful because they can do work that is too difficult or too dangerous for humans. However, 91% of respondents agree with the statement that robots "are a form of technology that requires careful management" (Special Eurobarometer 382, 2012). Concern that use of robots could lead to job losses was stated by 70% of respondents. Robots should therefore be used in particular when work is too tough or too dangerous for humans, for example in space exploration (52%), in manufacturing (50%), in military and security applications (41%), and in search and rescue (41%).

Whether the attitude toward robots changes after experience (actual experience or experience via media) with a robot is a controversial subject of debate. While some studies say that attitudes change through experience (Nomura, Suzuki, Kanda, & Kato, 2006; Nomura, Suzuki, Kanda, Yamada, & Kato, 2011; Weiss et al., 2009, p. 156), others say that they do not (de Graaf & Ben Allouch, 2013; Halpern & Katz, 2012).

When talking about the attitude toward technology or specifically about robots, it is reasonable not to neglect the demographic aspects as a control variable. The connection between attitude and age and gender is well documented.

In the European project "Keeping the Elderly Mobile", it was emphasized that the view that modern technology would make life more difficult was more pronounced in the group of people over 75 than in the group of people aged 55–75 (Marcellini, Mollenkopf, Spazzafumo, & Ruoppila, 2000). In this respect, age has an influence on the attitude toward technology. Also in the concrete context of attitudes toward robots, previous studies reported correlations between attitude and age (Dinet & Vivian, 2014; Scopelliti, Giuliani, & Fornara, 2005). However, age is not adequate as the sole means for measuring attitude. Previous experience with technology is important in this context (Claßen, 2013).

Studies of young people's attitudes toward technology have shown that boys use computers more often than girls; they have more experience in this way of dealing with technology and show greater interest (Claßen, 2013). Similar findings were obtained for the technological behavior of adults and older people. Furthermore, women were characterized by a greater phobia of technology and less self-confidence in their own abilities and greater fear when dealing with technology compared to men. In this respect, a gender difference can be observed in the use of technology (Broos, 2005; Sieverding, 2005). This difference could be explained, among other things, by the fact that boys are more motivated by their environment to deal with technology and take appropriate school courses. "Classical" role models and role distributions in the family, according to which the man is responsible for technical matters, promote the reported observations (Broos, 2005; Sieverding, 2005). Gender effects were also evident in attitudes toward robots (Nomura et al., 2004).

Technical experience plays a role insofar as it is positively related to the evaluation of technology (Melenhorst & Bouwhuis, 2004). It can be assumed that experience with technology is often gained in the context of professional activity. This is particularly true of men, despite the increasing number of women in employment. In addition, the level of development of the country in which the respective person lives plays an important role: Access to technology is much easier in industrialized countries (Czaja & Sharit, 1998).

These research examples show that both age and gender are strongly related to technical experience. In order to determine the effect of general previous experience with robotics and the effect of direct interaction with a collaborative robot, demographic aspects are included in the following study.

6.3 Study

In order to gain insight into the effect of experience on attitudes toward collaborative robots, an experiment was carried out. A between-subject experimental design was chosen, which means that two groups of participants were exposed to different experimental conditions. In the following, the effects of both groups were combined. In order to exclude the possibility that group differences in the results were based on differences that existed before the study, a pretest was carried out. The difference between pre- and post-measurement can provide additional information about the exact extent of change. This experimental design is referred to as an alternative-treatments design with a pretest (Shadish, Cook, & Campbell, 2002).

Whereas one group was provided with a positive human–robot experience where they could self-determine the work pace, the second group was provided with a more negative human–robot experience with a prescribed cycle time[1] (corresponding to the working speed of an experienced assembler). Participants were randomly assigned to the groups.

6.3.1 Task and Procedure

The subject of investigation was the robot Universal Robot 5 (UR-5). Collaborative work was approved according to EN ISO 13849 and EN ISO 10218-1 (Universal Robots, 2015).

The UR-5 was embedded in an assembly cell (width × height × depth: 170 cm × 100 cm × 190 cm). The assembly cell was equipped with storage boxes (including nuts, screws, spacers, and gear components), assembly tools (e.g., Allen key, torque screwdriver), a touch pad, holding devices, and a tray box (see Fig. 6.2). The UR-5 mainly operated in the left part of the cell; the human operated mainly in the right part.

The product to be built was a gear transmission with five gearwheels. The gear transmission consisted of 3D-printed plastic components, screws, nuts, and spacers (see Fig. 6.3).

Human and robot had to assemble three identical gear transmissions in a row in close cooperation and with partial collaboration, so their work steps were interdependent. While the human had to complete five work steps, the robot had to fulfill four work steps (see Table 6.1). Instructions were given on a touch pad where the

---

[1]Cycle time (sometimes also called takt time) is defined as the maximum time allowed for producing a product in order to meet the customer's demands (Schroer, 2004).

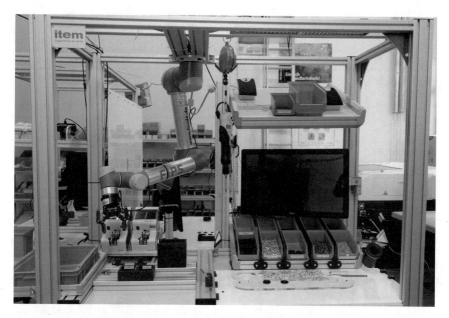

Fig. 6.2 Assembly cell with the UR-5 hanging from the ceiling

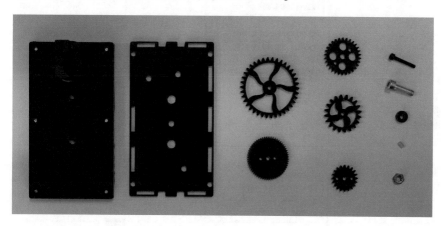

Fig. 6.3 Gear parts for assembly

human received pictorial and written instructions. The touch pad also explained the
robot's steps and gave an overview of all steps.

Table 6.1 Assembly steps for one gear transmission

Human		Robot	
Work step	Description	Work step	Description
Preparation 1	Position the base plate and the back plate according to the illustration into the front plate holder		
Preparation 2	Place two M5 nuts in the mounting device. For help you can use the positioning stick		
Assembly 1	Mount two gear sets. To achieve this, use two M3 grub screws. As an aid you can use the centering pins on the working surface and an Allen key	Positioning	The robot places the base plate in the assembly gadget
		Screws	The robot inserts four M3 socket screws. Subsequently it will position a back plate into the fixture
Assembly preparation	Wait until the robot has placed the back plate into the fixture. Then put two M5 construction screws into the back plate. Screw them with the torque screwdriver		
		Assistance	The robot presents the preassembled base plate
Assembly 2	Place the gears or gear sets and the spacer sleeve on the base plate as shown in the illustration. Subsequently, carry out a function test of the gear drive		
		Tray	The robot puts the product in the intended box

Table 6.2 Robot-related experiences

	M (SD)
Reading material	3.08 (1.87)
Watching media	2.94 (1.78)
Physical contact	2.62 (2.04)
Attending events	1.91 (1.96)
Built or programmed	1.79 (1.91)

six-point scale ($0 = $ *zero times* to $5 = $ *five or more times*)

6.3.2 Participants

The sample of the present experiment consisted of 80 participants. Due to the missing values or otherwise invalid data sets, lack of language skills of the participants, and technical difficulties with the UR-5, the sample size was reduced to 66 participants with an almost balanced gender distribution (female $= 34$, male $= 32$). The average age of the sample was $M = 25.58$ years (SD $= 5.77$, $n = 65$, 1 missing) with a range of 19–53 years.

The participants' previous experience with robots is shown in Table 6.2. The highest value was achieved by the item "How often did you read robot-related stories, comics [...] last year" with $M = 3.08$ ($SD = 1.87$) and the lowest by the item "How often have you built or programmed a robot so far in your life?" with $M = 1.79$ ($SD = 1.91$). The total average over the sum of the five items was $M = 12.43$ ($SD = 7.84$), with a range of 0–25.

6.3.3 Method

The experiment was framed with a pre- and post-test. Data were collected via a paper-and-pencil questionnaire.

The frequency with which the participants had read material, watched media, attended events concerning collaborative robots, had physical contact with them or built or programed a robot, was measured before the experiment with an adapted version of the scale by MacDorman, Vasudevan, and Ho (2009). Frequency was indicated on a six-point scale ($0 = $ *zero times* to $5 = $ *five or more times*).

The Special Eurobarometer (Special Eurobarometer 382, 2012) aims to capture public attitudes toward robots. The original scale comprised nine blocks, of which QA5 and QA6 were used and adapted for the present study. The first block asks for the subjectively perceived influence of robots on society through a five-point Likert scale ($1 = $ *I disagree* to $5 = $ *I agree*) and four items. An example item is "Collaborative robots destroy jobs". The second block asks for preferred areas of application of robots, in which participants must select a maximum of three areas out of 11. Derived from Thrun's observation (Thrun, 2004), it was assumed that a robot's key characteristic derives directly from its occupational category (industrial, professional service or personal service). Thus, we surmised that respondents could reasonably imagine an appropriate kind of collaborative robot for the job.

The "Negative Attitudes toward Robots Scale" (NARS; Nomura, Kanda, & Suzuki, 2006) asks for the participants' attitude to robots using a total of 14 items. On a five-point Likert scale ($1 = I \ disagree$ to $5 = I \ agree$), participants could communicate their consent. Again, "collaborative robots" complemented the wording. The items can be subdivided into three subscales: (S1) A negative attitude to situations of interaction with robots (six items), (S2) A negative attitude toward the social influence of robots (five items), and (S3) Negative attitudes about one's own emotions when interacting with robots (three items). An individual score on each subscale is calculated by summing the values of each subscale's items. Thus, the minimum and maximum points are 6 and 30 for S1, 5 and 20 for S2, and 3 and 15 for S3, respectively. In S3, all three items must be reversed.

The Special Eurobarometer (Special Eurobarometer 382, 2012) and NARS (Nomura et al., 2006) were measured both before and after the experiment.

6.4 Results

6.4.1 Robot-Related Experiences

Male participants ($M = 15.84, SD = 6.60$) had significantly more robot-related experience than female participants ($M = 9.12, SD = 7.61$), $t(63) = 3.801, p < 0.001$. The effect size was $r = 0.43$ and corresponded to a medium effect (Cohen, 1992). A comparison with the Japanese and US-American sample from MacDorman et al. (2009) with $n = 731$ showed that male participants in the present study had significantly more robot-related prior experience than male participants in the reference samples (Japan: $M = 11.8$; USA: $M = 7.0$). The average previous experience of female participants was significantly higher than that of the female reference sample from the USA ($M = 5.4$), but just below that of the Japanese reference sample ($M = 11.3$). In addition, there was a significant correlation of robot-related experience with age in this sample, $r = 0.307, p = 0.014$. According to Cohen (1992), this corresponds to a medium effect.

6.4.2 Negative Attitudes Toward Robots (NARS)

In order to find out whether there were systematic differences between participants with regard to their attitude toward robots before the actual experiment and whether correlations between certain characteristics and the attitude could be found, corresponding test procedures were carried out.

Table 6.3 Gender differences in NARS

	S1		S2		S3	
	M (SD)	t	M (SD)	t	M (SD)	t
Male	11.31 (3.49)	−3.728*	13.97 (3.68)	−1.65	9.06 (2.56)	−1.695
Female	15.15 (4.70)		15.52 (3.81)		10.12 (2.47)	

*p < 0 .001

Table 6.4 Experience-based differences in NARS

	S1		S2		S3	
	M (SD)	t	M (SD)	t	M (SD)	t
H (n = 31)	11.65 (3.05)	3.133*	14.27 (2.91)	1.072	9.03 (2.48)	1.058
L (n = 32)	15.03 (5.26)		15.29 (4.55)		10.23 (2.53)	

H group with much experience, *L* group with little experience
*p < 0.005

A t-test for independent samples showed a significant effect between the sexes. Women showed more negative attitudes toward situations of interaction with robots ($M = 15.15$, $SD = 3.49$) than men ($M = 11.31$, $SD = 3.49$), $t(63) = −3.728$, $p < 0.001$ (see Table 6.3). This corresponded to a mean effect where $r = 0.43$. With regard to the other dimensions of NARS, there were no significant differences.

There were no significant correlations with regard to age (S1: $r = −0.114$, $p = 0.368$, S2: $r = −0.112$, $p = 0.383$, S3: $r = −0.080$, $p = 0.531$).

In order to find out whether there was a difference in attitude between people with more or less robot-related experience, the participants were first divided into a group with much experience and a group with little experience by means of a median split. The subsequent t-test for independent groups showed significant differences. Participants with little previous experience ($M = 15.03$, $SD = 5.26$) showed more negative attitudes toward situations of interaction with robots than participants with much experience ($M = 11.65$, $SD = 3.05$), $t(47.203) = 3.133$, $p = 0.003$ ($r = 0.41$ corresponds to a mean effect). In S2 and S3, there were no significant differences (see Table 6.4).

As sex and robot-related experiences were correlated, a multiple regression was carried out to analyze the effect of both independent variables on S1. The analysis showed that both sex and robot-related experience predicted S1 significantly, $F(2.63) = 9.414$, $p < 0.0001$, $n = 64$. If the robot-related experiences increased by a factor of one within the same gender, the value of S1 decreased on average by −0.146. With the same previous experience, the value in S1 of women was 2.941 higher than the value of men. 21.1% of the value in S1 could be explained by the two independent variables, which corresponded to a medium effect.

To compare whether the NARS values before and after experience with the UR-5 differed, a *t*-test was performed for dependent samples. There was a significant difference in attitude to situations of interaction with robots, $t = 2.764$, $p = 0.008$, $n = 63$. Before the task, participants had more negative attitudes ($M = 13.44$, $SD = 4.49$) than after the task ($M = 12.59$, $SD = 4.75$). The effect size was $r = 0.33$ and thus corresponded to an average effect. The factors S2 and S3 showed no significant pre-post difference.

6.4.3 Eurobarometer

In both pre- and post-tests, the manufacturing industry was the preferred application area for collaborative robotics, closely followed by space research. Overall, frequencies in both pre- and post-tests showed that participants preferred collaborative robots in areas of application that were dangerous or physically demanding for humans. Only a few participants saw areas of application primarily in the social sphere, such as education, leisure or in care of people in need of help. The frequency distribution is shown in Fig. 6.4.

In order to test whether the interaction with the UR-5 led to a change in the preferred areas of collaborative robotics, a Pearson's chi-square test was conducted. The favored areas of application showed no connection with the time of measurement, $\chi^2 = 2.685$, $p = 0.988$.

When interviewed about their attitudes toward collaborative robots, the participants showed a positive opinion in the pretest. A majority agreed that "collaborative robots are necessary as they can do jobs that are too hard or too dangerous for peo-

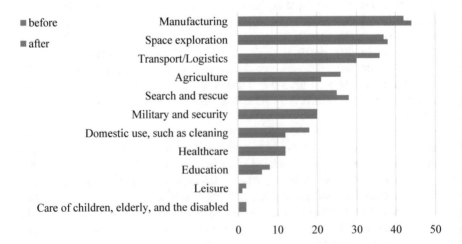

Fig. 6.4 Frequencies of the preferred areas of application of collaborative robots before and after the experiment

ple" (95.5% overall agreement) and that "they are good for society because they help humans" (87.9% overall agreement). On the other hand, there was no agreement on the statements concerning threat to jobs and benefits of collaborative robots for employment opportunities. While 34.8% of participants saw collaborative robots as a threat to jobs, 25.8% rejected this statement and 39.4% had a neutral opinion on this. The results showed that 48.5% agreed with the statement that "collaborative robots can boost job opportunities", while 19.7% rejected this statement and 31.8% evaluated it neutrally.

The evaluation of these items did not reveal any significant correlation with age. Only the item "Collaborative robots are good for society because they help people" showed a significant difference between the sexes, $t(64) = 2.909, p = 0.005$. Here women agreed less ($M = 3.97, SD = 0.67$) than men ($M = 4.47, SD = 0.72$). This item also showed that people with much robot-related previous experience agreed more with this statement ($M = 4.44, SD = 0.71$) than people with little previous experience ($M = 3.97, SD = 0.70$), $t(64) = -2.741, p = 0.008$. A multiple regression analysis showed that only sex influenced the evaluation of this item, $F(2.64) = 4.720, p = 0.011, n = 65$, as robot-related experience did not have a significant regression coefficient β. Women had an average value that was 0.401 lower than the men's value. It was found that 10.7% of the variance of this item could be explained by sex, which corresponded to a small effect. For the other items, there were no gender- or experience-based differences.

To compare the items before and after the task, a t-test for dependent samples was performed. It was found that the participants before the task considered collaborative robots to be a greater threat to jobs ($M = 3.15, SD = 1.08$) than after the task ($M = 2.80, SD = 1.14$), $t(64) = 3.486, p = 0.001$ ($r = 0.23$ corresponded to an average effect). In addition, it was shown that the participants considered the handling of collaborative robots more cautiously before the task ($M = 4.29, SD = 0.82$) than after the task ($M = 4.03, SD = 0.92$), $t(64) = 2.472, p = 0.016$). The effect sizes were $r = 0.40$ and $r = 0.30$, respectively, and thus corresponded to average effects. The other items did not differ significantly between the measurement times (see Table 6.5).

6.5 Conclusion

The results show that male participants had more general robot-related experience than female participants did. Age and general robot-related experiences were positively correlated. Regarding the attitude toward collaborative robots, women showed more negative attitudes than men did. Participants with more robot-related experience showed a more positive attitude toward collaborative robots. Both variables had their own significant explanatory contribution. There was no age effect regarding the attitude. The attitude toward situations of interaction with collaborative robots changed positively even after a short experience with the collaborative robot. Furthermore, after interaction with the robot, participants saw collaborative robots as

Table 6.5 Before and after comparison of Eurobarometer items

	Before	After	
	M (SD)	M (SD)	t
Collaborative robots are a good thing for society, because they help people	4.22 (0.74)	4.12 (0.78)	1.23
Robots steal peoples' jobs	3.15 (1.08)	2.80 (1.14)	3.486**
Collaborative robots are necessary as they can do jobs that are too hard or too dangerous for people	4.58 (0.53)	4.44 (0.66)	1.494
Collaborative robots are a form of technology that requires careful management	4.29 (0.82)	4.03 (0.92)	2.472*
Widespread use of collaborative robots can boost job opportunities in the EU	3.35 (0.98)	3.37 (0.98)	−0.173

*p < 0.05; **p < 0.005

less of a threat to their job and rated the difficulty of handling robots as less dangerous. The preferred areas of application did not change significantly between the measurements.

The results suggest that the experience with robots, regardless of whether the experience is abstract (e.g., through media) or direct through contact with a robot, has a positive effect on the attitude toward collaborative robots. This may be because people are better able to assess the robot with its strengths and weaknesses. They see, for example, that humans are still far superior to robots in terms of flexibility—which can reduce the fear that humans will soon become superfluous in the factory. The results of this study thus reinforce the results of Nomura et al. and Weiss et al. (see Sect. 6.2.2). Additionally, the study shows that a change of attitude is caused by interaction not only in humanoid but also in industrial robots.

For companies, this implies that the introduction of human–robot solutions should ideally be preceded by an opportunity for employees to acquire positive experiences with robots. Besides direct training, playful interaction during breaks can also be an alternative. As demonstrated at the Hanover Fair 2017, robots can be used to make cotton candy (e.g., Franka Emika) or to open bottles (e.g., KUKA), for example. As acceptance is key when implementing new technology, an early measurement of acceptance is important (Kummer, Schäfer, & Todorova, 2013), so the extent of (counter-) measures can be estimated. The lower the original acceptance, the more sensitization is necessary. If no real robot is available, abstract training and experience through videos, posters, etc. can be used to create a more positive attitude.

Even after the experience with the UR-5, the participants considered the areas of application to be only slightly changed. This indicates that even after positive experience, the preferred fields of application for robots are difficult or dangerous areas for humans. This is in accordance with the results of the original Eurobarometer study, but may also be because an industrial robot was used in this collaborative use case and an industrial robot would be unusual in other application areas such as care.

Since the study was a manufacturing scenario, it is no surprise that the application area of manufacturing was rated slightly higher than before the study.

6.5.1 Limitations and Future Work

The study has some limitations that should be considered when interpreting the results. First, the sample consisted mainly of students from a technical university. It is possible that construction workers would perceive human–robot interaction in a different way. Furthermore, it has been shown that age has an effect on attitudes toward technology (Marcellini et al., 2000) and the sample of this study was relatively young. However, age is not sufficient as the sole reason for measuring technology acceptance. Crucial in this context is the prior experience with technology (Claßen, 2013). The study's sample already had many robot-related experiences and cannot be seen as representative of a European or German population. Nevertheless, the sample of this study answered the Eurobarometer items similarly to the European sample. Second, habituation effects might occur, so that the described patterns might not remain over time. Third, it must be explored whether the findings are transferable to real production environments.

Future studies will look in a more detailed manner for the origins of positive or negative attitudes toward collaborative robots. Besides demographic factors, robot-related and environmental factors will be studied.

Acknowledgements This research and development project is funded by the German Federal Ministry of Education and Research (BMBF) within the "Innovations for Tomorrow's Production, Services, and Work" Program (funding number 02L14Z000) and implemented by the Project Management Agency Karlsruhe (PTKA). The authors are responsible for the contents of this publication.

References

Allport, G. W. (1935). Attitudes. In C. Murchison (Ed.), *Handbook of Social Psychology*. Worcester, Mass.: Clark University Press. Retrieved from http://web.comhem.se/u52239948/08/allport35.pdf.

Arndt, S. (2011). *Evaluierung der Akzeptanz von Fahrerassistenzsystemen: Modell zum Kaufverhalten von Endkunden*. Zugl.: Dresden, Techn. Univ., Diss., 2010 (1. Aufl.). *Verkehrspsychologie*. Wiesbaden: VS Verl. für Sozialwiss.

Bergman, M. M. (1998). A theoretical note on the differences between attitudes, opinions, and values. *Swiss Political Science Review, 4*(2), 81–93. https://doi.org/10.1002/j.1662-6370.1998.tb00239.x.

Broos, A. (2005). Gender and information and communication technologies (ICT) anxiety: Male self-assurance and female hesitation. *Cyberpsychology & Behavior: The Impact of the Internet, Multimedia and Virtual Reality on Behavior and Society, 8*(1), 21–31. https://doi.org/10.1089/cpb.2005.8.21.

Buche, M. W., Davis, L. R., & Vician, C. (2012). Does technology acceptance affect e-learning in a non-technology intensive course? *Journal of Information Systems Education, 23*(1), 41–50.

Buchenau, M.-W., Höpner, A., & Wocher, M. (2016). *Seite an Seite mit dem Menschen: Die Roboter kommen*. Retrieved from https://www.handelsblatt.com/technik/hannovermesse/seite-an-seite-mit-dem-menschen-die-roboter-kommen/13504490-all.html.

Chaplin, J. P. (1985). *Dictionary of psychology*. New York: Dell.

Cherubini, A., Passama, R., Crosnier, A., Lasnier, A., & Fraisse, P. (2016). Collaborative manufacturing with physical human–robot interaction. *Robotics and Computer-Integrated Manufacturing, 40,* 1–13. https://doi.org/10.1016/j.rcim.2015.12.007.

Claßen, K. (2013). *Zur Psychologie von Technikakzeptanz im höheren Lebensalter: Die Rolle von Technikgenerationen*. Dissertationsschrift, Heidelberg: Heidelberg University Library.

Cohen, J. (1992). A power primer. *Psychological Bulletin, 112*(1), 155–159. https://doi.org/10.1037//0033-2909.112.1.155.

Czaja, S. J., & Sharit, J. (1998). Age differences in attitudes toward computers. *Journals of Gerontology. Series B, Psychological Sciences and Social Sciences, 53*(5), 329–340.

de Graaf, M. M. A., & Ben Allouch, S. (2013). Exploring influencing variables for the acceptance of social robots. *Robotics and Autonomous Systems, 61*(12), 1476–1486. https://doi.org/10.1016/j.robot.2013.07.007.

DIN Deutsches Institut für Normung e. V. (2010). *Roboter und Robotikgeräte—Wörterbuch (ISO/DIS 8373:2010); Deutsche Fassung prEN ISO 8373:2010*. (DIN EN ISO, 8373). Berlin: Beuth Verlag GmbH.

Dinet, J., & Vivian, R. (2014). Exploratory investigation of attitudes towards assistive robots for future users. *Le Travail Humain, 77*(2), 105. https://doi.org/10.3917/th.772.0105.

Fishbein, M., & Ajzen, I. (1980). *Belief, attitude, intention, and behavior: An introduction to theory and research* (4. print). *Addison-Wesley series in social psychology*. Reading, Mass.: Addison-Wesley.

Fitts, P. M. (1951). *Human engineering for an effective air-navigation and traffic-control system*. Columbus Ohio.

Görke, M., Blankemeyer, S., Pischke, D., Oubari, A., Raatz, A., & Nyhuis, P. (2017). Sichere und akzeptierte Kollaboration von Mensch und Maschine. *ZWF Zeitschrift Für Wirtschaftlichen Fabrikbetrieb, 112*(1–2), 41–45. https://doi.org/10.3139/104.111668.

Halpern, D., & Katz, J. E. (2012). Unveiling robotophobia and cyber-dystopianism: The role of gender, technology and religion on attitudes towards robots. In H. Yanco (Ed.), *Proceedings of the Seventh Annual ACMIEEE International Conference on Human-Robot Interaction* (pp. 139–140). New York, NY: ACM.

Helms, E., & Meyer, C. (2005). Assistor—Mensch und Roboter rücken zusammen: Einsatz von sicheren und einfach bedienbaren Assistenzrobotern im Produktionsalltag. *Wt Werkstattstechnik Online, 95*(9), 677–683.

Kadir, B. A., Brodberg, O., & Conceicao, C. (2018). *Designing human-robot collaborations in Industry 4.0: Explorative case studies*.

Kummer, T.-F., Schäfer, K., & Todorova, N. (2013). Acceptance of hospital nurses toward sensor-based medication systems: A questionnaire survey. *International Journal of Nursing Studies, 50*(4), 508–517. https://doi.org/10.1016/j.ijnurstu.2012.11.010.

Louho, R., Kallioja, M., & Oittinen, P. (2006). Factors affecting the use of hybrid media applications. *Graphic Arts in Finland, 35*(3), 11–21. Retrieved from http://citeseerx.ist.psu.edu/viewdoc/download?doi=10.1.1.468.8752&rep=rep1&type=pdf.

MacDorman, K. F., Vasudevan, S. K., & Ho, C.-C. (2009). Does Japan really have robot mania? Comparing attitudes by implicit and explicit measures. *AI & Society, 23*(4), 485–510. https://doi.org/10.1007/s00146-008-0181-2.

Madhavan, P., & Wiegmann, D. A. (2007). Effects of information source, pedigree, and reliability on operator interaction with decision support systems. *Human Factors, 49*(5), 773–785. https://doi.org/10.1518/001872007X230154.

Marcellini, F., Mollenkopf, H., Spazzafumo, L., & Ruoppila, I. (2000). Akzeptanz und Nutzung öffentlicher Automaten durch Ältere: Ergebnisse einer europäischen Untersuchung. *Zeitschrift Für Gerontologie Und Geriatrie, 33*(3), 169–177. https://doi.org/10.1007/s003910070057.

Mayer, A. K., Fisk, A. D., & Rogers, W. A. (2009). *Understanding technology acceptance: Effects of user expectancies on human-automation interaction.* Atlanta, GA: Georgia Institute of Technology. Retrieved from https://smartech.gatech.edu/bitstream/1853/40567/1/HFA-TR-0907-UserExpectancies.pdf.

Melenhorst, A. S., & Bouwhuis, D. G. (2004). When do older adults consider the internet? An exploratory study of benefit perception. *Gerontechnology, 3*(2), 89–101. https://doi.org/10.4017/gt.2004.03.02.004.00.

Nomura, T., Kanda, T., & Suzuki, T. (2006a). Experimental investigation into influence of negative attitudes toward robots on human–robot interaction. *AI & SOCIETY, 20*(2), 138–150. https://doi.org/10.1007/s00146-005-0012-7.

Nomura, T., Kanda, T., Suzuki, T., & Kato, K. (2004). Psychology in human-robot communication: An attempt through investigation of negative attitudes and anxiety toward robots. In *Proceedings of the 13th IEEE International Workshop on Robot and Human Interactive Communication* (pp. 35–40). https://doi.org/10.1109/ROMAN.2004.1374726.

Nomura, T., Suzuki, T., Kanda, T., & Kato, K. (2006b). Measurement of negative attitudes toward robots. *Interaction Studies, 7*(3), 437–454. https://doi.org/10.1075/is.7.3.14nom.

Nomura, T., Suzuki, T., Kanda, T., Yamada, S., & Kato, K. (2011). Attitudes toward robots and factors influencing them. In K. Dautenhahn & J. Saunders (Eds.), *Advances in Interaction Studies. New Frontiers in Human–Robot Interaction* (Vol. 2, pp. 73–88). Amsterdam: John Benjamins Publishing Company. https://doi.org/10.1075/ais.2.06nom.

Onnasch, L., Maier, X., & Jürgensohn, T. (2016). *Mensch-Roboter-Interaktion—Eine Taxonomie für alle Anwendungsfälle.* Dortmund.

Roschelle, J., & Teasley, S. D. (1995). The construction of shared knowledge in collaborative problem solving. In C. O'Malley (Ed.), *Computer Supported Collaborative Learning* (pp. 69–97). Berlin, Heidelberg: Springer Berlin Heidelberg. https://doi.org/10.1007/978-3-642-85098-1_5.

Schmidtler, J., Knott, V., Hölzel, C., Bengler, K., Schlick, C. M., & Bützler, J. (2015). Human-centered assistance applications for the working environment of the future. *Occupational Ergonomics, 12*(3), 83–95. https://doi.org/10.3233/OER-150226.

Schroer, B. J. (2004). Simulation as a tool in understanding the concepts of lean manufacturing. *Simulation, 80*(3), 171–175. https://doi.org/10.1177/0037549704045049.

Scopelliti, M., Giuliani, M. V., & Fornara, F. (2005). Robots in a domestic setting: A psychological approach. *Universal Access in the Information Society, 4*(2), 146–155. https://doi.org/10.1007/s10209-005-0118-1.

Shadish, W. R., Cook, T. D., & Campbell, D. T. (2002). *Experimental and quasi-experimental designs for generalized causal inference* ([Nachdr.]). Belmont, CA: Wadsworth Cengage Learning. Retrieved from https://moodle2.units.it/pluginfile.php/132646/mod_resource/content/1/Estratto_ShadishCookCampbellExperimental2002.pdf.

Sieverding, M. (2005). Der "Gender Gap" in der Internetnutzung. In K.-H. Renner, A. Schütz, & F. Machilek (Eds.), *Internet und Psychologie: Vol. 8. Internet und Persönlichkeit: Differentiell-psychologische und diagnostische Aspekte der Internetnutzung* (159.172). Göttingen: HOGREFE.

Special Eurobarometer 382. (2012). *Public attitudes towards robots.* Retrieved from the European Commission website: http://ec.europa.eu/public_opinion/archives/ebs/ebs_382_en.pdf.

Thrun, S. (2004). Toward a framework for human-robot interaction. *Human-Computer Interaction, 19*(1–2), 9–24. https://doi.org/10.1080/07370024.2004.9667338.

Universal Robots. (2015). *Benutzerhandbuch UR5/CB3: Übersetzung der originalen Anleitungen (de).*

Weiss, A., Bernhaupt, R., Tscheligi, M., & Yoshida, E. (2009). Addressing user experience and societal impact in a user study with a humanoid robot. In *Proceedings of the Symposium on New Frontiers in Human-Robot Interaction* (pp. 150–157). Edinburgh, Scotland.

Wilson, T. D., Lindsey, S., & Schooler, T. Y. (2000). A model of dual attitudes. *Psychological Review, 107*(1), 101–126. https://doi.org/10.1037/0033-295X.107.1.101.

Chapter 7
A Social Robot in a Shopping Mall: Studies on Acceptance and Stakeholder Expectations

Marketta Niemelä, Päivi Heikkilä, Hanna Lammi and Virpi Oksman

Abstract Social robots are gradually being introduced in public places to perform various service tasks in which the robots interact with users in the service front line. The presence of social robots in stores and shopping malls is one noticeable aspect of this phenomenon. Customers tend to feel positive about such robots, but the long-term benefits and impact of social service robots are hard to estimate, especially from the business perspective. The MuMMER project has involved mall customers, store managers, and mall managers to study their expectations and concerns about a shopping mall robot. Pepper of SoftBank Robotics was used as the robot platform. All stakeholders showed mainly positive attitudes. Facilitating factors in the adoption of social robots in malls seem to be the capability of the robot to be both entertaining and useful; in particular, the robot requires advanced dialog capability in order to be able to serve customers and collaborate with personnel. Moreover, there needs to be a perceived potential of the robot to lead to increased sales or decreased costs in the mall. As part of the adoption of social robots in shopping malls, the mall and store staff should be involved in co-designing the robots' tasks, and roles, as their work will be influenced by the robots in many ways.

Keywords Social robots · Pepper · Shopping mall · Customer acceptance · Business perspective

M. Niemelä (✉) · P. Heikkilä · H. Lammi · V. Oksman
VTT Technical Research Center of Finland, Tampere, Finland
e-mail: marketta.niemela@vtt.fi

P. Heikkilä
e-mail: paivi.heikkila@vtt.fi

H. Lammi
e-mail: hanna.lammi@vtt.fi

V. Oksman
e-mail: virpi.oksman@vtt.fi

© Springer Nature Switzerland AG 2019
O. Korn (ed.), *Social Robots: Technological, Societal and Ethical Aspects of Human-Robot Interaction*, Human–Computer Interaction Series,
https://doi.org/10.1007/978-3-030-17107-0_7

7.1 Introduction

Social robotics as a branch of technology is taking leaps due to recent advances in many enabling technologies such as processors, sensors, wireless communication, and algorithms related to face and expression recognition and natural language processing. It is foreseen that socially interactive robots—i.e., robots for which social interaction plays a key role and that exhibit various "human social" characteristics (Fong, Nourbakhsh & Dautenhahn, 2003)—will have significant applications in many areas in which it is beneficial that the interaction with technology is intuitive and easy to learn. Social robots might help aging people to live independently at home for longer, people with cognitive disabilities (e.g., dementia or autism) to manage their everyday lives, and children to learn cognitive and communicative skills. In these contexts, social robots may take both (physically) assistive roles and more emotional roles as companions (Korn, Bieber, & Fron, 2018).

Social robots are also entering public places. Robots have been developed and tested, in some cases for months and years, e.g., in scientific centers, museums, stores, and shopping malls to serve customers, and typically have received positive feedback (Chen et al., 2015; Gross et al., 2009; Huang, Iio, Satake, & Kanda, 2014; Kanda, Shiomi, Miyashita, Ishiguro, & Hagita, 2010; Shiomi, Kanda, Ishiguro, & Hagita, 2006; see also Sect. 7.2, Related work). And they are not just used for research: a humanoid robot, Pepper, is a commercial device that has been taken into use in many stores to present products and services and entertain customers, although mainly in Japan (Pandey & Gelin, 2018). In Europe, the Care-O-Bot robot has been serving customers in a chain of German electronics stores since 2016.[1]

This trend of social service robots becoming gradually more common is supported by the International Federation of Robotics (IFR) in their robot sales estimations. According to IFR (2018), almost 10,400 units of "public relation robots", i.e., robots that provide consumer-facing guidance and delivery services in retail stores, hotels, museums and other public spaces,[2] were sold in 2017. This represents more than 50% growth compared to the year before. In 2018, the sales will be 15,900 and in 2019–2021, 93,400, so roughly 30,000 units per year. Furthermore, according to the estimation of Loup Ventures (n.d.), sales of social and entertainment robots are expected to grow from 1.7 million units in 2015 to 7.43 million units by 2025. While these latter numbers do not tell us directly about the march of advanced social robots into public places, as the category includes a wide variety of robotic devices from remote-controlled cars and drones to conversational artificial intelligence with no living-like physical embodiment (e.g., the Amazon Echo and Google Home), they indicate that consumers will be encountering developing robotic devices and improving robotic capabilities in various contexts of life more and more.

[1] https://www.ipa.fraunhofer.de/en/press/2016-11-06_Care-O-bot-4-celebrates-its-premiere-as-shopping-assistant.html.

[2] https://www.robotics.org/content-detail.cfm/Industrial-Robotics-Industry-Insights/Service-Robots-on-the-World-Stage/content_id/7061.

Overall, we are witnessing a gradual but predictable accelerating introduction of robotics in general and social robots in particular in service environments and tasks. This trend is certainly driven by a sheer technology push and needs for business renewal, but also—if not driven, at least shaped—by societal needs, as robots in public places are of special importance considering the aging population. For instance, Kobayashi, Yamazaki, Takahashi, Fukuda, and Kuno (2019) have experimented with a robotic shopping trolley that facilitates the shopping activity of an elderly person and would possibly reduce the need for a caregiver. The robotic trolley would serve as a technology to support the everyday activities of the elderly person, but also as a technology of rehabilitation for walking and cognitive functions, both needed and practiced in shopping. In the longer term, another societally significant role of robots might be to replace part of the shrinking workforce, as with the aging population the number of people of working age decreases. This trend is particularly strong in Japan but is foreseen in European countries as well (Schneider, Hee Hong, & Van Le, 2018).

MuMMER (Multimodal Mall Entertainment Robot) is one of the research projects advancing public relations robots. MuMMER is an EU-funded project that is developing a humanoid robot that is able to operate autonomously and naturally in a public shopping mall (Foster et al., 2016). The robot platform deployed in the project is

Fig. 7.1 A pepper robot in a shopping mall

Pepper (SoftBank Robotics), a 120 cm high humanoid on wheels, with expressive gesturing and dialog capabilities (Fig. 7.1). The project develops both human-aware navigation and natural language interaction for the social robot to perform its interactive tasks independently and naturally in a dynamic shopping mall environment and with several users simultaneously. Importantly, MuMMER also focuses on the social acceptability of such robots: although in experiments the responses from customers toward such robots have been reported to be positive, is the general trend of robotization in public places and customer service perceived as positive as well? What are the expectations of customers about the tasks and roles a shopping mall robot should take, and are there constraints in that regard? And even if customers would welcome shopping mall robots, would malls and store managers show willingness to adopt them and for what kinds of uses? We have also studied the business perspective on shopping mall robots: what potential and challenges do mall store managers and mall managers perceive with regard to social robots as part of their daily business and longer-term strategies.

In this chapter, we present MuMMER research on the perspectives, expectations, and acceptance of a shopping mall robot from the three main stakeholders: mall customers, store managers and mall managers, and personnel. This research provides an overview of the stakeholder perspectives and implications for further design of shopping mall robots as well as their wide-scale adoption in public places.

7.2   Related Work

A number of studies on shopping mall robots have been reported, but here we concentrate on those that have collected feedback about the robot from customers and possibly from other stakeholders. Many of these studies have deployed robot platforms *TOOMAS*, *Robovie* or Pepper.

The TOOMAS shopping assistant robot was developed in the period 2005–2009, and ten robots were employed in three large home improvement stores for long-term everyday use (Doering et al., 2015). The 1.5-m high robot was based on SCITOS A5 and worked in fully autonomous mode to greet customers, to provide a search system for articles and product and location information and to guide the customer to the target location (Gross et al., 2009). In user evaluations, the acceptability of the robot was rather high, and most of the participants reported an intention to use the shopping robot in the future, especially for article searches. Robot-assisted shopping was found to be as effective and satisfactory as conventional shopping, although it was still slower with the robot (Doering et al., 2015).

Robovie is a robot tested in a Japanese shopping mall in order to develop its navigational abilities in crowds. Usually, Robovie is deployed semi-autonomously, using a human operator to carry out speech recognition, to monitor and override its behavior selection, and to provide additional domain knowledge when needed. Robovie research has reported various interesting results: those customers who used the robot visited more in shops and bought more, compared to those who used the traditional info screen (Kanda et al., 2010). In particular, a small Robovie robot encouraged people to try to print coupons for shops (Shiomi et al., 2013). Customers felt positive about personal greetings that the robot learns during longer-term use (Glas et al., 2017). Customers also liked when Robovie recognized that they might be "lost" in the mall and approached them to provide guidance (Brscic, Ikeda, & Kanda, 2017). Finally, people took more flyers from the robot that mimicked the behavior of a human flyer distributer, than from a human distributer, although this may be at least partly explained by the novelty of such robots—people want to see what the robot does after taking the flyer (Shi, Satake, Kanda, & Ishiguro, 2018).

Another, recently widely introduced robot platform for shopping assistive tasks is Pepper. Unlike many other humanoid robots, Pepper was developed for B2B purposes for SoftBank, to help reduce the workload of store staff and attract more customers (Pandey & Gelin, 2018). However, as many as 7000 Pepper robots are with consumers, from all 10,000 robots sold globally (but mainly in Japan). The other 3000 Peppers are being used in SoftBank shops, sushi bars, clothing stores, and Nespresso boutiques. In Europe, there have been trials with Pepper in railway stations, supermarkets, health and elderly care facilities, and cruise ships (Pandey & Gelin, 2018). Compared to Robovie, Pepper is used as an autonomous robot capable of moving and speech. The robot's interactive capabilities in service tasks have been tested by using it as a "salesperson", and it was found that the robot should start the interaction dialog with a short, easy-to-answer speech that helps people to engage with the robot for longer interaction (Iwasaki, Zhou, Ikeda, Kawamura, & Nakanishi, 2018). In front of a chocolate shop, the robot was able to create a more positive customer experience than when the customers interacted with a tablet kiosk (De Gauquier et al., 2018).

Entertainment is one part of the design of Pepper, and it is expected to increase the societal acceptability of the robot (Pandey & Gelin, 2018). Also the MuMMER project perceives it as central to develop the entertaining functions for the shopping mall robot (Foster et al., 2016). However, in observations during demonstrations of Pepper in a mall, we found that while children like to play with the robot, adults ask for practical information (Aaltonen, Arvola, Heikkilä, & Lammi, 2017). With Robovie, it has been found that customers tend to perceive the Japanese mall robot as a "mascot", which may make it challenging to become aware that the robot is able to provide useful services as well (Sabelli & Kanda, 2016), and to, e.g., save the customer time and effort. The perceived usefulness of the robot will probably be significant for its long-term use, at least for the adult customers, in addition to entertainment.

The success of shopping mall robots also depends greatly on business issues and how business actors such as store managers and mall managers can take advantage of the robot. In an interview study of several stakeholders of a restaurant (Lai & Tsai, 2018), it was found that in the short term, the restaurant would use a Pepper robot to greet and welcome customers and take care of reservations. In the longer term, customers could order food and give feedback through the robot, and at some point in the future, the robot would replace restaurant staff. In another study in a shopping mall, the interviewed store managers said that they perceived robots as providing a

Table 7.1 Summary of the three study settings

Study #	Perspective	Purpose of study	Method	Participants	N
1	Customers	Perceptions, expectations and worries	Workshop	Mall customers	10
2	Customers	Attitude, intention to use, beliefs/impressions of the robot and its interaction	Acceptance questionnaire survey, 3 × 2 rounds	Mall customers	252 + 254
3	Business	Expectations, business ideas and strategies	Interviews	Store managers in the mall; Managers of shopping malls	8 + 3

cheap workforce and unique value that humans cannot provide (Shi, Satake, Kanda, & Ishiguro, 2016). Both these studies were Asian. Our MuMMER study results provide illumination of the business stakeholders' perspective on shopping mall robots in the European (Finnish) context.

7.3 Studies in MuMMER

In MuMMER, we studied the perceptions and acceptance of the shopping mall robot, Pepper, from both customer and business perspectives. Customers took part in a workshop and a three-round acceptance survey, and store managers and mall managers were individually interviewed. The study settings are summarized in Table 7.1, and we give details of the studies and results in the following sub-chapters. A major part of the data was collected in Finland in a large shopping mall that consisted of approximately 200 shops and stores of mainly fashion and leisure equipment. The customers in the mall were of all ages: families with children, teenagers, and elderly persons.

Previously, Study 1 (customer workshop) was partially reported in (Niemelä, Heikkilä, & Lammi, 2017) and the first round of Study 2 (acceptance survey) was reported in (Niemelä, Arvola, & Aaltonen, 2017). The results of Study 3 were published in (Niemelä, Heikkilä, Lammi, & Oksman, 2017). This article is the first that draws the extended results of the three studies together to give a richer insight into the introduction of social robots in shopping malls.

7.3.1 Study 1. Customer Workshop

We invited customers of the mall to a workshop session to discuss their impressions, expectations, use ideas, and concerns about social robots and Pepper in particular in the shopping mall context. The participants were recruited via announcements in the mall's magazine and social media channels.

7.3.1.1 Method and Participants

The two-hour workshop was arranged in a meeting room of the mall. Pepper was physically present, having its "autonomous life" mode turned on in the session. First, the robot was introduced to the participants and they were able to have simple interactions with it. Then we illustrated the wider functionality and human–Pepper interaction through three short demonstration videos: Pepper as a general attraction in a Japanese shop, as providing marketing information on coffee machines for a customer, and as interactively serving a customer in a clothes store (giving recommendations and providing information).

After a short discussion of general thoughts about shopping mall robots and Pepper in particular, the participants individually wrote down on post-it notes potential use ideas for the robot in the mall and then presented their ideas one by one for general discussion. The same procedure was repeated for potential concerns and risks related to using the robot at the mall. After the workshop, we analyzed the ideas and concerns in a data-driven way by grouping them according to their affinity.

The participants were ten regular consumers (four women, six men), quite evenly distributed in the age range 26–66. Their attitudes toward social robots were more positive than negative, and a few participants knew about such robots beforehand.

7.3.1.2 Results

The participants produced 97 use ideas and 58 potential problems as risks for the robot in the mall. They were grouped into the following categories of tasks or roles for a shopping mall robot (Table 7.2), and concerns (Table 7.3).

The appearance and the interaction of the robot raised discussion. The appearance was characterized as "sympathetic" and positive because the robot was not too human-like. For interaction, the capability of Pepper to engage in social interaction by looking at the eyes (the robot turns its head and body to follow the face of the human by its "gaze") was experienced as impressive from the very beginning. The participants were positive about the robot's capability to speak clearly and wait for the user's response, thus mimicking natural turn-taking in human conversation. Their expectation was that the robot should also be able to undisturbedly engage in speech interaction with the user in the mall, when there would be many people and much talking around, background noise, and people speaking in dialects, and in free form.

Table 7.2 Tasks and roles for a shopping mall robot

Task/role	Description
Information and guidance	Informing about locations and providing directions and routes to stores and services in the mall, and answering questions about stores, products, and services (e.g., opening times, where to find a certain product to buy).In addition, it was hoped the robot would be able to answer complex, open questions such as what to buy as a present for a 14-year-old boy
Entertainment and attraction	Greeting and welcoming customers, saying good-bye/thank you; providing a show by telling jokes and stories and dancing; entertaining children; hosting events in the mall, crowd-puller in the mall
Advertising and sales	Drawing customers to stores; advertising special offers and campaigns, distributing coupons; giving recommendations about restaurants; advertising the local area to tourists
Shopping company	Having entertaining chats with customers in stores and cafes or when queueing or idle in the mall corridors, keeping up a good mood; adjusting its behavior to the mood of the customer (requiring emotion recognition: the robot should not be cheerful for sad people, and it should turn away from irritated people)
Shopping assistant	Assistant for elderly and disabled customers in particular; carrying bags, calling a taxi, giving directions, and accompanying the customer to target locations; translating languages; assisting in emergency situations
Robot for children	Entertaining and playing, babysitting; helping lost (and crying) children to find their parents, providing a safe place for lost children to wait for their parents

In spite of high expectations for the human-like speech interaction, the robot should be capable of, the participants preferred both a robot-like appearance and behavior: "*It is good to keep its behavior 'robot', not to pretend to be human, [...], it looks like robot, so it should be a robot.*"

This comment raised discussion about how the robot would be not pleasant to talk to if its communication was robot-like (i.e., mechanical). Overall, the discussion indicated a difference between how the participants judged human-like interaction and a human-like robot. The interaction should be human-like (i.e., natural, intuitive to a human user), but the robot should not behave as if it were in the position of a human: the robot should not in speech refer to itself as human, for instance, saying "in my view" or "personally I think", since the robot was not perceived to have a personal view or personhood. Instead, the robot could respond by referring to what people (in general) think.

The participants thought that for a shopping mall, having a social robot to serve customers would be an advantage. A mall having a robot was characterized as forward-looking, modern, pursuing speciality, enticing customers, experiential, and willing to experiment.

Table 7.3 Concerns related to a shopping mall robot

Task/role	Description
Privacy	Are photographs, voice, or video recorded in customer interactions? Where are the records are stored, who can use them and for what purposes?
Reliability	How can customers trust that the information the robot gives is up-to-date and correct?
Physical safety	Overheating of the robot; people accidentally bumping into the robot; can children's fingers be stuck in the joints of the robot?
Data security	Can hackers take over the robot's video camera and microphones for eavesdropping?
Vandalism, mistreatment	Customers or children pushing or twisting the robot on purpose
Liability	If something bad happens, who takes responsibility?
Lack of proper use in the mall	Limited or dysfunctional interaction or too simple entertainment may get boring for the customers; the robot is perceived only as a toy
The robot is too human or too strange	Vulnerable persons (children, mentally disabled) may be scared of the robot or misperceive the robot as a real friend

7.3.2   Study 2. Customer Acceptance Survey

To complement the qualitative data collected in the customer workshop with quantitative data, we arranged a customer acceptance survey. The three-round survey was carried out to investigate mall customers' attitudes to shopping mall robots, their intention to use such a robot in the mall, and their beliefs and impressions concerning the robot and interaction with it.

7.3.2.1   Method and Participants

The acceptance survey was conducted as part of the Pepper demonstrations in the mall, which took place in autumn 2016, spring 2017, and autumn 2017. During the demonstrations, the robot provided simple English or Finnish dialog and applications such as greetings, hand-shaking and dance, and gave out mall-related information. In each round, a compact paper form questionnaire including items on *Attitude* (one item), impressions of *Interaction* (three items), and *Intention to use* (one item) was given to mall visitors who came to interact with the robot or who monitored others interacting with it. There were other items as well, for instance, concerning the impressions of the robot. The thorough analysis of the survey data is ongoing, and for the purposes of this chapter, we limit discussion to the five items mentioned.

128

M. Niemelä et al.

Fig. 7.2 (Left and middle) A limited version of the questionnaire was implemented on feedback kiosks in the shopping mall. (Right) Paper form questionnaires were filled in as part of the demonstrations of pepper in the mall

Before each robot demonstration, a limited version of the acceptance survey was carried out in the mall. The purpose of this "pre-survey" was to survey attitudes toward the idea of a shopping mall robot without the robot present. The survey was implemented on four feedback tablet kiosks with a short description of the robot and a picture of the robot's head. The limited "Without robot" questionnaire also included the items of *Attitude, Interaction* (as beliefs, not impressions), and *Intention to use* (in case the robot was available in the mall with services interesting to the customers). The questionnaire was activated for one or two weeks at a time and it was openly available for visitors (also children) to answer. The feedback kiosks and a demonstration situation are illustrated in Fig. 7.2.

Both questionnaires, the longer one provided on paper and the shorter one on the feedback kiosk, included the demographic items of gender and age group of the respondent, and whether the robot was familiar to the respondent. For the two later rounds, it was asked whether the person had responded in the earlier survey rounds (for this analysis, the item was used to exclude multi-respondents).

Overall, acceptance survey data with 506 respondents in total has been collected, as shown in Table 7.4. Respondents under 16-years old (or age not given), those who answered only the first question on attitude, and those who reported answering earlier questionnaires were excluded. *Attitude* and *Intention to use* were measured on a 1–5 point Likert scale and beliefs/impressions as five-step semantic differentials. The results based on the feedback kiosk data are presented here for comparison, but should be interpreted with caution, as the method clearly includes challenges with regard to the quality of the data due to lack of control over respondents in the mall. Although an electronic survey method as such should not be more prone to lying by the respondents than a paper survey (Lazar, Feng, & Hochheiser, 2010, p. 117), the feedback kiosks in the mall were easily accessible to children, who may have randomly responded to the questions and produced data points that were impossible to differentiate from the responses of an adult, cognizant participant.

The most common age groups in the "Without robot" data were 21–30 (23%) and 31–40 (18%) years old. For "With robot", most of the respondents were either

Table 7.4 Customer acceptance survey

Condition of data collection	Survey round			Total
	1: September–October 2016	2: February–April 2017	3: August–September 2017	
"Without robot": on tablet feedback kiosks	$N = 102$	$N = 72$	$N = 78$	$N = 252$ (117 F, 107 M, 28 unknown)
"With robot": during pepper demonstrations, on paper forms	$N = 88$	$N = 77$	$N = 89$	$N = 254$ (145 F, 105 M, 4 unknown)

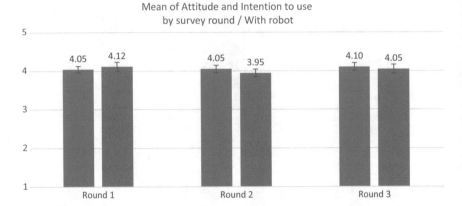

Fig. 7.3 Mean *attitude* (blue columns) and *intention to use* (red columns) by survey round in the "With robot" condition. Attitude scale: 1 = negative, ..., 5 = positive; Intention to use scale: 1 = very unlikely, ..., 5 = very likely. The vertical lines on top of the bars refer to standard error (SE)

31–40 (27%) or 41–50 (18%) years old. All age groups (from 16–20 to 70+) were represented in both sets of data, at least with a 7% share.

7.3.2.2 Results

The "With robot" condition. The mean value of *Attitude* was 4.1 (5 being the most positive); a clear majority (77%) of all "With robot" respondents perceived the idea of a shopping mall robot positively and only 4% felt it would be a bad idea. The mean value of *Intention to use* was 4.0. Almost four out of five (79%) would probably or very probably use the robot, if the robot was available in the mall with services interesting to them. However, 9% indicated they were unlikely to use the robot. The *Attitude* and *Intention to use* were similarly high, irrespective of the survey round ($F(2) = 0.120, p = 0.887$ for *Attitude*; $F(2) = 0.543, p = 0.587$ for *Intention to use*) as shown in Fig. 7.3.

There was no significant effect of age group on *Attitude* or *Intention to use*. Female respondents were slightly less positive in their *Attitude* ($M = 3.96$) than male respondents ($M = 4.18$) ($F(1) = 3.25, p = 0.073$). There was no such effect for *Intention to use* ($F(1) = 1.00, p > 0.1$).

The "Without robot" condition. The "Without robot" data collected on the electronic feedback kiosks in the shopping mall must be treated very cautiously because with this method there is no control on how the questionnaire is responded to and by whom. Anyway, we were interested to investigate whether this kind of effortless data collection brings added value for the field study and whether it can shed light on perceptions of mall visitors not in contact with the robot or researchers. We paid extra attention to the cleaning of the data by manually removing respondents

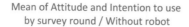

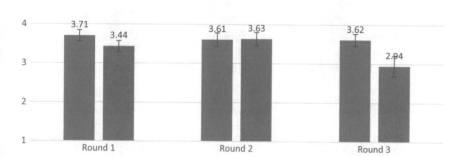

Fig. 7.4 Mean attitude (blue columns) and intention to use (red columns) by survey round in the "Without robot" condition. Attitude scale: 1 = negative,..., 5 = positive; Intention to use scale: 1 = very unlikely,..., 5 = very likely. The vertical lines on top of the bars refer to SE

whose data appeared random or included free comments clearly out of context, or who reported their age as under 16 or no age at all.

The mean values in this condition were 3.7 for *Attitude* and 3.5 for *Intention to use*. Approximately two-thirds (65%) of these respondents perceived the idea of the shopping mall robot positively, and half of them (53%) said they would use the robot in the mall. One-fifth (21%) evaluated the mall robot negatively and one-fourth (26%) thought they probably would not use the robot, even if its services were interesting.

The means of *Attitude* and *Intention to use* were at the same level irrespective of the survey round ($F(2) = 0.127, p = 0.881$; and $F(2) = 1.553, p = 0.215$, respectively). In Fig. 7.4, the third survey round shows exceptionally low *Intention to use*, only 2.9. This data contains only 18 respondents after excluding the invalid responses, so the reliability of these results is low.

All three "Without robot" datasets showed a systematic tendency toward a less positive attitude and lower intention to use than in the "With robot" condition. Statistically, the respondents in the "Without robot" data were significantly more negative than those who saw or used the robot themselves and took the paper questionnaire (for *Attitude*, $F(1) = 15.77, p = 0.000$; for *Intention to use* $F(1) = 22.98, p = 0.000$).

In the "Without robot" data, there was no significant effect of age on *Attitude* (One-way Anova $F(6) = 1.59, p > 0.05$) or *Intention to use* ($F(6) = 1.78, p > 0.05$). Similarly to the "With robot" condition, female respondents were a little more negative in their *Attitude* ($M = 3.61$) than male respondents ($M = 3.80$) (One-way Anova $F(1) = 3.25, p = 0.045$). There was no such effect for *Intention to use* ($F(1) = 0.63, p > 0.1$).

Beliefs/impressions concerning the robot interaction. Beliefs/impressions regarding the interaction with the robot were measured with three semantic differen-

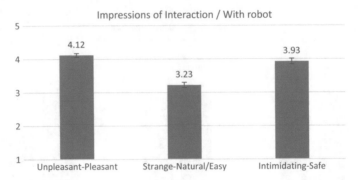

Fig. 7.5 Mean values of impressions (the "With robot" data pooled) regarding the interaction with the robot, with standard error (SE)

tial items: *Unpleasant-Pleasant, Strange-Natural/Easy*[3] *and Intimidating-Safe*. (The *Unpleasant-Pleasant* differential was not asked in the first round of the "Without robot" condition.)

In the "With robot" condition, 75% of respondents rated the robot's interaction as *Pleasant*, and only 4% as *Unpleasant*; 45% as *Natural/Easy* and 28% as *Strange*; and 64% as *Safe* and 12% as *Intimidating*. The mean values of the impressions are presented in Fig. 7.5.

In the "Without robot" data, there were considerable amounts of missing data for these items, especially in the third survey (this might have been due to technical problems with the feedback system). In addition, the *Pleasant-Unpleasant* item was not included in the first "Without robot" survey at all, as mentioned above. Therefore, for the first survey round the actual $N = 89$, second round $N = 62$, and third round only $N = 15$–24, depending on the item. We report here N per item and valid percentages only (excluding missing data).

There were 86 responses to the *Unpleasant-Pleasant* item: 49% believed the interaction with the shopping mall robot to be *Pleasant*, and 36% *Unpleasant*. Of 168 respondents, 40% believed the interaction to be *Natural/Easy* and 47% *Strange*. Of 166 respondents, 54% believed the interaction felt *Safe*, and 27% *Intimidating*. The mean values of the beliefs are presented in Fig. 7.6.

The study showed a fairly high positive attitude, intention to use, and impressions of the interaction with the social shopping mall robot providing customer services. Without seeing and personally experiencing the robot, the responses tended to be more negative. Potential factors underlying this difference are (1) presence of the robot, (2) presence of the researcher, (3) method of data collection, and (4) unintended selection of respondents interested in the robot in the "With robot" condition (we can expect mostly interested people to come to look at the robot and then fill in the

---

[3]The original Finnish word *luonteva* can be translated to "natural" in this sense: "marked by easy simplicity and freedom from artificiality, affectation, or constraint" (https://www.merriam-webster.com/dictionary/natural, def. 13b, accessed November 1, 2018). To emphasize this, we use the translation Natural/Easy.

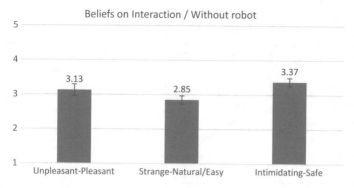

Fig. 7.6 Mean values of beliefs (the "Without robot" data pooled) regarding the interaction with the robot, with standard error (SE)

questionnaire). The "Without robot" condition gives us insight into the robot-related perceptions of those mall visitors in general who are not affected by the robot's or researchers' direct presence.

7.3.3 Study 3. Store Managers' and Mall Managers' Perspective

In order to get understanding from the business perspective in addition to the customers' perspective, we interviewed retailers (store managers) and shopping mall managers. The focus in these interviews was the store managers' and mall managers' business-oriented perspective on the expected tasks and roles of a shopping mall robot, their related concerns, the requirements that stores and malls have as a robot deployment environment, the related business models, and expected measurable business impact of the robot. This study has been previously published in (Niemelä, Heikkilä, Lammi, & Oksman, 2017); here we report the results as part of the multi-stakeholder perspective analysis of shopping mall robots.

7.3.3.1 Method and Participants

Eight store managers from one shopping mall and three shopping mall managers from different malls were interviewed (Table 7.5). All participants were from southern Finland. The retailer interviewees all represented different branches of retail or service businesses, and none of them had interacted with a social robot before (three retailers had seen them on TV, the Internet, or movies). Of the three shopping mall managers, one was the manager of the mall with the eight interviewed retailers (with approximately 200 stores). One of the other malls consists of approximately

Table 7.5 Interviewees

ID	Interviewee	Store type	Familiarity[a]	Gender
1	Store manager	Bedroom furniture and accessories store	1	F
2	Store manager	Digital games shop	2	M
3	Restaurant manager	Family restaurant	3	M
4	Store manager	Jeweler shop	3	F
5	Store manager	Shoe shop	3	F
6	Store manager	Department store	1	M
7	Store manager	Hobby/arts and crafts shop	1	F
8	Managing director	Cultural center for children	2	F
A	Shopping mall manager	Large shopping center in metropolitan area	1	M
B	Shopping mall manager	Middle-sized shopping center in province	3	F
C	Shopping mall manager	Large shopping center in province	4	M

[a]Familiarity with "shopping mall robots": 1-not at all familiar,..., 5-very familiar

70 shops and stores, and the third mall approximately 220 with also fashion and leisure as main retail types. None of the mall managers had interacted with a social robot before, but two of them had some background knowledge of mall robots. Ten of the 11 interviewees described their attitude toward shopping mall robots as "quite positive"; one participant (a store manager) had a "quite negative" attitude.

The retailers were interviewed one by one in a meeting room in their mall. The Pepper robot was physically present with the autonomous life function activated. The mall managers were interviewed in their offices, and Pepper was not physically present due to travel challenges. For them, the robot was demonstrated on two videos: Pepper as a "receptionist" at the entrance of a restaurant, welcoming customers and telling them about the menu of the day, and Pepper as a salesperson in a fashion shop, discussing clothes with the customer.

As a data collection method, we applied a semi-structured, open-ended interview with certain key questions such as: How would you apply a social service robot such as Pepper in your shop/store/mall? How do you expect your customers to respond to the robot? What kind of customer service can a robot provide? How would you measure the success of the robot in your business? What kind of concerns or risks might there be? The mall managers were also asked about potential business models for the robot. Each interview session lasted about 1–1.5 h. The data was collected as written notes and as audio records.

7.3.3.2   Results

Uses of a shopping mall robot in stores and mall

Retailers. The most frequently stated uses for social robots and Pepper in particular were attracting potential customers' attention, welcoming visitors to the store, and creating a pleasant atmosphere. Pepper as a robot was perceived as funny and approachable, "clearly designed to appear likable and harmless", which was considered as a special benefit for customer service tasks. Pepper was expected to be especially attractive to children, which was found an asset as the robot would "free" the parents to concentrate on shopping without distractions. The robot might also defuse tension in cases where children started misbehaving due to hunger or tiredness in the middle of shopping. The retailers considered these kinds of tasks significant but found the personnel too busy to do them. Another important emotional characteristic of the robot was it capability to stay calm and positive in all situations, even if the customer was rude or the situation was stressful. A robot cannot be insulted like a salesperson and that can have a concrete impact on sales: "Sometimes the product is not sold if the salesperson gets upset because of something that the customer says."

The robot was also seen as suitable for assistive sales tasks: providing product information, guidance, advertising new products, and demonstrating their uses to customers. The robot should be connected to the stock system to be able to provide appropriate and up-to-date information. The robot could remind customers to leave feedback and even persuade them to provide their data for the customer register. Two retailers said that they would like to try Pepper as a cashier or as a mobile payment terminal. In addition, two retailers found that Pepper with its tablet screen attached to the chest could be used to integrate online services with the physical shop, e.g., through providing access to social media or the online store.

A robot in a store should not be idle. In quiet times, it should do cleaning or similar tasks. In the shopping mall, the stores are obliged to stay open until the closing time of the whole mall. At late hours, they are usually quite empty of customers.

The interviewed retailers expected that most of the customers would have a positive attitude toward Pepper. Even though there might be some people who were reluctant toward using new technology in general or who currently found the robots too strange for social interaction, these were not considered as problems.

Shopping mall managers. The three shopping mall managers suggested that the most critical service provided by a shopping mall robot is to guide customers inside the shopping mall and accompany them to places they want to go. A social robot would provide a more feasible or human-friendly option to information boards due to its human-like behavior: talking to customers personally and indicating directions. Person identification (of regular customers) was seen as having business potential to enable personalized marketing and advertising. The robot should assist customers by finding products and providing product information, promote campaigns by sharing flyers, and telling customers about offers in shops located on other floors or locations in the mall, thereby helping people to better utilize the services of the whole mall. The robot would also be suitable for new or improved services concerning the shopping mall as a whole, such as selling gift cards that are valid in all stores and supporting maintenance functions by monitoring the mall area to detect defects in cleaning, maintenance or guarding.

When asked whether the mall owners would be interested in providing the robot as a service or marketing platform for service providers outside of the mall, they were interested but only if the services did not compete with the ones provided in the mall. For example, public services such as making appointments with the local healthcare unit, providing access to the (public) digital library system, or mail, banking and insurance services were seen as having potential for further development, as they bring clear added value to the customers.

Business aspects and risks of a shopping mall robot

Retailers. All the interviewed retailers were willing to adopt the robot in their business. They expected the robot to bring positive experiences to the customers and increase the value of the retail brand. The robot would—and should—attract more customers and increase sales of products and services. The robot's value for business would also directly show in the number and quality of feedback it received or collected itself.

The main obstacle for the interviewees was the cost of the robot. Instead of purchasing the robot, five of the eight retailers said they would prefer to borrow or rent (leasing) it just for campaigns or for new product releases. However, three retailers (a department store, a shoe shop, and a restaurant) considered taking the robot into permanent use, especially to provide information and guidance for the customers in their shop premises. Pepper was foreseen to replace salespersons at the info desk of the department store.

Another challenge in adopting the robot for their store was that some retailers in the mall were dependent on the approval of their higher management, e.g., in the case of retail chains. The ownership structure of the retailer's business determined "where" the decision to take a robot into use in a single shop was made.

If there was just a single robot or a few in a mall, a business model for short-period lending or renting the robot would lead to some kind of competition between the retailers, as many stores would be interested in having the robot on weekends when the mall was more crowded. One proposal was that all stores could use it systematically in turn or that the bigger stores would have the privilege of using the robot.

Considering the actual implementation of the robot, the retailers identified the following challenges: the robot was expected to cost too much considering its (current) value in customer service; the presence and moving of the robot required more space that many shops have; the robot's speech interaction capability was not yet advanced enough to reliably serve naïve users in shops; and the liability issues in cases of breakage or stealing should be clear. In addition, there were shops that required a certain delicacy in their services, such as in selling wedding rings or a special mattress for a person suffering from sleep problems, and the robot would be disturbing in these sensitive situations. In these cases, however, Pepper could entertain or inform the customers waiting for the service. The robot's physical limitations were also noted: "Even though the robot could sell shoes, it cannot do the packaging."

Shopping mall managers. Although the managers were broadly interested in the opportunities offered by service robots in the malls, none of them had any plans or strategy to utilize them in the future business yet. For the moment, the shopping mall managers found leasing to be the most feasible business model: the mall would lease the robot for a certain time, e.g., two years, with a maintenance contract, and would rent it further to the retailers for their campaigns. The main challenge the managers perceived was not the capabilities of the robot as such, but rather the general conservativeness of *retailers* in utilizing new digital technologies in their business or advertising.

The mall managers expected service robots to bring more customers to the mall, cut the costs of equipment (e.g. information screens) or personnel, and improve the overall brand value of their malls or the mall chain. The late, empty hours in the malls were also of concern for the managers: one of them wondered whether a robot could take care of a shop alone or with just one salesperson.

Pepper's (marketed) capability to recognize human emotions aroused special interest. It was seen as a potentially useful feature for collecting customer feedback and other customer-related information. For example, the robot could monitor the feelings of a customer who had taken offense, to estimate in situ how to optimally compensate the offended customer to their satisfaction in a cost-efficient way.

The managers foresaw that the robot's novelty effect would wear off, and they expected the improved social capabilities and useful information and guidance services of the robot to become more and more crucial for customers. Otherwise, the robot might end up being perceived as useless. Currently, they found it difficult to measure its real value for business, although they admitted that it had some unique business potential. Without experience, it is challenging to justify the investment price for retailers and in shopping mall business in general. However, the shopping mall managers envisioned that the robots' social interaction skills would become smoother in five years and robots would have a role in the future shopping mall business. The excellent usability of the robot, data security in all functions, and small consumer-delighting gestures in the human–robot interaction would be of importance then.

7.4   Summary and Discussion

To understand the expectations and acceptance issues that relate to the gradual introduction of social services robots in public places and in shopping malls in particular, we engaged with both end users and robot service providers to discover their perspectives. We had discussions with customers in a workshop, carried out a three-round acceptance survey for customers in the mall, and interviewed store managers and the mall manager. Here we summarize the results and draw them together.

The results of the first study, the customer workshop, gave us an initial under-standing of the consumer expectations and concerns toward a shopping mall robot and Pepper in particular. The participants expected Pepper to give information and guidance as well as entertain customers and create a pleasant atmosphere by sim-ple gestures, such as greeting customers when they entered a shop. Other use ideas included advertising and giving recommendations, serving as a shopping companion, assisting elderly and disabled customers and entertaining and babysitting children as well as helping in situations where they children had lost their parents. The concerns were related to issues with privacy, reliability, physical safety, data security, mis-treatment, and liability. In addition, the participants mentioned that the robot should be used in an interesting and appropriate way to avoid people getting bored with it.

In general, the number and variety of ideas, positive comments about the robot and the lively discussion around it indicated clear consumer interest toward a shop-ping mall robot. The Pepper robot was considered sympathetic and easy to approach and robot-like enough not to create an impression that it was mimicking or replacing humans. In development of a shopping mall robot, this aspect is important. The work-shop participants thought that the robot should not be giving an impression of having real personhood and person-like agency. It can still be beneficial to model human behavior for certain tasks of the robot, such as giving intuitive and effective guidance (as described e.g. in Heikkilä, Lammi, & Belhassein, 2018). In some tasks, Pepper could be capable of creating a unique customer experience by utilizing appropriate robot-like behavior and capabilities.

The second study, the customer acceptance survey, supported and extended the findings of the consumer workshop. The respondents had fairly positive attitudes and a high intention to use the robot, especially when they experienced the robot themselves through interacting with it or seeing others communicating with it. The robot's capability to follow the face of the human with its own "gaze" combined with the perception of the robot as a generally sympathetic creature seems to make Pepper a robot that is able to make a solid positive impression among different people. In the case of the "Without robot" survey, the respondents answered more based on their preconceptions and personal ideas about shopping mall robots (although a picture of Pepper's head was provided with the questionnaire). Still, even in that group, the attitude and acceptance were moderately high: two-thirds of the respondents had a positive attitude and half of them said they would use the robot in the mall.

Apart from one unsuccessful data collection (round 3 on the feedback kiosks), the level of attitude and intention to use stayed stable throughout the three survey rounds during a year. The female respondents were somewhat more negative than the male respondents in their attitude toward shopping mall robots but not in their intention to use. This was found in both datasets, thus irrespective of the robot's presence during the data collection.

With regard to impressions, the robot interaction was perceived as highly pleasant, mostly safe, and reasonably natural/easy. On the other hand, it is worth remembering that more than one-fourth of the respondents perceived the robot interaction closer to "strange" than to natural/easy. However, the perceived "unpleasantness" of the robot interaction was very low. These results indicate that the design of the robot services should be considered to minimize the feeling of strangeness of the robot interaction. One aspect of that, at least before people become generally familiar with social robots, may be that the robot should communicate as a robot, not as if it were a person, as discussed in the consumer workshop.

The interviews in the third study from the business perspective confirmed that the currently expected roles of the robot are consistently as a greeter, an attractor, and a guide and information provider in the mall. Other often mentioned tasks were entertaining, promoting, collecting feedback, and also working as a cashier. Many store managers and all three mall managers saw special potential in the human-likeness of the robot: the robot would be more "human" in guiding than info screens, which customers often find hard to interpret, or the robot would ease tension in challenging customer service situations.

On the other hand, the limits of the robot's social capabilities were identified as three retailers discussed whether the robot was able to manage the required sensitivity and complexity in more challenging customer interaction situations, such as when selling wedding rings or special equipment for people suffering from sleep problems.

All but one of the store managers raised issues related to breakage or malfunction of the robot or vandalism and mentioned that the robot cannot be watched over all the time by staff. Many also pointed out negative attitudes of people or that the elderly or children might be scared. However, these were not considered prominent issues that would stop the managers considering social robots as part of their services and business.

The mall managers were concerned about the cost, the cost-efficiency, and the hard-to-prove value of the robot, and that it may be challenging to encourage stores to take new marketing methods and technologies into use. The most promising business model appeared to be that the malls lease the robot with full maintenance and an update service—often malls do not have suitable technical expertise in house—and further rent the robot to the stores, which could use the central information system of each mall for renting. Mall managers were also interested in using the robot as a business platform to provide external services complementing those supplied by the mall, e.g., services of banking, post, and insurance, calling taxis and making appointments to barbers or health centers. The business impact should be measured by assessing how the robot affected the number of customers, sales, services used, customer satisfaction, and cost savings, for instance, savings in guidance costs (the info screens might be expensive) or personnel costs. The robot should also increase the brand value of malls. At least in our mall customer workshop, the participants' comments were in line with this, as a mall providing social robot services was appraised as forward-looking and as attracting customers.

Overall, the perception of Pepper-like social robots and their introduction into mall contexts was shown to be constantly positive among all stakeholder groups. The expected tasks for the robot were very consistent as well. For instance, the simple greeting function that the robot is able to do was often mentioned by both customers and store managers (in particular, people may like personalized greetings) (Glas et al., 2017).

All stakeholder groups also discussed the robot's novelty effect and its inevitable wearing off after a period of usage. As a counter effect, the customers called for useful services (not just entertainment), and the managers called for the robot to be capable of fluent dialog with customers and so be able to serve in more complex and sensitive customer service tasks. Without these improvements, the robot may be perceived as just an entertainer for children or as a "mascot" in the mall, which may even hinder finding useful functions for the robot (Sabelli & Kanda, 2016). It is a clear challenge to develop a robot to understand natural language and to produce it in a shopping mall environment—especially if the language is Finnish.

Conversational AI in English has been greatly advancing lately with new algorithms and tools, using (deep) learning techniques over large public datasets and training on real user feedback (Papaioannou et al., 2017). For instance, a dialog system solution called Alana was able to keep up a coherent free-speech conversation for 2–3 min on average in the Alexa Challenge 2017 (an international competition for such systems, evaluated by actual users; ibid.). The MuMMER project team has also tested using the Google Translate cloud service to test Alana in German. However, automated translating software does not necessarily work well with smaller language groups. For instance, English, German, and Dutch all belong to the main group of Indo-European Germanic languages. Finnish is part of a distinct Uralic group, spoken by about 5.5 million people, and both its vocabulary and grammar are very different to the Germanic group (Ager, 1997).

A related but different challenge concerns the hearing ability of the robot in a noisy shopping mall. As part of the MuMMER project, we carried out acoustical measurements in the mall. The "silent" time during a weekday morning was around 50 dB, while the peak noise on Saturday afternoon reached 63 dB. Although acoustically the mall is good, one problem for the robot is the various sources of speech around it, as mall visitors talk near the robot and mall announcements and advertisement are given from time to time.

The most central concern impeding the introduction of social robots into malls seems to be the costs of the robot and its maintenance, and the difficulty to discern what the robot will bring in return as a business value. The robot should increase the number of customers, sales, or the value of the business brand of the mall or store. Alternatively, the robot should bring savings in guidance costs (e.g., through making info screens unnecessary) or in personnel costs. Especially the latter indicates that in the future, business stakeholders could and probably will use social robots to replace human work, and the same observation has been made by Shi et al. (2016) and Lai and Tsai (2018). This is not to say human workers would be superseded by social robots, but robots could be assigned to take care of customer service tasks when workers are not available, or when it is just not profitable to hire a person to do the task. The empty evening hours in stores are an example of that. How this would influence customer behavior is a question for further research.

The introduction of social robots in customer service tasks in public places could also be seen as a possibility to increase the value of human work: the robot could work as a collaborative partner with humans such as workers in the info booth in the mall, and support their work, thereby promoting more efficiency or better quality.

The collaborative role of a shopping mall robot was touched upon by some store managers when discussing the robot as a means to help the staff in challenging service situations. With the improving social capabilities of the robots entering malls and shopping environments, the collaborative roles and tasks of the robot would ideally be co-designed in consultation with the staff themselves.

7.5 Conclusions, Limitations and Future Work

In the MuMMER project, we studied the perspectives, expectations, and acceptance of three central stakeholder groups—consumer customers in the mall, store managers in the mall, and mall managers in three shopping malls—as part of the co-design process of a social robot and its introduction in a shopping mall in Finland. The perception of Pepper-like social robots and their introduction into mall contexts was shown to be constantly fairly positive among all stakeholder groups, and the expected roles of the robot were very constant among the different stakeholders.

The research presented in this chapter has limitations which may affect the validity and reliability of the results and which we have to consider in order to generalize from them. First, the study was almost fully limited to one shopping mall, in one country. As a consequence, the results are representative of stakeholders in this mall but possibly not of shoppers or shopkeepers in other malls in Finland, not to mention other European countries. There is, however, some supporting research already carried out in Europe (De Gauquier et al., 2018).

Second, the participants in the studies (especially the customer workshop and the customer acceptance survey during the robot demonstrations) were found based on convenience sampling or were self-selecting volunteers; thus, it is likely that persons that have interest in technology in general or robots in particular are over-represented in the participants. Therefore, the results may be biased toward more positive and possibly more technology-savvy responses, than they would be when taking a representative sample from the customer population of the mall.

On the other hand, there is no such baseline customer data that could be used as a base for representative sampling in the mall. Self-selection may be the most natural data collection method when investigating new user populations or new phenomena of usage (Lazar et al., 2010, p. 109). From this perspective, we can consider our data to be valid and reliable for the purposes of this study: to build and consolidate the understanding of attitudes and expectations of different shopping mall stakeholders for social robots in the mall. Furthermore, the study results are very much in line with earlier research (see Sect. 7.2, Related work), and the results from the three stakeholder groups are consistent with each other in terms of the perceived challenges and potential of shopping mall robots.

Third, although not a limitation but rather a matter of the focus of our studies, this research does not include the perspective of a major stakeholder group, the mall workers, who have a high probability of being influenced by the introduction of social robots in service front line tasks. For instance, the workers may have to operate the mall robot, collaborate with it, change their work practices in order to gain benefit or provide customers with benefit from it, or they may even be replaced by it. In future work, we aim to systematically involve mall workers to study their expectations and have them participate in co-designing roles and tasks of social robots, so that the integration of the shopping mall robot with the practices and business in the mall would continue to be seen as positive to all stakeholders.

Acknowledgements This research is part of the project MuMMER, which has received funding from the European Union's Horizon 2020 research and innovation program under grant agreement No. 688147.

References

Aaltonen, I., Arvola, A., Heikkilä, P., & Lammi, H. (2017). Hello pepper, may I tickle you? Children's and adults' responses to an entertainment robot at a shopping mall. In *Proceedings of the Companion of the 2017 ACM/IEEE International Conference on Human-Robot Interaction (HRI'17)* (pp. 53–54).

Ager, D. (1997). *Language, community and the state*. Exeter, England: Intellect.

Brscic, D., Ikeda, T., & Kanda, T. (2017). Do you need help? A robot providing information to people who behave atypically. *IEEE Transactions on Robotics, 33*(2), 500–506. https://doi.org/10.1109/TRO.2016.2645206.

Chen, Y., Wu, F., Shuai, W., Wang, N., Chen, R., & Chen, X. (2015). KeJia robot? An attractive shopping mall guider. In A. Tapus et al. (Ed.), *ICSR2015, LNAI 9388* (pp. 145–154). Cham: Springer. http://doi.org/10.1007/978-3-319-25554-5_15.

De Gauquier, L., Cao, H.-L., Gomez Esteban, P., De Beir, A., van de Sanden, S., Willems, K. … Vanderborght, B. (2018). Humanoid robot pepper at a Belgian chocolate shop. In *Companion of the 2018 ACM/IEEE International Conference on Human-Robot Interaction—HRI'18* (p. 373). New York, NY, USA: ACM Press. http://doi.org/10.1145/3173386.3177535.

Doering, N., Poeschl, S., Gross, H.-M., Bley, A., Martin, C., & Boehme, H.-J. (2015). User-centered design and evaluation of a mobile shopping robot. *International Journal of Social Robotics, 7*(2), 203–225. https://doi.org/10.1007/s12369-014-0257-8.

Fong, T., Nourbakhsh, I., & Dautenhahn, K. (2003). A survey of socially interactive robots. *Robotics and Autonomous Systems, 42*(3–4), 143–166. https://doi.org/10.1016/S0921-8890(02)00372-X.

Foster, M. E., Alami, R., Gestranius, O., Lemon, O., Niemelä, M., Odobez, J.-M., & Pandey, A. K. (2016). The MuMMER project: engaging human-robot interaction in real-world public spaces. In A. Agah, J. Cabibihan, A. Howard, M. Salichs, & H. He (Eds.), *Social robotics. ICSR 2016. Lecture notes in computer science* (Vol. 9979 LNAI, pp. 753–763). Cham: Springer. http://doi.org/10.1007/978-3-319-47437-3_74.

Glas, D. F., Wada, K., Shiomi, M., Kanda, T., Ishiguro, H., & Hagita, N. (2017). Personal greetings: Personalizing robot utterances based on novelty of observed behavior. *International Journal of Social Robotics, 9*(2), 181–198. https://doi.org/10.1007/s12369-016-0385-4.

Gross, H. M., Boehme, H., Schroeter, C., Mueller, S., Koenig, A., Einhorn, E. … Bley, A. (2009). TOOMAS: Interactive shopping guide robots in everyday use—Final implementation and experiences from long-term field trials. In *IEEE/RSJ International Conference on Intelligent Robots and Systems (IROS) 2009* (pp. 2005–2012).

Heikkilä, P., Lammi, H., & Belhassein, K. (2018). Where can I find a pharmacy? Human-driven design of a service robot's guidance behavior. In *4th Workshop on Public Space Human-Robot Interaction (PubRob 2018), International Conference on Human-Computer Interaction with Mobile Devices and Services (MobileHCI 2018), Barcelona, Spain.*

Huang, C.-M., Iio, T., Satake, S., & Kanda, T. (2014). Modeling and controlling friendliness for an interactive museum robot. In *Robotics: Science and Systems 2014*. Berkeley, CA, USA, July 12–16, 2014.

International Federation of Robotics. (2018). *Executive Summary World Robotics 2018 Service Robots*.

Iwasaki, M., Zhou, J., Ikeda, M., Kawamura, T., & Nakanishi, H. (2018). A customer's attitude to a robotic salesperson depends on their initial interaction. In *27th IEEE International Conference on Robot and Human Interactive Communication (RO-MAN2018)*. Nanjing and Tai'an, China.

Kanda, T., Shiomi, M., Miyashita, Z., Ishiguro, H., & Hagita, N. (2010). A communication robot in a shopping mall. *IEEE Transactions on Robotics, 26*(5), 897–913.

Kobayashi, Y., Yamazaki, S., Takahashi, H., Fukuda, H., & Kuno, Y. (2019). Robotic shopping trolley for supporting the elderly. In N. Lightner (ed), *Advances in Human Factors and Ergonomics in Healthcare and Medical Devices. AHFE 2018. Advances in Intelligent Systems and Computing* (Vol. 779, pp. 344–353). Cham: Springer. http://doi.org/10.1007/978-3-319-94373-2_38.

Korn, O., Bieber, G., & Fron, C. (2018). Perspectives on social robots: From the historic background to an experts' view on future developments. In *Proceedings of the 11th Pervasive Technologies Related to Assistive Environments Conference* (pp. 186–193). http://doi.org/10.1145/3197768. 3197774.

Lai, C.-J., & Tsai, C.-P. (2018). Design of introducing service robot into catering services. In *Proceedings of the 2018 International Conference on Service Robotics Technologies-ICSRT '18 - ICSRT '18* (pp. 62–66). New York, NY, USA: ACM Press. http://doi.org/10.1145/3208833. 3208837.

Lazar, J., Feng, J. H., & Hochheiser, H. (2010). *Research methods in human-computer interaction*. Hoboken: Wiley.

Loup Ventures. (n.d.). *Unit sales of social and entertainment robots worldwide from 2015 to 2025 (in millions). In Statista—The Statistics Portal*. Retrieved November 30, 2018, from https://www.statista.com/statistics/755677/social-and-entertainment-robot-sales-worldwide/.

Niemelä, M., Arvola, A., & Aaltonen, I. (2017). Monitoring the acceptance of a social service robot in a shopping mall: First results. In *Proceedings of the Companion of the 2017 ACM/IEEE International Conference on Human-Robot Interaction (HRI'17)* (pp. 225–226). Vienna, March 6–9. http://doi.org/10.1145/3029798.3038333.

Niemelä, M., Heikkilä, P., & Lammi, H. (2017). A social service robot in a shopping mall—Expectations of the management, retailers and consumers. In *Proceedings of the Companion of the 2017 ACM/IEEE International Conference on Human-Robot Interaction (HRI'17)* (pp. 227–228). Vienna, March 6–9. http://doi.org/10.1145/3029798.3038301.

Niemelä, M., Heikkilä, P., Lammi, H., & Oksman, V. (2017). Shopping mall robots—Opportunities and constraints from the retailer and manager perspective. In A. Kheddar (Ed.), *Lecture Notes in Computer Science (Including Subseries Lecture Notes in Artificial Intelligence and Lecture Notes in Bioinformatics)* (Vol. 10652, pp. 485–494). Springer International Publishing AG 2017.

Pandey, A. K., & Gelin, R. (2018). A mass-produced sociable humanoid robot: Pepper: The first machine of its kind. *IEEE Robotics and Automation Magazine, 25*(3), 40–48. https://doi.org/10.1109/MRA.2018.2833157.

Papaioannou, I., Curry, A. C., Part, J. L., Shalyminov, I., Xu, X., Yu, Y. … Lemon, O. (2017). Alana: Social dialogue using an ensemble model and a ranker trained on user feedback. In *1st Proceedings of Alexa Prize (Alexa Prize 2017)*. https://s3.amazonaws.com/alexaprize/2017/technical-article/alana.pdf.

Sabelli, A. M., & Kanda, T. (2016). Robovie as a mascot: A qualitative study for long-term presence of robots in a shopping mall. *International Journal of Social Robotics, 8*(2), 211–221. https://doi.org/10.1007/s12369-015-0332-9.

Schneider, T., Hee Hong, G., & Van Le, A. (2018). Managing Japan's shrinking labor force with AI and robots. *IMF F&D Magazine—June 2018*.

Shi, C., Satake, S., Kanda, T., & Ishiguro, H. (2016). How would store managers employ social robots? In *ACM/IEEE International Conference on Human-Robot Interaction* (Vol. 2016–April, pp. 519–520). IEEE. http://doi.org/10.1109/HRI.2016.7451835.

Shi, C., Satake, S., Kanda, T., & Ishiguro, H. (2018). A robot that distributes flyers to pedestrians in a shopping mall. *International Journal of Social Robotics, 10*(4), 421–437. https://doi.org/10.1007/s12369-017-0442-7.

Shiomi, M., Kanda, T., Ishiguro, H., & Hagita, N. (2006). Interactive humanoid robots for a science museum. In *Proceeding of the 1st ACM SIGCHI/SIGART Conference on Human-Robot Interaction - HRI '06*. http://doi.org/10.1145/1121241.1121293.

Shiomi, M., Shinozawa, K., Nakagawa, Y., Miyashita, T., Sakamoto, T., Terakubo, T. … Hagita, N. (2013). Recommendation effects of a social robot for advertisement-use context in a shopping mall. *International Journal of Social Robotics, 5*(2), 251–262. http://doi.org/10.1007/s12369-013-0180-4.

Chapter 8
Multi-party Interaction in Public Spaces: Cross-Cultural Variations in Parental and Nonparental Response to Robots' Adaptive Strategies

Saida Mussakhojayeva and Anara Sandygulova

Abstract Social environments are often complex and ambiguous: Many queries to the robot are collaborative and do not have an assigned addressee, for example, a family. In contrast, in the case of conflicting queries, social robots need to participate in value decisions and negotiate multi-party interactions. With the aim of investigating who robots should adapt to (children or adults) in multi-party negotiations within human–robot interactions in public spaces, this chapter presents two studies: a real-world study conducted in a shopping mall and a follow-up cross-cultural study conducted online. The results include a number of interesting findings based on people's relationship with a child and their parental status. In addition, a number of cross-cultural differences were identified in respondents' attitudes toward robot's multi-party adaptation in various public settings.

Keywords Human–robot interaction · Multi-party interaction · Public spaces · Cross-cultural robotics · Social robotics

8.1 Introduction

Recent progress in the area of service robotics, including research which seeks to integrate speech, sensing, acting, and networking, has resulted in increasingly versatile and reliable service robots. One of the most promising application domains for service robots is to be deployed in public spaces, for example, as reception and information desk attendants (Makatchev, Simmons, Sakr, & Ziadee, 2013), museum and city guides (Shiomi, Kanda, Ishiguro, & Hagita, 2006) servants in bars and restaurants (Foster et al., 2012), health care, rehabilitation, and therapy assistants in hospitals (Korn, Bieber, & Fron, 2018), and educators and learning companions in

S. Mussakhojayeva · A. Sandygulova (✉)
Nazarbayev University, Nur-Sultan, Kazakhstan
e-mail: anara.sandygulova@nu.edu.kz

S. Mussakhojayeva
e-mail: saida.mussakhojayeva@nu.edu.kz

© Springer Nature Switzerland AG 2019
O. Korn (ed.), *Social Robots: Technological, Societal and Ethical Aspects of Human-Robot Interaction*, Human–Computer Interaction Series,
https://doi.org/10.1007/978-3-030-17107-0_8

educational institutions (Hood, Lemaignan, & Dillenbourg, 2015). In such public environments, robots must necessarily deal with situations that demand they engage humans in a socially appropriate manner. Similar to human–human communication, if a robot does not adjust its communication style to the interlocutor or situation at hand, this can lead to confusion and misunderstanding. It can also cause annoyance, displeasure, and dissatisfaction with the service, ultimately leaving a feeling of disengagement with a robot. Adaptive human behavior reflects an individual's social and practical skills in meeting the demands of society. Thus, a socially competent robot needs to behave according to the contemporary conventional norms regularly accepted within the society, social class, or user group (Sandygulova, Dragone, & O'Hare, 2016).

Social environments are complex and ambiguous in terms of task ownership, responsibilities, and accountability (Davidoff, Lee, Zimmerman, & Dey, 2006). Many queries to the robot are collaborative and do not have an assigned addressee, for example, a family (Mussakhojayeva, Zhanbyrtayev, Agzhanov & Sandygulova, 2016; Mussakhojayeva, Kalidola, & Sandygulova, 2017). In contrast, in cases of conflicting queries, social robots need to participate in value decisions and in negotiating multi-party settings (Mussakhojayeva & Sandygulova, 2017). This chapter presents two studies: a field study and an online study that aim to explore what people's thoughts are in such situations, and what is the most appropriate solution to mediate such conflicts.

Past work (Torrey, Powers, Marge, Fussell, & Kiesler, 2006) suggests that adaptive robots not only have advantages in terms of information exchange and efficient communication, but also add social advantages as well. Our work aims to address a minimally explored issue of who robots should adapt to within multi-party interactions. In particular, the focus of this work is to investigate what adults have to say about a robot's adaptivity and to create such an adaptive robot that can address these requests.

This chapter firstly presents an HRI study conducted at a local shopping mall with a group of adult–child participants. The goal of the study was (a) to investigate whether people preferred the robot to adapt to children or to adults within multi-party interactions and (b) to analyze whether these responses were different according to the relationship of an adult–child pair. It is followed by a description of the conducted online study that aimed (a) to further investigate these research questions and (b) to compare the responses of local people with the opinions of the US participants on this matter.

8.2   Related Work

This section presents the existing state of the art relevant to the work of this chapter, which is divided into three subsections. A broad set of different perspectives constitutes the interdisciplinary nature of HRI field. Thus, the related work is organized accordingly in each subsection of this related work. Each subsection establishes the context and situates this chapter in relation to each of these related work topics.

The first subsection introduces related work on robots designed for public environments, then discusses their capabilities and the challenges associated with public human-populated settings. Secondly, this section presents a brief theoretical background dedicated to the theories on cross-cultural differences, which motivates the research study of this chapter and sets hypotheses. The final part of this section reviews a number of related cross-cultural studies conducted with social robots. This part introduces the reader to the current state of the art in cross-cultural research in HRI.

8.2.1 Robots in Public Spaces

Research and development of robots to be deployed in public spaces need to account for a wide range of challenges depending on the application domain. Indeed, each public environment sets common technical challenges; however, HRI requirements may be different. This section provides a review of the related work in the domain of robots for public spaces and examines each work's requirements, challenges, and contributions. The "network robot system" (NRS) framework (Glas et al., 2012) is the result of several years of research in the domain of public environments and has been utilized in a number of field studies in train stations (Shiomi et al., 2011), science museums (Shiomi, Kanda, Ishiguro, & Hagita, 2006), and shopping malls (Kanda, Shiomi, Miyashita, Ishiguro, & Hagita, 2009) in Japan. In the NRS framework, four mobile robots (two Robovie humanoid robots and two cart robots), sensors embedded in the environment, and planning servers are integrated to provide robot services such as guiding and carrying shopping bags for people in social contexts. Features of NRS include recognition and anticipation of people's behavior, identification of individuals, coordination of services and navigation paths between robots, and support for human operator supervision. The NRS framework performed successfully during the experiment, and the participants responded in a positive way, indicating they would like to use these services in the future. Similar challenges presented to mobile robots have been addressed by the research teams working on city guide robots such as FROG, the Fun Robotic Outdoor Guide, (Evers et al., 2014) and Autonomous City Explorer (ACE) (Lidoris et al., 2007), as well as the airport guide robot SPENCER (Spencer, 2015), a fully autonomous mobile robot for smart passenger flow management.

Robots deployed at information or reception desks are often stationary and do not need to deal with navigation and socially aware mapping. However, such robots need to behave according to the social norms accepted within the society they are deployed in. One such robot is utilized as a receptionist, the Hala robot (Simmons et al., 2011), which consists of a human-like stationary torso with an LCD mounted on a pan-tilt unit. The LCD "head" allows rendering of character faces, appearance cues, verbal, and non-verbal behaviors of ethnicity that can be controlled for the ethnic similarity to a local population of Doha, Qatar. An HRI experiment was conducted in Education City, Doha, with 30 participants: adult native speakers of Arabic (fluent in English) and native speakers of American English. The results show that the Hala robot with a relatively low human likeness could evoke associations between the robot's verbal and non-verbal behaviors and its attributed ethnicity. However, the results of this experiment did not find evidence of ethic homophily (Makatchev, Simmons, Sakr, & Ziadee, 2013). Another study by Salem (Salem, Ziadee, & Sakr, 2014) exploring culture-specific variations of HRI between Arabic and English native speakers highlighted the importance of addressing and exploiting cultural differences when designing multilingual and cross-cultural service robots.

Challenges present in multi-party interactions have been addressed by a robot bartender JAMES (Foster et al., 2012) that is designed to work in dynamic, multi-party social situations. The JAMES system incorporates state-of-the-art components for computer vision, linguistic processing, state management, high-level reasoning, and robot control. The bartender robot consists of two manipulator arms with humanoid hands mounted in a position to resemble human arms, along with Microsoft Kinect and an animatronic talking head, iCat. During a study conducted in laboratory settings, the system performed successfully with 31 university participants (Foster et al., 2012). The specific domain of public exhibitions has only been addressed by a few research groups. One example is a gesture-centric android system (Kondo, Takemura, Takamatsu, & Ogasawara, 2013), which is able to adjust gestures and facial expressions based on a speaker's location or situation, for multi-party communication. The speaker location is identified by face recognition and microphone position. An experiment was conducted with 1662 subjects interacting with the Actroid-SIT android in a shopping mall in Japan. Another field experiment was conducted during a six-day Fleet Week in New York with 202 subjects (Martinson, Lawson, & Trafton, 2013). Groups of three people firstly trained the Octavia robot to memorize their soft biometrics information (complexity, height, and clothes), then although the people tried to trick Octavia by changing their location, she could successfully identify them 90% of the time. Both systems successfully address multi-party HRI relying on multi-modal recognition of people through sound and vision.

8.2.2   Cultural Differences by Hofstede

We believe that culture might have an effect on how people negotiate multi-party settings. Geert Hofstede (Hofstede, 1986) defines the term "culture" as "collective programming of mind that distinguishes the members of one group or category of people from others." The social psychologist and his followers (Wursten, Jacobs, & Magazine, 2013) argue that every society is represented by a certain set of fundamental values, and these values influence the educational system and the way children are brought up.

One comprehensive study (Hofstede, 2011) described the impact of culture on the values of its members, and how the values reflected on their behavior. The authors conducted a cross-cultural study across 50 countries and identified that the values associated with one culture may not be applicable to another one. In the same manner, the behavior and attitude typical of one culture might not hold true for another one.

8.2.3 Cross-Cultural HRI

O'Neill-Brown (1997) emphasized that the culture shapes the flow of interaction, and intelligent agents should consider users' cultural background. The number of works published recently reaffirms the importance of culture in HRI. Li, Rau, and Li (2010) found cultural differences in the perception and likability of a robot which acted as a companion: Korean and Chinese participants found the robot more enjoyable and satisfactory compared to the Germans. The authors tried to integrate Hofstede's dimensions into the context of their study and their result showed a positive response between the dimensions and the perception of the robot: Germans showed a negative attitude toward the companion robot, as they would prefer more control over the robot because of their masculine culture and individualism, as suggested by Hofstede. There are also different concerns and interests in the robot design from culture to culture. Japanese subjects generally want the robot to be more engaging and socially appropriate, whereas UK subjects like the robot as long as it is not too good at simulating humans (Syrdal, Nomura, & Dautenhahn, 2013). The differences between Eastern and Western cultures were also identified in a study by Wang (Wang, Rau, Evers, Robinson, & Hinds, 2010). An in-group (peer) robot had a more positive response among Chinese participants compared with US subjects when the robot was speaking in a culture-based audience-oriented form. The authors identified that, compared with US subjects, the relationship of the Chinese to robots was mediated to a greater extent by trust, and the Chinese were more likely to take the robot's advice if the robot communicated in more culturally normative ways. These results suggest the importance of testing the hypothesis in a culturally diverse manner, since the cultural background has a reflection on a robot's effective adaptation techniques. There is also a correlation between Hofstede's cultural dimensions and parental disciplinary methods (Schwab, 2013). For instance, the USA was quantified as a highly individualistic and a low-power distance (i.e., people relate to each other more as equals regardless of formal positions) country. Consequently, the study indicated that parental discipline is liberal: Children are encouraged for good behavior and punished for poor. During the nighttime, children are expected to sleep separately from parents. However, the study had a limited number of cultures analyzed and did not cover Kazakhstan, which is essential for proper comparison. The absence of the Soviet Union/Russia is also a limitation, since Kazakhstan could be related to Russia as an ex-part of USSR. The majority of parents in Kazakhstan nowadays were born when the country was part of the USSR and share similar values to those held in the USSR. Thus, any conclusion based on social behavior in one cultural environment should be studied in a different cultural context before making any extrapolations.

8.3 Studies

8.3.1 Robots' Adaptive Scenarios for Self-introductions

We developed two adaptive strategies for two humanoid NAO robots' self-introductions (child-friendly and adult-friendly) with the aim to convey to the general public that robots could easily adapt to people in real time by dynamically changing only their utterances. The robot's appearance, synthesized voice, and behaviors were exactly the same for both robots. We modified an existing self-introduction application, which is available at NAOStore. The videos of robots' self-demonstrations can be found at the link.[1]

NAO's default synthesized voice was used for English and Russian versions of the demonstrations. The differences between conditions were in the wording of the robot's utterances such as greetings, its self-introduction, and goodbyes. The self-introduction script of the robot's speech included a demonstration of the NAO's technical features and functionality. For example, the child-friendly robot described its vision capabilities as "I see the world around me with my eyes" in contrast to the adult-friendly robot which stated that "my vision consists of two embedded cameras." Verbal content delivered by the two robots was the only difference manipulated for adaptation. All non-verbal behaviors such as waving, gesticulating, eye gaze and other robot movements were the same in both conditions. The child-friendly version of the script is provided below:

Hi! I am a robot and my name is NAO. I come from France and I can speak eight languages. I have more than 5000 brothers and sisters all around the world in universities and research laboratories. I am learning new things every day and I need a lot of practice. For example, I have learned to recognize people's faces, answer questions, sing, grab objects, and even play soccer like a pro. Moving was very difficult to learn. But I learned it quite well and I am able to walk and dance. The device located on my torso allows me to detect large objects in front of me. With my eyes I can see the world around me. Thanks to them, I can look right in front of me or at my feet. And I can speak and hear with the help of my mouth and my two ears on my head. Well, I guess that's enough about me. It is time to show you some of my moves. Thank you for your time. Bye-bye!

The adult-friendly version of the script is provided below:

Hello! I am a humanoid robot and my name is NAO. I was created and manufactured by Aldebaran Robotics in France. Counting every articulation in my body, I have 25 degrees of freedom. I am easily programmable in my own software in a number of programming languages. For example, with my face and object detection and recognition software I am programmed to recognize people, grab objects and play soccer like a pro. To help me keep my balance, there are four sensitive resistors under each of my feet combined with my inner self sensor. The sonar located on my torso allows me to detect obstacles that I might encounter.

In addition, my vision is due to two embedded cameras in my face. Thanks to them I can look right in front of me or at my feet. The voice synthesizer, two speakers, and four microphones allow me to interact with people and the outside world. Well, I guess that is enough about me. It is time to demonstrate one of my performances. Thank you for your time. Good bye!

---

[1] https://goo.gl/forms/JCDQ4hW4zPP5EGWx1

Fig. 8.1 Experimental setup

8.3.2 Study 1: Shopping Mall

The first study took place at the food court of a shopping mall on Sunday throughout the day, where people were approached and asked to participate in the experiment with the robots. A parent–child group of participants were asked to sit in front of the robots at a specifically allocated area for about 10–15 min and watch the robots' self-presentations. The setup of the experiment is depicted in Fig. 8.1. There were two humanoid Aldebaran NAO robots of the same blue color: a child-friendly robot and an adult-friendly robot. The difference between the robots is described in the previous section.

8.3.3 Method

We used a mixed-subject design, in which all of the participants experienced two robot demonstrations. In an attempt to prevent order effects, we counterbalanced the order of the demonstrations. The child-friendly robot was presented first for half of the participants followed by the adult-friendly robot. The other half of the participants interacted with an adult-friendly robot first followed by the child-friendly robot. At the end of the interaction, the adult and child participants were interviewed by the researchers.

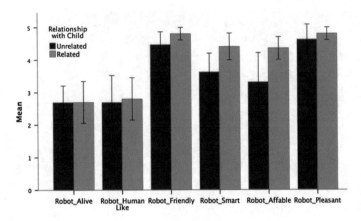

Fig. 8.2 Results

Adult–child paired participants were invited to take part in the experiment, where the adults' gender as well as relation to the child were varied. A total of 33 groups participated in the study: 14 male adult–child and 19 female adult–child groups. Among these groups, 20 were related and 13 were unrelated. In parent–child groups, there were cases where a parent had more than one child with them, and cases where some of the children had both parents with them. In the first case, multiple children were mapped to one parent, and in the second case, both parents were interviewed individually, thus creating two distinct groups. There were also children without their parents that watched the robots' presentations multiple times. In these cases, random adults were asked to participate in the study by joining these children. These cases were considered as unrelated adult–child groups. Demographic information about the gender and age of the adults and children were collected during the post-interview sessions.

8.3.4   Results and Discussion

First it should be noted that all the children responded positively to the robots and the majority of them correctly replied to the post-interview about the robot. Since there were no differences in how the children rated the robots, the adults' responses are discussed in detail (Fig. 8.2).

We conducted a series of one-way ANOVA tests on the collected data of adults' responses. First we tested how the adults rated the robots from 0 to 100 using a Funometer scale (Read, MacFarlane, & Casey, 2002). There was a statistically significant difference between the ratings of parents and strangers $F(1.31) = 5.982, p = 0.020$. When comparing unrelated adults (70.23 ± 22.5) to related adults (88.20 ± 19.3), the Funometer scale values were significantly lower in the interaction of unrelated adults. These findings suggest that parents rated the robots much more positively after seeing their child(ren)'s enjoyment. Similarly, Pearson's chi-square test revealed a significant difference ($p = 0.022$) in what adults compared the robots to: Options were toy, electrical appliance, computer, pet, or human. Unrelated adults were significantly more likely to compare the robot to a toy in contrast to parents whose most popular comparison was to a computer.

Finally, Pearson's chi-square test was conducted to compare the opinions of related and unrelated adults on whether robots should adapt to adults or to children in cases of multi-party situations. The majority of related and unrelated adults replied that the robot should adapt to children in such situations. Then, the question was whether they would change their opinion if it were a different setting such as a hospital, bank, or police station. For this question, some adults did change their opinion and claimed that the robot should ignore the child and adapt its verbal content to the adult. However, the change in opinion significantly differed according to the relationship with the child. Pearson's chi-square test showed statistically significant ($p = 0.023$) differences: Strangers did not want the robot to adapt to the unknown child in other public environments in service/business related domains. In contrast, the majority of parents said they would still prefer the robot to adapt itself and its content to their children (i.e., they did not change their opinion).

8.3.5 Study II: Online Survey

To further investigate the above questions about rules for robots' adaptation in public spaces, we developed two versions of an online survey that could probe people's attitudes and intuition toward robots in public spaces. We used Amazon Mechanical Turk (AMT) in order to recruit participants in the USA, and we distributed the local version of the survey through local social media.

The participants were informed that the purpose of the study was to investigate their views on robots in public spaces. The scenarios described in Sect. 8.3.1 were video-recorded. Then, these videos were randomly counterbalanced to avoid order effects.

8.3.6 Structure

The survey consisted of four parts. Part 1 consisted of 10 background questions about people's age, sex, education, and experience with robots, and whether they had children or lived with children younger than 12 years old.

Part 2 consisted of two videos and 24 questions (12 questions about each video). The videos showed the self-demonstrations of the two robot conditions: child-friendly and adult-friendly robots.

Each video was followed by 12 questions that were exactly the same for both conditions. The questions were related to people's perception of the robot's age, gender and identity. We used the Godspeed questionnaire (Bartneck, Croft, & Kulic, 2009) measuring anthropomorphism and perceived intelligence on a five-point scale. Table 8.1 presents the questions that were utilized in our questionnaire.

Part 3 was composed of nine questions regarding the preferences for the particular robot, its age and gender. The participants were also asked to name the features of the robot based on what gender they thought it had. The final questions of this section asked about the participant's preferred robot type under different conditions, i.e., if they were with a child/accompanied by a child and if this child was related/unrelated to them.

Part 4 consisted of six questions designed to identify the robot's adaptation algorithm. The participants were presented with a scenario where the robot detected more than one person in front of it and were asked whether the robot should be adapted toward the child or adult. The question was also asked under varying conditions—whether the detected adult–child group was related/known or unrelated/unknown—to identify the differences in responses.

Part 5 asked the same questions as Part 4 but under different settings. The user was now presented with the robot in the role of receptionist in a bank, hospital, and police station. Exactly the same questions were asked, i.e., what adaptation strategy the participant would prefer under the two different relationship conditions. Finally, Part 6 consisted of three open-ended questions where the participants were asked to provide examples where their opinion expressed in Parts 4 and 5 changed.

8.3.7 Participants

A total of 60 subjects were involved in the study. Thirty of them were US participants recruited through AMT and 30 were from Kazakhstan (KAZ) recruited through local social media. There were 35 male and 25 female participants. The age of all participants ranged from 18 to 65 years ($M = 34.05$, SD $= 11.74$).

The age of the US participants ranged from 25 to 65 years ($M = 39.33$, SD $= 11.49$), where the number of males was 20 ($M = 38.30$, SD $= 9.93$), and the number of females was 10 ($M = 41.40$, SD $= 14.50$). Nine out of the 30 US subjects had at least one child, among which, four had a child under 12 y.o. and shared a home with them. The age of the KAZ participants ranged from 18 to 60 years ($M = 28.77$, SD $= 9.52$), with the number of males being 15 ($M = 30.00$, SD $= 11.36$) and the number of females being 15 ($M = 27.53$, SD $= 7.43$). Ten out of the 30 subjects from KAZ had at least one child, among which seven had and lived with a child under 12 y.o.

Table 8.1 Mean ratings for the adult-friendly and child-friendly robots

Robot	Unintelligent—Intelligent		Machinelike—Human-like		Unconscious—Conscious		Artificial—Lifelike		Useless—Useful	
	Adult	Child	Adult	Child	Adult	Child	Adult	Child	Adult	Child
KAZ	**3.43**	**3.77**	2.87	3.00	2.57	2.67	2.53	2.63	**3.33**	**3.60**
USA	3.77	3.77	2.83	3.00	2.40	2.33	2.50	2.53	3.97	3.93

Results in bold have a significant difference

8.3.8 Results

We found that 78.3% of participants from KAZ ($M = 12.87$, SD $= 5.30$), and the USA ($M = 10.21$, SD $= 3.21$) judged the child-friendly robot to have the age of a child. However, there were significant differences ($x2$ (2, $N = 60$) $= 34.737$, $p < 0.001$) between cross-country responses on the perception of the robot's gender. While the voice tone was the same for both languages, all the KAZ participants identified the robot as male, whereas 53.3% of the US participants perceived the robot as female.

The adult-friendly version of the robot was perceived by the KAZ respondents to be aged $M = 14.37$, SD $= 4.82$. The US participants estimated the robot's age to be $M = 15.24$, SD $= 7.818$ with no significant differences between the two countries.

However, we found statistically significant results for the child-friendly robot's perceived gender: $x2$ (2, $N = 60$) $= 25.578$, $p < 0.001$. In total, 93.3% of the KAZ participants perceived the robot as male, while only 50% of the US subjects thought the same.

In the following data analysis, we use *age group, gender, home country, parental status, parent of a child younger than 12 y.o.* and *living with a child younger than 12 y.o.* as the main independent variables. Age group is split by the mean age of the participants: "younger than 34" and "34 and over." What is more, each participant was assigned to the following categories: "Male/Female," "KAZ/US," "Parent/Not a parent," "Parent of a young child/Not a parent of a young child," and "Lives with a young child/Does not live with a young child."

People's Ratings of the Robots: A repeated measures ANOVA with a Greenhouse-Geisser correction determined that ratings differed statistically significantly on the five-point Likert scale for Intelligent ($F(1, 29) = 4.677$, $P = 0.039$) and Useful ($F(1, 29) = 7.864$, $P = 0.009$) between the robot types. The KAZ participants rated the child-friendly robot as significantly more intelligent ($M = 3.77$, SD $= 1$ vs. $M = 3.43$, SD $= 1.19$) and more useful ($M = 3.60$, SD $= 1.19$ vs. $M = 3.33$, SD $= 1.21$). Interestingly, for the US participants the two robots were not significantly different according to their ratings.

Child-friendly versus Adult-friendly Robot: A series of Pearson's chi-square tests was conducted to find whether the KAZ and US participants had significant differences in their preferences for a particular robot type in the answers to the questions from Part 3 presented in the following Table 8.2.

Apart from cross-cultural differences, the differences in peoples' responses for the questions of this part between most independent variables such as gender, parental status, age group, and others were not significant. Thus, we report only statistically significant results between countries. There was a statistically significant difference between the KAZ and US participants: $x2$ (2, $N = 60$) $= 12.94$, $p = 0.002$ for the third question of this part: *Would you prefer the robot to be...?* with three options: *child, adult* or *it does not matter*. We found that 73.3% of the KAZ respondents reported that they preferred the robot to be an adult in contrast to only 40% of the US respondents. The main difference was that the majority of the US participants

Table 8.2 Questions from Part 3

Questions	Answers	USA (%)	KAZ (%)
Which robot did you like more? (from 1st video vs. 2nd video)	Child-friendly	40	60
	Adult-friendly	60	40
What robot would you prefer to interact with? (from 1st video vs. 2nd video)	Child-friendly	40	53.3
	Adult-friendly	60	46.7
Would you prefer the robot to be...? (child, adult vs. does not matter)	Child	**26.7**	**26.7**
	Adult	**40**	**73.3**
	Does not Matter	**33.3**	
Would you prefer the robot to be...? (male, female vs. does not matter)	Male	33.3	10
	Female	30	40
	Does not matter	36.7	50
If you had to choose to watch only one robot speak about itself, would you choose the first or the second video? (1 vs. 2)	Child-friendly	43.3	56.7
	Adult-friendly	56.7	43.3
If you had to choose to watch only one robot speak about itself, what video of the robot would you choose if there was a child (younger than 12 y.o.) sitting next to you that had to watch the same video chosen by you? (1 vs. 2)	Child-friendly	93.3	76.7
	Adult-friendly	6.7	23.3
What video of the robot would you choose if this child was related/known to you? (1 vs. 2)	Child-friendly	90	76.7
	Adult-friendly	10	23.3
What video of the robot would you choose if this child was unrelated/unknown to you? (1 vs. 2)	Child-friendly	**93.3**	**73.3**
	Adult-friendly	**6.7**	**26.7**

Results in bold have a significant difference

reported that the age group of the robot did not matter to them, in contrast to the KAZ respondents where no one selected that option, i.e., the remaining participants (26.7%) chose the child robot.

And the last question of this part—*What video of the robot would you choose if this child was unrelated/unknown to you?*—had a significant difference in the responses of the KAZ and US participants: $x2 (1, N = 60) = 4.32, p = 0.038$. We found that 73.3% of the KAZ and 93.3% of the US respondents chose to watch the child-friendly robot. There was also a significant difference between gender groups: $x2 (1, N = 60) = 3.963, p < 0.046$. 91.4% of male participants reported choosing to watch the child-friendly robot while 72.0% of the female participants would watch the video with the child-friendly robot. In addition, there was a statistically significant difference between age groups: $x2 (1, N = 60) = 3.675, p < 0.05$. The adult-friendly robot would be watched by 23.7% of people younger than 34 but only by 4.5% of people older than 34.

Rules for Robot Adaptation: The next part of the survey had a question on the rules for the robot's adaptation that are detailed in Table 8.3. We only report the significant differences.

For the question, *When the robot sees an adult and a child approach the robot together (i.e. they are related/known to each other), should this robot adapt its speech, behaviors, services,* etc. *to the adult or to the child?* There was a similar pattern in the responses from both countries: People thought that the robot needed to adapt to the child: 66.7% of the KAZ and 63.3% of the US respondents. However, the KAZ parents with children under 12 years old responded differently to the KAZ non-parents: $x2 (1, N = 12) = 5.182, p = 0.023$: 85.7% of those that had a child said that the robot should adapt to the child, in contrast to only 20% of adults who did not have a child.

So when such robot sees an adult and a child approach the robot from different directions (i.e. they are unrelated/unknown to each other), should this robot adapt its speech, behaviors, services, etc. *to an adult or to a child?* This is where there was a statistically significant difference in people's opinions: $x2 (1, N = 60) = 13.416, p < 0.001$. We found that 56.7% of the KAZ and 96.7% of the US participants believed that such a robot should adapt to the adult. It should be noted that the US participants still thought that such a robot needed to adapt to an adult in either case while 43.3% of the KAZ respondents believed that such a robot needed to adapt to a child in the unrelated/unknown adult–child scenario. Again, parental or living status did not have a significant effect on people's opinion.

And who should such a robot look after first in an unrelated scenario? People could choose between the following options: *child, adult, FCFS* and the *robot should listen to both queries and decide which one is of higher priority*. A statistically significant difference was found between the KAZ and the US responses: $x2 (1, N = 60) = 26.211, p < 0.001$. While 60% of the KAZ respondents thought that the robot should look after the people in the order in which they arrived, 60% of the US respondents believed that the robot should look after the adults first. We found that 20% of participants from both countries chose the last option.

Table 8.3 Questions from Part 5

Question	Answer	USA (%)	KAZ (%)
When the robot sees more than one person in front of it (e.g. an adult and a child), should this robot adapt its speech, behaviors, services, etc. to be suitable to the adult or to the child?	Child	53.3	63.3
	Adult	46.7	36.7
When the robot sees an adult and a child approach the robot together (i.e. they are related/known to each other), should this robot adapt its speech, behaviors, services, etc. to the adult or to the child?	Child	63.3	66.7
	Adult	36.7	33.3
When the robot sees an adult and a child approach the robot from different directions (i.e. they are unrelated/unknown to each other), should this robot adapt its speech, behaviors, services, etc. to the adult or to the child?	Child	53.3	56.7
	Adult	46.7	43.3
And who should such a robot look after first in an unrelated scenario? (child vs. adult vs. first-come, first-served (FCFS) vs. the robot should decide given the query priority)	Child	26.7	26.7
	Adult	20	6.7
	FCFS	26.7	26.7
	Robot	27.6	40
Imagine a situation when the robot works at the information desk in a Bank and ..., should this robot adapt its speech, behaviors, services, etc. to an adult or to a child?			
...sees an adult and a child approach the robot together (i.e. they are related/known to each other)...	Child	3.3	10
	Adult	96.7	90
...sees an adult and a child approach the robot from different directions (i.e. they are unrelated/unknown to each other)...	Child	**3.3**	**43.3**
	Adult	**96.7**	**56.7**
And who should such a robot look after first in an unrelated scenario? (child vs. adult vs. first-come, first-served vs. the robot should decide given the query priority)	Child	**0**	**16.7**
	Adult	**60**	**3.3**
	FCFS	**20**	**60**
	Robot	**20**	**20**
A robot working at a hospital reception desk			
Adult and a child approach the robot together (i.e. they are related/known)	Child	10.0	3.3
	Adult	90	96.7
Adult and a child approach the robot from different directions (i.e. they are unrelated/unknown)	Child	**23.3**	**50.0**
	Adult	**76.7**	**50**
And who should such a robot look after first in an unrelated scenario?(child vs. adult vs. first-come, first-served vs. the robot should decide given the query priority)	Child	**6.7**	**16.7,**
	Adult	**56.7**	**10**
	FCFS	**20**	**46.7**
	Robot	**16.7**	**23.3**

(continued)

Table 8.3 (continued)

Question	Answer	USA (%)	KAZ (%)
A robot working at the police station			
Adult and a child approach the robot together (i.e. they are related/known)	Child	6.7	6.7
	Adult	93.3	93.3
Adult and a child approach the robot from different directions (i.e. they are unrelated/unknown)	Child	**16.7**	**53.3**
	Adult	**83.3**	**46.7**
And who should such robot look after first in an unrelated scenario? (child vs. adult vs. first-come, first-served vs. the robot should decide given the query priority)	Child	**3.3**	**26.7**
	Adult	**50**	**3.3**
	FCFS	**16.7**	**36.7**
	Robot	**30**	**33.3**

Significant results are in bold

So when such a robot sees an adult and a child approach the robot from different directions (i.e. they are unrelated/unknown to each other), should this robot adapt its speech, behaviors, services, etc. *to the adult or to the child?* Similar to a bank robot setting, a statistically significant difference was found between the two countries: $x2$ $(1, N = 60) = 4.593, p = 0.032$. We found that 50% of the KAZ and 23% of the US participants thought that such a robot should adapt to an unrelated/unknown child in a hospital setting. In addition, there was a statistically significant difference between gender groups: $x2$ $(1, N = 60) = 6.898, p < 0.009$. The majority of men (77.1%) reported that such a robot should adapt to adults, in contrast to 44.0% of female respondents who thought so. Interestingly, 81.8% of adults older than 34 years old also reported that such a robot should adapt to adults, in contrast to 52.6% of adults younger than 34 years old who thought so. This difference is statistically significant: $x2$ $(1, N = 60) = 5.11, p < 0.024$.

And who should such a robot look after first in unrelated scenario? Again, a statistically significant difference was identified: $x2$ $(1, N = 60) = 15.619, p = 0.004$. The majority of the KAZ participants (46.7%) believed that the robot should follow the FCFS logic, in contrast to the majority of the US participants (56.7%) who believed that such robot should look after adults first. Interestingly, there was a significant difference between the responses of those people that had children aged younger than 12 years old: $x2$ $(4, N = 60) = 10.521, p < 0.033$. We found that 41.7% of those that had children younger than 12 years old answered that such a robot should decide which query was of higher priority. The second most popular answer (33.3%) of the parents of children younger than 12 years old was looking after people in the order in which they arrive, i.e., FCFS. In contrast, the majority of non-parents (37.5%) reported that such a robot should look after adults first. Moreover, a statistically significant result was obtained for people that reported living with children younger than 12 years old: $x2$ $(4, N = 60) = 11.644, p < 0.020$.

So when such robot sees an adult and a child approach the robot from different directions (i.e. they are unrelated/unknown to each other), should this robot adapt its speech, behaviors, services, etc. *to the adult or to the child?* Similar to the bank and hospital robot settings, there was a statistically significant difference between the two countries: $x2$ $(1, N = 60) = 8.864, p < 0.003$. We found that 53.3% of the KAZ and 16.7% of the US respondents believed that such a robot should adapt to the unrelated/unknown child. In addition, 81.8% of adults older than 34 years old also reported that such a robot should adapt to adults, in contrast to 55.3% of adults younger than 34 years old. This difference is statistically significant: $x2$ $(1, N = 60) = 4.319, p < 0.038$.

And who should such a robot look after first in an unrelated scenario? The differences in the responses from the two countries were again statistically significant: $x2$ $(3, N = 60) = 19.997, p < 0.001$. The responses of the KAZ participants were as follows: FCFS (36.7%), the robot should decide (33.3%), the child (26.7%), and the adult (3.3%). In contrast, the responses of the US participants were as follows: the adult (50%), the robot should decide (30%), FCFS (16.7%) and the child (3.3%). However, those people that lived with children younger than 12 years old had a significant effect on people's choices: $x2$ $(3, N = 60) = 9.810, p < 0.020$. The majority of those (53.8%) who shared a home with children reported that such a robot should decide who to look after first. The second most popular option was to look after the unrelated/unknown child (30.8%). For people that did not live with children younger than 12 years old the two most popular options were: an adult (31.9%) and FCFS (31.9%).

8.4 Discussion and Conclusion

The main cross-cultural differences were found in people's opinions on adaptation in cases when the robot was approached by unrelated/unknown adult–child pairs. The US participants believed that robots in banks, hospital, and police settings should always adapt to adults. On the contrary, the KAZ participants believed that only in cases when children are accompanied by related/known adults, should the robot adapt to an adult. In other cases (i.e., when a child is alone), the robot should adapt to the child in bank, hospital, and police settings. Similarly, there was a clear cross-country difference in people's opinion on the order in which adults and children should be looked after. The KAZ participants believed that the FCFS strategy should be followed by the robots in unrelated scenarios, while the US participants believed that the priority for the robots should be looking after adults first.

Taken together, the results from the two studies suggest that the relationship within the multi-party group of participants had a significant effect on people's opinion about the robot's need for adaptation in both countries. Specifically, the KAZ parents expressed an opinion that regardless of the environment, robots should always adapt to their children apart from business/emergency settings such as banks, hospitals, and police stations. On the other hand, in situations of high urgency, children within unrelated groups should be looked after first and adapted to according to the parents and adults who shared a home with a child younger than 12 years old. They justified their opinion with the fact that it is important that the child understands the content of the robot's speech regardless of whether it is an entertaining or a service robot delivering important instructions.

In contrast, although unrelated adults in the USA agreed that robots should adapt to children within the entertainment settings, they believed that adults should be addressed and the robot's speech should be adapted to adults in non-entertainment settings (i.e., hospitals, banks, and police stations) regardless of whether the nearby child is related or unrelated to them. In summary, adaptation of the robot should be context-specific and should take into the account the relationship between children and adults interacting with the same robot.

8.4.1 Future Work

Although this research would have benefitted from collecting more responses, these findings show the need to create a dynamically adaptive robot, which is able to detect whether a particular group of people is related to each other or not. The robot would need to estimate age and gender groups of children and adults [e.g., based on 3D body metrics (Sandygulova, Dragone, & O'Hare, 2014)], detect their relationship/non-relationship status (e.g., based on social cues such as proxemics), and should be able to switch to child-friendly language once a non-parent adult is detected in a non-entertainment context.

References

Bartneck, C., Croft, E., & Kulic, D. (2009). Measurement instruments for the anthropomorphism, animacy, likeability, perceived intelligence, and perceived safety of robots. *International Journal of Social Robotics, 1*(1), 71–81.

Davidoff, S., Lee, M. K., Zimmerman, J., & Dey, A. (2006). Socially-aware requirements for a smart home. In *Proceedings of the International Symposium on Intelligent Environments* (pp. 41–44).

Evers, V., Menezes, N., Merino, L., Gavrila, D., Nabais, F., Pantic, M., et al. (2014). The development and real-world deployment of frog, the fun robotic outdoor guide. In *Proceedings of the 2014 ACM/IEEE International Conference on Human-Robot Interaction* (p. 100). ACM.

Foster, M. E., Gaschler, A., Giuliani, M., Isard, A., Pateraki, M., & Petrick, R. P. (2012). Two people walk into a bar: Dynamic multi-party social interaction with a robot agent. In *Proceedings of the 14th ACM International Conference on Multimodal Interaction* (pp. 3–10), ICMI '12, New York, NY, USA. ACM.

Glas, D. F., Satake, S., Ferreri, F., Kanda, T., Hagita, N., & Ishiguro, H. (2012). The network robot system: Enabling social human-robot interaction in public spaces. *Journal of Human-Robot Interaction, 1*(2), 5.

Hofstede, G. (1986). Cultural differences in teaching and learning. *International Journal of inter-cultural relations, 10*(3), 301–320.

Hofstede, G. (2011). Dimensionalizing cultures: The Hofstede model in context. *Online Readings in Psychology and Culture, 2*(1), 8.

Hood, D., Lemaignan, S., & Dillenbourg, P. (2015). The cowriter project: Teaching a robot how to write. In *Proceedings of the Tenth Annual ACM/IEEE International Conference on Human-Robot Interaction Extended Abstracts* (pp. 269–269). ACM.

Kanda, T., Shiomi, T., Miyashita, Z., Ishiguro, H., & Hagita, N. (2009). An affective guide robot in a shopping mall. In *Proceedings of the 4th ACM/IEEE International Conference on Human Robot Interaction* (pp. 173–180), HRI '09, New York, NY, USA. ACM.

Kondo, Y., Takemura, K., Takamatsu, J., & Ogasawara, T. (2013). A gesture-centric android system for multi-party human-robot interaction. *Journal of Human-Robot Interaction, 2*(1), 133–151.

Korn, O., Bieber, G., & Fron, G. (2018). Perspectives on social robots: From the historic background to an experts' view on future developments. In *Proceedings of the 11th Pervasive Technologies Related to Assistive Environments Conference* (pp. 186–193), PETRA '18, New York, NY, USA. ACM.

Li, D., Rau, P. L. P., & Li, Y. (2010). A cross-cultural study: Effect of robot appearance and task. *International Journal of Social Robotics, 2*(2), 175–186.

Lidoris, G., Bauer, K. K. A., Xu, T., Kuhnlenz, K., Wollherr, D., & Buss, M. (2007). The autonomous city explorer project: Aims and system overview. In *2007 IEEE/RSJ International Conference on Intelligent Robots and Systems* (pp. 560–565).

Makatchev, M., Simmons, R., Sakr, M., & Ziadee, M. (2013). Expressing ethnicity through behaviors of a robot character. In *Proceedings of the 8th ACM/IEEE International Conference on Human-robot Interaction* (pp. 357–364), HRI '13, Piscataway, NJ, USA. IEEE Press.

Martinson, E., Lawson, W., & Trafton, J. (2013). Identifying people with soft-biometrics at fleet week. In *2013 8th ACM/IEEE International Conference on Human-Robot Interaction (HRI)* (pp. 49–56).

Mussakhojayeva, S., Kalidolda, N., & Sandygulova, A. (2007). Adaptive strategies for multi-party interactions with robots in public spaces. In A. Kheddar, E. Yoshida, S. S. Ge, K. Suzuki, J.-J. Cabibihan, F. Eyssel, & H. He (Eds.), *Social Robotics* (pp. 749–758). Cham: Springer International Publishing.

Mussakhojayeva, S., Zhanbyrtayev, M., Agzhanov, Y., & Sandygulova, A. (2016). Who should robots adapt to within a multi-party interaction in a public space? In *The Eleventh ACM/IEEE International Conference on Human Robot Interaction* (pp. 483–484). IEEE Press.

Mussakhojayeva, S. & Sandygulova, A. (2017). Cross-cultural differences for adaptive strategies of robots in public spaces. In *2017 26th IEEE International Symposium on Robot and Human Interactive Communication (RO-MAN)* (pp. 573–578).

O'Neill-Brown, P. (1997). Setting the stage for the culturally adaptive agent. In: *1997 AAAI Fall Symposium* (pp. 93–97).

Read, J., MacFarlane, S., & Casey, C. (2002). Endurability, engagement and expectations: Measuring children's fun. *Interaction design and children* (Vol. 2, pp. 1–23). Eindhoven: Shaker Publishing.

Salem, M., Ziadee, M., & Sakr, M. (2014). Marhaba, how may I help you? Effects of politeness and culture on robot acceptance and anthropomorphization. In *Proceedings of the 2014 ACM/IEEE International Conference on Human-robot Interaction* (pp. 74–81), HRI '14, New York, NY, USA. ACM.

Sandygulova, A., Dragone, M., & O'Hare, G. M. (2014). Real-time adaptive child-robot interaction: Age and gender determination of children based on 3d body metrics. In *2014 RO-MAN: The 23rd IEEE International Symposium on Robot and Human Interactive Communication* (pp. 826–831).

Sandygulova, A., Dragone, M., & O'Hare, G. M. (2016). Privet–a portable ubiquitous robotics testbed for adaptive human-robot interaction. *Journal of Ambient Intelligence and Smart Environments, 8*(1), 5–19.

Schwab, K.-W. (2013). Individualism-collectivism and power distance cultural dimensions: How each influences parental disciplinary methods. *Journal of International Education and Leadership, 3*(3).

Shiomi, M., T. Kanda, H. Ishiguro, & Hagita, N. (2006). Interactive humanoid robots for a science museum. In *Proceedings of the 1st ACM SIGCHI/SIGART Conference on Human-robot Interaction* (pp. 305–312), HRI '06, New York, NY, USA. ACM.

Shiomi, M., Sakamoto, D., Kanda, T., Ishi, C., Ishiguro, H., & Hagita, N. (2011). Field trial of a networked robot at a train station. *International Journal of Social Robotics, 3*(1), 27–40.

Simmons, R., Makatchev, M., Kirby, R., Lee, M. K., Fanaswala, I., Browning, B., et al. (2011), Believable robot characters. *AI Magazine 32*(4).

Spencer. (2015). Social situation-aware perception and action for cognitive robots. Retrieved from www.spencer.eu.

Syrdal, S., Nomura, T., & Dautenhahn, K. (2013). *The Frankenstein syndrome questionnaire—Results from a quantitative cross-cultural survey* (pp. 270–279). Springer International Publishing, Berlin.

Torrey, C., Powers, A., Marge, M., Fussell, S. R., & Kiesler, S. (2006). Effects of adaptive robot dialogue on information exchange and social relations. In *Proceedings of the 1st ACM SIGCHI/SIGART Conference on Human-robot Interaction* (pp. 126–133), HRI '06, New York, NY, USA. ACM.

Wang, L., Rau, P.-L. P., Evers, V., Robinson, B. K., & Hinds, P. (2010). When in Rome: The role of culture & context in adherence to robot recommendations. In *Proceedings of the 5th ACM/IEEE International Conference on Human-robot Interaction* (pp. 359–366), HRI '10, Piscataway, NJ, USA. IEEE Press.

Wursten, H., Jacobs, C., & Magazine, E. (2013). The impact of culture on education. *The Hofstede Centre, Itim Internation.*

Chapter 9
Cross-Collaborative Approach to Socially-Assistive Robotics: A Case Study of Humanoid Robots in a Therapeutic Intervention for Autistic Children

David Silvera-Tawil and Scott Andrew Brown

Abstract Autism is a developmental condition that can cause significant social, communication, and behavioral challenges. Children on the autism spectrum may have difficulties developing social and communication skills. A recent trend in robotics is the design and implementation of robots to assist during the therapy and education of children with learning difficulties. In this chapter, we reflect on lessons learned from a cross-collaborative research project involving a socially-assistive robot, KASPAR, as a tool to support therapy with autistic children. We provide experimental results from a small study using this humanoid robot in combination with Social Stories ^{TM}. We point to the strengths and challenges of our approach and discuss how others might use our experience as a guide to improving experimental and therapeutic outcomes.

Keywords Socially-assistive robotics · Autism spectrum disorder · Social stories · Applied behavioral analysis

9.1 Introduction

Autism spectrum disorder (ASD) is an ongoing neurodevelopmental condition that results in deficits in communication, social interaction, and behavior. The degree of impairment related to ASD varies significantly across the spectrum, ranging from severe to near-typical social functioning. For some, there are significant impacts on quality of life and independence, affecting education, employment, and social rela-

D. Silvera-Tawil (✉)
Australian e-Health Research Centre, Commonwealth Scientific and Industrial Research Organisation (CSIRO), Sydney, Australia
e-mail: david.silvera-tawil@csiro.au

S. A. Brown
Art and Design, University of New South Wales (UNSW), Sydney, Australia
e-mail: scott.brown@unsw.edu.au

© Springer Nature Switzerland AG 2019
O. Korn (ed.), *Social Robots: Technological, Societal and Ethical Aspects of Human-Robot Interaction*, Human–Computer Interaction Series,
https://doi.org/10.1007/978-3-030-17107-0_9

tionships (Farley et al., 2009). Relevant evidence-based practices and research-based therapies seek to improve the individual's social and communication skills, while at the same time promoting engagement in interpersonal interactions. Early intervention programs have been shown to be particularly beneficial and can lead to long-term gains in cognitive, social, emotional, and motor functioning, providing considerable improvements to the individual's quality of life and independence (Bennett, 2012).

Traditional education for children on the autism spectrum is often supplemented with digital technology—including apps and computer games—to help students acquire the skills necessary to navigate the world outside the classroom (Bauminger-Zviely, Eden, Zancanaro, Weiss, & Gal, 2013; Newbutt, Sung, Kuo, & Leahy, 2017). The strengths of these technologies are in providing a teaching environment that allows for expectation management, self-paced learning, and immediate feedback, while minimizing the need for 'real-world' social interactions during the learning process—a common source of anxiety for autistic people (Golan & Baron-Cohen, 2006). While considered generally safe and effective, there are concerns that a child who is taught to communicate using interactive technology may become dependent on the virtual world and its rewards, at the expense of developing interpersonal skills (Bauminger-Zviely et al., 2013).

For over a decade, researchers have explored the use of social robots as tools to supplement traditional therapy and education (Scassellati, Admoni, Matarić, & Admoni, 2012). Autonomous and remotely operated robots have shown that social robots can promote, among other skills, facial expression recognition (Vanderborght et al., 2012), shared attention (Warren et al., 2013), imitative free-form play (Robins & Kerstin, 2005), and turn-taking (Robins & Kerstin, 2010). The most effective approaches to date are those that use robots in free or semi-structured interactions (Boccanfuso et al., 2016; Costa, Lehmann, Dautenhahn, Robins, & Soares, 2015; Pennisi et al., 2016). However, outcomes have varied according to the intervention method, the robot being used, and the severity of the child's symptoms. Although many different robots have been developed, simple anthropomorphic shapes with limited expressivity and basic human-like behavior seem to offer the most promise for therapy and education, providing enhanced generalization of skills (Sartorato, Leon, & Sarko, 2017).

In this chapter, we demonstrate that a humanoid robot, KASPAR, can be used to facilitate the Social Stories™ intervention within an Applied Behavioral Analysis (ABA) framework, supporting the acquisition of social interaction and communication skills of autistic children. Through qualitative research based on interviews and observations, our project describes promising experiences for the three children participating in the study. We explore the impact of humanoid robots used as 'peers' during therapy and point to future work that could strengthen the cross-collaborative potential of research in this area.

9.2 Literature Review

9.2.1 Social Robots in Therapy and Education

Play is an important element in the development of language skills, cognitive skills, and opportunities for social interaction (Pierucci, Barber, Gilpin, Crisler, & Klinger, 2015). Pierucci et al. (2015) recommend that if a child does not typically engage in social reciprocal play, object play should be used to support the child in achieving social communication goals.

It is now generally agreed that children on the autism spectrum enjoy playing with mobile apps, computer games, and virtual reality devices (Bauminger-Zviely et al., 2013; DiGennaro Reed, Hyman, & Hirst, 2011; Grynszpan, Weiss, Perez-Diaz, & Gal, 2013). These technologies offer realistic-looking scenarios that can be built to depict everyday social situations, providing environments that allow for safe, self-paced learning and immediate feedback, while minimizing the need for 'real-world' social interactions during the learning process, a common source of anxiety for many people on the autism spectrum (Golan & Baron-Cohen, 2006).

By using first-person, realistic-looking, computer-generated environments, autistic individuals can develop a functional range of daily living skills (e.g., social and communication skills) that increase their opportunities for a more independent life (Bozgeyikli, Bozgeyikli, Raij, & Katkoori, 2016; Newbutt, Sung, Kuo Hung, & Leahy, 2016; Newbutt et al., 2017; Rajendran, 2013). It has been argued that the realism of computer-simulated environments, as well as the increased sense of presence provided by immersive virtual environments, can help promote learning and increase the probability that a person will generalize newly learned skills into everyday living (Miller & Bugnariu, 2016; Newbutt et al., 2016). Although these technologies appear to be effective, a significant concern is that the large gap between the safe and structured environment of computer-based interventions and the complexity and ambiguity of real-world social behavior may result in poor transfer of skills to real-world interactions (Bauminger-Zviely et al., 2013).

Over the last decade, researchers have explored the use of social robots as tools to supplement traditional therapy and education (Scassellati et al., 2012). These robots are often presented as toys. These objects are novel, animated, appear autonomous, and set themselves apart from traditional toys, thereby further maintaining a child's interest. The three-dimensional presence of robots, furthermore, provides a compromise between the virtual world—available through digital technologies—and the real world, by promoting an embodied experience on the part of the child. A robot can provide complex behavior patterns, such as those available in interpersonal interactions, and evoke social behaviors and perceptions in the people they interact with, while appearing less intimidating and more predictable than humans (Michaud & Theberge-Turmel, 2002).

There are a number of applications for social robots in the therapy and education of young people on the autism spectrum. In assessment, for example, a protocol to assist during the diagnosis of autism was proposed by Petric (2014). This protocol is based on four tasks extracted from the Autism Diagnostic Observation Schedule and modified to be implemented using humanoid robots. By using this approach, Petric (2014) believes that diagnosis of ASD can be standardized, improving accuracy and consistency.

In terms of treatment, existing research has focused on three main areas: the use of robots to (a) increase engagement and motivation; (b) elicit behaviors; and (c) model, teach, and/or practice skills (Diehl, Schmitt, Villano, & Crowell, 2012). For example, a 24-month trial conducted using the humanoid robot NAO to support autistic students with their academic learning, daily living, social and communication skills (Silvera-Tawil & Roberts-Yates, 2018; Silvera-Tawil, Roberts-Yates, & Bradford, 2018). Other research examples include: a nine-month trial with the anthropomorphic robot, 'Lucy,' aimed at enhancing sensory enrichment through free interaction in a home-based care environment (Khosla, Nguyen, & Chu, 2016); semi-structured interactions with the humanoid KASPAR, to increase body awareness in children with autism (Costa et al., 2015); and semi-structured play with the robot CHARLIE, to promote communication and social skills (Boccanfuso et al., 2016). Recent reviews of the research in the area have been presented by Begum, Serna, and Yanco (2016), Bodine et al. (2017) and Pennisi et al. (2016).

9.2.2 Social Stories^{TM}

Social Stories was devised as a tool to help individuals on the autism spectrum better understand the nuances of interpersonal communication, so that they could interact in an effective and appropriate manner (Gray, 2000). The stories are tailored to each person and can be used to teach behaviors, routines, and a curriculum, as well as to increase independence. Social Stories are commonly used in combination with other evidence-based practices, such as reinforcement, prompting, priming, and corrective feedback (Reynhout & Carter, 2007), and have been successfully used during intervention programs to teach autistic children appropriate social skills and behaviors across various situations and environments (Barry & Burlew, 2004; Reynhout & Carter, 2007).

Social Stories typically comprise short text and are accompanied by pictures or illustrations (Fig. 9.1). The use of illustrations depends on the age, level of cognitive ability, and learning preferences of the person that the story is being developed for. Four types of sentences are used when creating a Social Stories narrative: descriptive sentences, directive statements, affirmative sentences, and partial sentences (Hall, 2009). Suggested elements for the development of effective Social Stories include: (a) age-appropriate vocabulary, (b) use of the present tense, (c) a story written from the individual's perspective, and (d) the use of and adherence to a recommended ratio of directive and perspective sentences (Gray, 2000).

Social Stories can be written by different stakeholders, such as parents/carers, teachers, and therapists. They can be used to teach appropriate behaviors, social skills, routines, and a curriculum, as well as to increase independence. They are

Fig. 9.1 Social Stories example: shaking hands. Images in the top left and bottom right squares were obtained from Adobe Stocks

commonly used in combination with other evidence-based practices, such as reinforcement, prompting, priming, and corrective feedback (Reynhout & Carter, 2007). It is important to introduce a social story in a setting that is comfortable for the autistic individual, and to read it often, especially prior to a situation in which the targeted skill can be practiced.

Social Stories are usually read by a human carer or therapist. Recent research, however, argues that the social performance of autistic children can be improved when a robot is used as a storyteller instead of a human (Gillesen, Barakova, Huskens, & Feijs, 2011; Vanderborght et al., 2012). Gillesen et al. (2011) suggest that substituting a human with a robot can be preferable for a child, as it is more consistent and predictable. Furthermore, Simut et al. (2012) used the robot 'Probo' when using the Social Stories intervention to address the social skills deficits of four preschool-aged children with ASD. The social skills included sharing toys, saying hello, and saying thank you. The results of this study demonstrated that the social story told by the robot led to less prompting from the therapist in order for the children to perform the appropriate social response.

9.2.3 Applied Behavior Analysis

ABA is an evidence-based, intensive education therapy that has been shown to be highly effective in educating children and youths with disabilities, particularly those on the autism spectrum (Kearney, 2007). ABA relies on the collection of behavioral and/or academic data, examining the interaction between antecedent variables and consequences, and using this information to systematically plan desired learning and behavior change programs (Alberto & Troutman, 2009). ABA is a personalized program designed to meet the individual needs of each child to help support the development of their social, academic, and behavioral skills. It is dependent upon strengthening and maintaining desired behaviors through the application of positive reinforcement and modeling (Kerr & Nelson, 2010).

ABA has its roots in behaviorism and as such is dependent upon data collection. A functional behavioral assessment (FBA) is the first step in this process. The FBA consists of observations and interviews with the child and those closest to him/her (e.g., parents/carers, teachers). The results of the assessment are used to identify target behaviors to change and strategies and resources that can be used in the design of the intervention plan (Alberto & Troutman, 2009). Both the FBA and the behavior plan are highly individualized, as problem behavior serves different functions for different children. Social skills training programs, such as social narratives and Social Stories, are often included in the behavior plans of children who exhibit deficits in social understanding and communication (Reynhout & Carter, 2007).

An important component of the implementation of any ABA strategy is reinforcement. Positive reinforcement for desired responses will strengthen the response and has been shown to be effective when paired with several ABA-based strategies, including discrete trial training, incidental teaching, and pivotal response training (Hall, 2009). Discrete trial training is an adult-initiated contingency strategy that has been used to teach communication, language, and many other skills to autistic children (Lovaas, 2003). Incidental teaching focuses on child-initiated interactions to increase language skills (Hall, 2009). This strategy uses carefully planned environmental cues to elicit initiation from the child, such as putting a preferred item out of reach and using access to the item to motivate the child for performing a specific skill. Pivotal response training is a child-initiated teaching interaction with choice-making and turn-taking embedded. Important elements of this strategy on the part of the child include motivation and responsivity to multiple cues. The adult's focus is on: (a) following the child's lead, (b) giving the child choices, (c) reinforcing attempts as well as successes, (d) providing natural consequences, and (e) mixing maintenance tasks in with new tasks. Researchers have suggested that integrating robots into these and other ABA teaching strategies may increase their effectiveness (Diehl et al., 2012; Silvera-Tawil, Strnadová, & Cumming, 2016; Tang, Jheng, Chien, Lin, & Chen, 2013).

In this vein, Barakova and Lourens (2013), Gillesen et al. (2011) and Yun et al. (2014) designed three ABA-based robot interventions to promote self-initiated social behaviors in children. Yun et al. (2014) used a discrete trial teaching protocol with three task modes of therapy, encouragement, and interruption to improve the children's social capabilities, while Barakova and Lourens (2013) and Gillesen et al. (2011) used a pivotal response treatment approach and tailored interactive scenarios to promote self-initiated social behaviors, language and communication, problem-solving, and appropriate social responses.

9.3   KASPAR the Robot

The base platform used during this study was the humanoid robot KASPAR (Fig. 9.2). Developed by the Adaptive Systems Research Group at the University of Hertford-shire, KASPAR comprises 17 actuated degrees of freedom with motors located in its face, neck, arms and torso, one video camera in each eye, and a microphone and speaker for audio/verbal communication. Position sensors are incorporated in each of the robot's joints. KASPAR can produce pre-recorded speech and upper-body movement to engage children in play and conversation and provide feedback about their actions. All hardware is wirelessly connected to a host computer that controls all sensors and actuators.

Fig. 9.2 KASPAR interacting with a child while controlled by a researcher

9.4 Pilot Study

Prior to the main part of this project, a pilot study was conducted with a single participant, referred to here as Liam. This pilot study was designed to incrementally develop Liam's social skills by conducting five daily sessions with KASPAR, over a period of one week. To generate appropriate intervention goals and a Social Stories narrative specific to his interests and the aspirations of his parent, a pre-study interview was carried out with Liam's mother in late 2016.

At the time of the study, Liam was a very active five-year-old; lively and sensory seeking. He knew about 200 words and had better receptive language than expressive language. He liked books, computers, and interactive toys, but not music or singing.

During the pre-study interview, the main areas identified as needing improvement before Liam moved toward mainstream schooling were focus and attentiveness. As a result, the Social Stories and intervention goals for Liam's study focused on the idea of 'staying connected' through eye contact, listening to others, and expressing this intent through bodily orientation. The Social Stories was supported by text and images on a computer screen.

The intervention sessions were structured as a cause-and-effect game, as suggested by Silvera-Tawil et al. (2016), and were conducted by a researcher with a background in autism but no therapeutic experience. Controlled by the researcher, KASPAR would lead and direct the conversation with Liam using the pre-programmed Social Stories and provided positive reinforcement and reward through speech, movement, and music. The content of the Social Stories included interactions such as the following:

KASPAR: *Connecting means looking at people, facing them and listening closely.*
KASPAR: *Connecting with people is a great way to learn… learning new things is fun!*
KASPAR: *Now it is your turn Liam. To show you are connected with KASPAR, face me with your arms, legs, and body.*
[Waits for Liam to respond]
KASPAR: *'Well done!' or 'Let's try again.'*

Due to the nature of this interaction, Liam might do what KASPAR asked or might choose to do something different. The limited set of responses pre-programmed into the robot led to a disjointed or unnatural interaction between Liam and KASPAR, as the researcher tried to choose the 'best fit' from this set. It was also problematic for the researcher to be regularly focused on a control interface (keyboard), which drew Liam's attention away from KASPAR. When Liam went 'off script' from the pre-planned Social Stories, the researcher did not have the skills to redirect him back toward KASPAR, leading to missed opportunities to reflect on the interactions between child and robot.

As a child that is interested in computers and interactive toys, Liam quite quickly turned his attention to the keyboard interface and lost interest in KASPAR (in later discussions with his mother, she described Liam's disappointment that KASPAR did not look like the '60s sci-fi' robot he had expected before the study). Equally distracting for Liam was the study space chosen: a large room divided by panels (the Creative Robotics Lab at UNSW Art & Design). Behind the panels, there was a range of technical equipment such as cables, electronics, and televisions. When Liam discovered the equipment behind the panels, this became more interesting to him than the deliberately bland space presented for interacting with KASPAR. The difficulty in trying to keep Liam engaged persisted throughout the five sessions across the week.

While the pilot study was challenging and unsuccessful, key lessons were learned that led to the study design that is described below. Central to this is the importance of a trained professional and appropriate space to conduct the sessions. In the remainder of this chapter, we look at how we leveraged KASPAR's strength as a tool to enhance therapy.

9.5   Study Design

This study aims to contribute to the evidence base of using robots and Social Stories within an ABA framework to enhance the social communication abilities of autistic children. The potential benefits, challenges, limitations, and opportunities of the suggested approach were explored through observational research using a single-subject design, supported by qualitative insights collected from parents. Ethics approval was sought and obtained in February 2016 (UNSW HC15655). Semi-structured interviews, intervention sessions, and participant questionnaires were implemented between January and March 2018.

The recruitment of participants was carried out through autism clinics and service providers in Australia. The researchers had no direct contact with any parents/carers or children during this process. All intervention sessions and interviews were video- and audio-recorded with the participants' informed consent.

9.5.1   Wizard of Oz Technique

During the current study, KASPAR was controlled using the Wizard of Oz (WoZ) technique (Kelley, 1983). WoZ is used to simulate intelligent systems and interfaces by having a 'wizard'—typically a researcher—teleoperating the robots to simulate a portion of the system's purported functionality, in this case, the ability of KASPAR to understand and respond to the child's questions. Dahlbäck et al. (1993) showed that this technique is suitable when high-quality empirical data is required, but gathering the data is not a simple task.

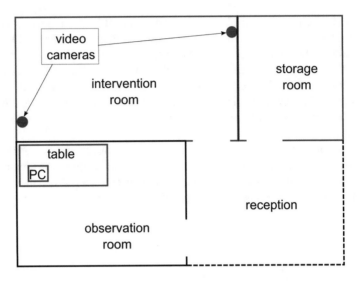

Fig. 9.3 Study space

9.5.2 Space

The study was carried out at a dedicated ASD clinic with access to a range of specialized toys and activities. All sessions were performed in a private therapy room within the clinic (Fig. 9.3). Only the child, a psychologist (therapist), and KASPAR were allowed into the main intervention room during the sessions. The parent and researchers were located in an adjacent observation room where they were able to control the robot and observe the sessions from two video cameras positioned at opposite corners of the main room.

9.5.3 Procedure

This study was divided into three stages, which are detailed in the following sub-sections. The sessions were distributed over seven weeks (Fig. 9.4).

Pre-intervention survey and interview
The aim of this stage was to collect information regarding the parents'/carers' perceptions of the children's ability to understand a current situation and to respond to it appropriately. This stage informed the personalized intervention goals and Social Stories for each child.

   The pre-intervention survey was an adapted version of the assessment questionnaire presented by Vicker (2003), including questions related to demographic information (e.g., gender, age, diagnosis, etc.), language comprehension, language

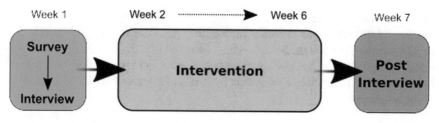

Fig. 9.4 Study procedure

expression, interest, communication effectiveness, play behaviors, and experience with Social Stories. Following the survey, additional information about the parents' expectations and the child's skills, routine, and motivations were collected through a semi-structured interview with the parents. These interviews were conducted by the researchers and therapists by means of a teleconference. The survey and interview were performed within seven days of the first session.

Intervention

Each child took part in five weekly intervention sessions at the clinic (Fig. 9.5). All sessions were conducted by a clinical psychologist with experience in ASD child therapy within an ABA framework, with two other researchers (including the 'wizard') and the child's parent viewing the session from an adjacent room. Each session lasted approximately 45 minutes. For all participants, KASPAR and paper-based Social Stories were introduced by the therapist during the second session. In all cases, KASPAR was introduced as a 'peer' for interaction. Sessions were planned weekly by the researchers and therapist using the outcomes from the previous session as a guideline. KASPAR's behaviors were re-programmed every week, according to the identified needs for each session.

Managed by the therapists, the intervention sessions increased interaction between KASPAR and the child over sessions two to four. During the last session (Session 5), interaction with KASPAR was reduced, providing an opportunity for the therapist to test the child's responses to the Social Stories without KASPAR's presence.

Fig. 9.5 Video from an intervention session. Therapist, child, and KASPAR interacting with each other with Social Stories displayed on the wall and table

Post-intervention interview

The aim of this stage was to collect information regarding the parents'/carers' perceptions of the robot and the behaviors they observed in their child during and immediately after the study (both within the sessions and at home). The post-intervention interview was performed by a researcher by means of a teleconference, within a week of the last session.

9.6   Results

Two parents (and their children) agreed to participate and will be presented here as case studies. The parents expressed interest in participating, as they believed their children would enjoy interacting/playing with robots (i.e., KASPAR) and additional therapy would also be beneficial. To protect their privacy, we will refer to the children as Jack and Carl. Both children had a previous diagnosis of ASD and were involved with different styles of therapy. Neither of them had interacted with humanoid robots before the study. While both participants were familiar with the clinic, neither child had attended therapy at that location for at least 12 months before the study.

9.6.1   Case Study 1: Jack Develops Social Communication Skills

Background

Jack was (at the time of the study) a five-year-old male student with a diagnosis of autism and global developmental delay. He was the only child of divorced parents. He had good verbal abilities but lacked communication skills. He spoke in a low voice, did not look at people or attracted their attention during a conversation. His attention span was short (between one and five minutes); he found it difficult to stay on topic, and excessive noise or activity caused him to lose concentration, get confused, and stop listening.

Jack was good at following instructions when sentences were kept short and simple, had a good memory, and recognized words and familiar people. He could understand when someone was upset as a result of his actions and was starting to understand concepts such as 'how,' 'what,' and 'why' and to ask for clarification when needed. He could link small pieces of information together (e.g., a person and their place of work) and bring that information into a conversation.

He was interested in books, songs, and movies, particularly books on spiders, making friends, families, and space. He enjoyed playing with toys and had great imagination. Jack liked physical play, being chased, tickled, and wrestled. He could not read or write.

According to his mother, Jack's vocabulary and communication skills 'rocketed in the last year or so,' as he had more opportunities for connection with peers through social activities (e.g., judo), play dates, and swimming lessons. Outside one hour per week of speech therapy, Jack did not take part in any further therapeutic activities during the study.

Social Stories were often used to support Jack. At home, Social Stories were used as visual itineraries, when new activities were expected, or when mom/dad needed to be away for an extended period of time. They allowed him to understand what was happening during each day and who was he staying with. Most of the time, Social Stories were created using a combination of text and images. When needed, they were read multiple times a day.

Intervention goal and Social Stories

Jack's Social Stories were designed to improve his social communication skills, including conversational skills (e.g., starting and finishing conversations; maintaining a topic of conversation; taking turns in a conversation) and non-verbal communication skills (e.g., judging proximity and distance). To support this goal, a Social Stories narrative was created to explore how to talk to family and friends, with particular attention given to turn-taking, personal distance during interaction (i.e., close but not too close), speaking clearly and loudly (but not too loud), and staying on topic. The stories were supported by text and images.

KASPAR was programmed to support the sessions by including a number of questions and answers around Jack's favorite topics, including food, drinks, books, movies, and songs. Additional phrases were introduced every week as needed, through reflection on session outcomes and in consultation with the therapist.

Intervention

KASPAR was introduced as a conversational 'peer' to give Jack the opportunity to practice new skills. Jack was aware that a robot would be part of the study and was looking forward to it. KASPAR was introduced during the second session.

Jack was immediately attracted to the technology and was willing to interact with it. Encouraged by the therapist, he would ask questions and answer KASPAR's questions in return. Throughout this session, however, his speech was low and difficult to hear.

Social Stories were introduced during the same session. Particular attention was given to turn-taking, speech volume, and interpersonal distance during interaction. Motivated by the therapist, Jack would practice the new skills while interacting with KASPAR, giving him the opportunity to learn through repetition. Throughout the remaining sessions, it became clear that KASPAR would keep Jack engaged and motivated, serving not only as a conversational 'peer' but also as a positive reinforcer. Jack enjoyed not only talking to KASPAR, but also singing and dancing while it played musical recordings and performed simple dance routines. Jack liked hugging KASPAR and shaking hands when saying 'hello' and 'good bye.'

KASPAR was only introduced during the second half of each session. With KASPAR in the room, Jack would engage in conversation, asking and answering questions about both their interests (e.g., favorite food, drink, movie, song, etc.) and practicing new social communication skills introduced through the Social Stories. Jack would show KASPAR his drawings, books, and photographs, and KASPAR would provide positive reinforcement through phrases such as 'great,' 'sounds nice,' and 'tell me more.' When not around, KASPAR served as a reference and topic of conversation, improving the communication between the therapist and Jack. According to the therapist, however, Jack appeared to answer more of his questions and managed to stay on topic for longer periods of time when KASPAR was in the room.

Parent interview

Jack's mother believed the therapy sessions were beneficial for her child. She highlighted that Jack engaged with KASPAR more than he normally would with 'other children,' as he did not normally volunteer to interact with his peers but he did with KASPAR, not only because 'he was fascinated by the robot,' but because he had a good connection with the therapist. She highlighted that the opportunity to interact with KASPAR and practice care was a great opportunity for Jack to learn about being gentle with others. Jack's mother also mentioned that the location and structure of the therapy sessions were adequate and comfortable for her child. It was a place where he had been before, so it was easy for him to adapt.

In terms of the changes observed during the study, she mentioned that since the first session with KASPAR Jack had started showing great recall skills in reference to what KASPAR likes, and that was not something he would normally do. Furthermore, he seemed to be able to recall more information about his own peers than previously and would also consider other's needs and interests more than before. For example, Jack would now offer his mother food when he was eating—something that never happened before.

Although Jack's mother acknowledged that it was hard to quantify how many of these new skills could be attributed to the robot-assisted therapy, she believed that every new experience delivered a reward, and the opportunity to interact with KASPAR afforded Jack the ability to talk and show more interest in peers at school. She said that the exposure to this 'new and cool' experience was making Jack more confident and open to people and experiences.

Finally, Jack's mother mentioned that KASPAR could be very helpful for practicing skills such as touch, because it is hard to learn these boundaries with people. She also suggested that an agenda during each session could be used to increase structure to better manage the child's expectations.

9.6.2   Case Study 2: Carl Develops Focus and Attentiveness

Background

Carl was (at the time of the study) a four-year-old male with a diagnosis of autism. He was the younger child in a family of two children. His most developed skills were in the domains of vocal and visual perceptual and matching to sample. Skills in all other domains were emerging at different rates. His language and communication skills were limited. He could follow simple instructions, but sentences needed to be kept short (less than three words). Comprehension was harder if the content was unfamiliar, or if there was background noise. He only communicated to satisfy a need, such as food or play. When he communicated, he used simple language, short sentences (two to three words), and gestures. He spoke softly and quickly, making it hard to understand what he meant. He seemed to recognize simple words, such as names, vocabulary describing family members, and food. Occasionally, he also recognized pictures of common items. He did not ask questions, made comments, shared information, or engaged in conversation. He appeared to understand more than he could express.

Carl liked cars and physical play (e.g., the trampoline, swing, climbing, jumping, etc.). He enjoyed listening to songs and sometimes sang and danced. He only looked at books very briefly. In general, he was more interested in interacting with adults than children. He played around other children (parallel play), but not with them (social play).

At the time of the study, Carl was attending approximately 30–35 hours of weekly ABA therapy. His therapy goals included: improvements in functional communication skills, receptive language, expressive language, social interaction, emotion regulation, and daily living skills. He had no previous experience of using Social Stories.

Intervention goals and Social Stories

The aim of Carl's Social Stories included improving receptive language (the ability to understand spoken language), matching skills (e.g., match pictures and objects/activities based on shared features), and engagement with others. To support these goals, the Social Stories were developed to match his comprehension skills and were presented as a 'visual agenda' using pictures and words to represent different tasks and activities such as playing, jumping, eating, and group time.

Intervention

The first session with Carl was challenging. He only engaged with the therapist during physical play and had no interest in sitting down, talking, dancing, or singing. On multiple occasions, he looked upset, and as a coping mechanism, he would look out of the windows of the intervention room. He repeated this behavior multiple times. He would not follow instructions and would only communicate to ask for food. When food was provided, he would sit down, eat, and calm down until he had finished eating.

KASPAR was introduced during the second session. Carl's behavior was unchanged, and he did not engage with either KASPAR or the therapist. He looked tired, hungry, and overwhelmed. There was no noticeable difference to the first session. At the end of this session, however, the therapist and Carl's mother encouraged him to say good-bye to KASPAR and touch its hands and face.

In the next two sessions (three and four), there was a marked change in Carl's demeanor. He appeared to be well rested and in a good mood. Physical play was more effective, and for the first time, he laughed and enjoyed the session. He still used the window as a coping mechanism, but not as often as before. The time he spent looking out the window decreased from 17 and 19% in sessions one and two to 10 and 5% in sessions three and four respectively. During these sessions, Carl appeared to be more comfortable with the therapist and the activities requested of him. He responded well to the Social Stories (introduced in session two) and was able to use the images provided by the therapist to choose his preferred activity.

Carl was also more interested in KASPAR. He was willing to sit near it for short periods of time, touch its face and hands, and participate during singing (e.g., singing single words from a song). While his main reinforcers were still food and physical play, interacting with KASPAR became part of the learning time, or 'group time.' His engagement and interest in group time increased from session to session, while his need for food and physical play decreased. During session four, he would occasionally choose group time over physical play. By the end of these sessions, Carl was constantly responding to the visual agenda and was willing to participate in both play time (with the therapist) and group time (with KASPAR).

During the last session, Carl seemed tired and distracted. During the first half of this session, he displayed similar behaviors to those of the first two sessions. It was hard to engage him in group time or physical play. He would get overwhelmed, go to the window and cover his ears. KASPAR was brought in halfway through this session. Motivated by the therapist, Carl seemed to be interested in interacting with KASPAR. He would sit to interact with it, would touch its hands and face, and would look at its body very closely. During this time, his engagement with the robot and therapists increased. Unlike previous sessions, he did not ask for food.

Parent interview

Carl's mother believed that the experience with KASPAR was very successful. She mentioned that Carl was a very active child, and in general, he did not sit and interact with anyone for extended periods of time. She highlighted that during the last few sessions Carl was interacting with the robot 'quite well' and sang the words of songs played by KASPAR. Surprisingly, she mentioned, Carl actually wanted to sit and interact with KASPAR and would do so without being asked; something that he had not done with anyone before. According to her, 'sitting with other children is not natural for him.'

Carl's mother posited that at first Carl probably thought KASPAR was another toy, and he only got interested in it when he was encouraged to touch it. She also acknowledged that Carl's routine was off during sessions two and five: 'He was tired and hungry.' If there was more time, she believed, Carl would have interacted with KASPAR for longer. She also believed the therapist was instrumental in Carl's progress.

It is hard to tell if any of the learning from interacting with KASPAR or the therapist was transferred to outside the sessions. However, Carl's mother believed that for the last few weeks Carl started to engage with more toys and started sitting down more with 'things that are moving.' She also mentioned that in the last few weeks of the study Carl went into the toy room and for the first time explored on his own. Additionally, it appeared that he participated more in songs and repeated more words than he used to. Carl's mother believed there were a few things that may have contributed to this improvement, given that he had daily therapy, but was confident that the sessions with KASPAR were definitely one of the contributing factors.

9.7 Discussion

Over a period of five weeks, the humanoid robot KASPAR was used as a tool to assist during therapy for autistic children. KASPAR is anthropomorphic, communicates via speech and movement, and can be programmed to act and respond according to the needs and context of therapy sessions. During the five-week intervention, KASPAR was controlled by a researcher via a wireless interface. Therapy sessions were supported by Social Stories and followed an ABA framework that relied on personalization, reinforcement, and repetition.

Two case studies were presented. In both cases, the therapy and activities were tailored to the child's needs and interests. The therapy was guided by a clinical psychologist, with KASPAR used as a 'peer' to practice social and communication skills. One of the strengths of this approach was the therapist's experience with autistic children. Unlike the original pilot study, this allowed us to use the specialist knowledge of the therapist to ease the child into the intervention over a series of weeks and respond to the child's feedback and behavior during each session. These case studies demonstrate the ability of the KASPAR robot to engage children during therapy, irrespective of the child's social and communication skills. It is difficult to quantify the benefits of the robot in isolation from the therapist, or the impact of the intervention in isolation from other activities that happened during the five-week period (e.g., external therapy). In line with previous research, however, our data suggests that social robots (i.e., KASPAR) can increase engagement and motivation; improving the child's focus, concentration, and ability to acquire new skills (Bauminger-Zviely et al., 2013; Scassellati et al., 2018; Silvera-Tawil & Roberts-Yates, 2018; Winkle & Paul, 2018).

As mentioned before, the intervention was guided by Social Stories—a tool to help individuals on the autism spectrum to better understand the nuances of interpersonal communication. In this study, the Social Stories provided structure and clear guidelines within each session, while KASPAR provided the opportunity to practice the new skills introduced through the Social Stories. Together, they enriched the learning experiences of the children. With Jack, the robot became a rewarding social partner, facilitating intrinsic interest through various levels of social communication and interaction. Motivated by the interactions with KASPAR, Jack would talk, play, sing, and dance as long as KASPAR was involved. As a result, his ability to stay focused, recall information, and respond to others' needs improved during the study.

For Carl, on the other hand, the main reinforcers were food and physical play. His interest in KASPAR was not immediate. Motivated by the therapist, he would sit with, touch, and sing with the robot. By the third session, his interest increased. Eventually, he started interacting with KASPAR in ways not seen before by his mother. Carl was very tactile, and while he did not often play with other children, he enjoyed object-based play. This suggests that he could have classified KASPAR as an object with which he was willing to interact. It is not clear, however, if he was able to identify KASPAR's attributes as anthropomorphic. If not, does it really matter? By interacting with KASPAR, he was able to interact with the therapist, respond to the therapy, and practice newly introduced skills better than without the robot's presence. Perhaps the anthropomorphic appearance was not key to increasing engagement, but it is possible that it enabled behaviors that Carl could identify and would eventually transfer to interpersonal interactions.

Overall, these case studies demonstrate that the framing of KASPAR as a facilitating object during therapy, supported by Social Stories, can provide an environment where new skills can be introduced and practiced with the therapist as a mediator. Furthermore, when KASPAR was not present, it became a point of reference to improve the relationship and communication between the child and therapist. However, using a robot during therapy supported by Social Stories is not an easy task. In the following section, we point to the strengths and challenges of our approach as a guide to improving experimental and therapeutic outcomes.

Lessons learned

1. Children are unpredictable. Working closely with a therapist allowed us to personalize the activities to meet the goals of each session and prepare KASPAR accordingly. When needed, the therapist would change the direction of each session and use the robots to improve the acquisition and practice of skills.
2. Every session counts. The success of a session is not purely dependent on the therapist's skills or the robot's ability to motivate and engage the children, but also depends on the child's emotional and physical state. As seen in this study, when a child is having a bad day (e.g., hungry or tired), they can respond very differently to the same session compared to when they feel well. Furthermore, the connection with the robot and therapist can vary depending on the personality and cognitive abilities of the children. For these reasons, longitudinal experiments with multiple sessions are recommended.
3. Building the connections between the child, the therapist, and the robot is important. In this study, the therapist was introduced to the child first, establishing a social relationship to which the robot could be added. Framing the sessions in this way afforded an opportunity to respond naturally to an unpredictable population. The therapist was able to 'fall back' on his training and existing relationship with the child if interactions with the robot did not go to plan, or if the child was less interested in engaging due to external factors (e.g., being tired or hungry).
4. The experimental space is key. It is well known that autistic people can have difficulty with unfamiliar situations, and therefore, it was expected that a study location known by both parents and children would reduce stress and anxiety during the child's participation in the study. Related to this, the use of a dedicated therapy space, where design considerations such as the amount of distraction visible to a child, was useful for focusing attention on both the robot and therapist.

Additionally, presenting the Social Stories on paper instead of a computer screen helped reduce unnecessary distractions within the sessions.

5. Using a robot for engendering conversation and social skills can be challenging. Even when teleoperated, KASPAR lacks the level of flexibility and adaptability that humans have. While in some cases this lack of flexibility could affect flow during an interaction, limiting responses and behaviors provides a level of predictability and repeatability that is needed during therapy. Given that the robot was only used as a tool to assist the therapist, we believe this limitation also served a purpose in that the therapist had to intervene and 'assist' during the interaction, providing additional opportunities for the child and therapist to interact with each other.

6. Technical issues should be anticipated. Issues with sound, such as delays and broken sound, made interaction via the robot complex through teleoperation. This further strengthens the case for having an expert (i.e., a therapist) present to keep the session progressing. In future research, more robot autonomy might be beneficial to reduce the need for a 'wizard' (robot teleoperator) during each session. It is important to remember, however, that inappropriate responses could confuse the children, and the system should be reliable enough to improve and not adversely affect the intervention.

7. Involve parents and carers. The 'observation room' setup proved comforting for parents who liked being able to see the session in real time. That not only gave them great confidence, but it provided the researchers with immediate and insightful feedback on behaviors that were new and beneficial for their child.

8. Keep the goals in perspective. Limited movement by KASPAR when the child moved around the room appeared to reduce engagement between the robot and the child. For example, Jack wanted to dance or play with KASPAR (e.g., hide and seek). The goal of the session, however, was to increase interpersonal interaction and communication. When it was impossible for the robot to participate, the therapist took a mediation role to help the child and robot interact as needed. As a result, this improved the connection, communication, and engagement between Jack and the therapist.

9.8   Conclusions

The two case studies presented in this chapter demonstrate the potential of using robots in combination with Social Stories as tools to support therapy with autistic children. While it is difficult to quantify the impact of the robot and Social Stories in isolation from the therapist and other activities external to the study, our data suggests that the Social Stories provided structure to the sessions, while KASPAR increased engagement and motivation. Together, they improved the focus and concentration of the children as well as their ability to acquire new skills. For some children, progress could be more marked than for others.

Future work should be considered to evaluate this pilot study using pre- and post-intervention measures that can be quantitatively compared in similar situations during longitudinal interventions with and without the use of robots.

Acknowledgements We would like to thank the Adaptive Systems Research Group at the University of Hertfordshire for providing the KASPAR robot used during this study. A very special thank you to Chris, our therapist, for all his help with planning and managing the sessions.

References

Alberto, P. A., & Troutman, A. C. (2009). *Applied behavior analysis for teachers* (8th ed.). Upper Saddle River, NJ: Merrill.

Barakova, E., & Lourens, T. (2013). Interplay between natural and artificial intelligence in training autistic children with robots. *Natural and Artificial Models in Computation and Biology*, 161–170.

Barry, L. M., & Burlew, S. B. (2004). Using social stories to teach choice and play skills to children with autism. *Focus on Autism and Other Developmental Disabilities, 19*(1), 45–51.

Bauminger-Zviely, N., Eden, S., Zancanaro, M., Weiss, P. L., & Gal, E. (2013). Increasing social engagement in children with high-functioning autism spectrum disorder using collaborative technologies in the school environment. *Autism, 17*(3), 317–339.

Begum, M., Serna, R. W., & Yanco, H. A. (2016). Are robots ready to deliver autism interventions? A comprehensive review. *International Journal of Social Robotics, 8*(2), 157–181.

Bennett, A. (2012). *Parental involvement in early intervention programs for children with autism.* Master's Thesis, St. Catherine University.

Boccanfuso, L., Scarborough, S., Abramson, R. K., Hall, A. V., Wright, H. H., & O'Kane, J. M. (2016). A low-cost socially assistive robot and robot-assisted intervention for children with autism spectrum disorder: Field trials and lessons learned. *Autonomous Robots, 1–19.*

Bodine, C., Sliker, L., Marquez, M., Clark, C., Burne, B., & Sandstrum, J. (2017). Social assistive robots for children with complex disabilities. *Robotic Assistive Technologies: Principles and Practice, 261–308.*

Bozgeyikli, E., Bozgeyikli, L., Raij, A., & Katkoori, S. (2016). Virtual reality interaction techniques for individuals with autism spectrum disorder: Design considerations and preliminary results. *Proceedings of the International Conference on Human-Computer Interaction, 4551, 127–137.*

Costa, S., Lehmann, H., Dautenhahn, K., Robins, B., & Soares, F. (2015). Using a humanoid robot to elicit body awareness and appropriate physical interaction in children with autism. *International Journal of Social Robotics, 7*(2), 265–278.

Dahlback, N., Jonsson, A., & Ahrenberg, L. (1993). Wizard of Oz studies—Why and how. In *Proceedings of the International Conference on Intelligent User Interfaces*, pp. 193–200.

Diehl, J. J., Schmitt, L. M., Villano, M., & Crowell, C. R. (2012). The clinical use of robots for individuals with autism spectrum disorders: A critical review. *Research in Autism Spectrum Disorders, 6*(1), 249–262.

DiGennaro Reed, F. D., Hyman, S. R., & Hirst, J. M. (2011). Applications of technology to teach social skills to children with autism. *Research in Autism Spectrum Disorders, 5*(3), 1003–1010.

Farley, M., McMahon, W., Fombonne, E., Jenson, W., Miller, J., Gardner, M., ... Coon, H. (2009). Twenty-year outcome for individuals with autism and average or near-average cognitive abilities. *Autism Research, 2,* 109–118.

Gillesen, J. C., Barakova, E. I., Huskens, B., & Feijs, L. M. (2011). From training to robot behavior: Towards custom scenarios for robotics in training programs for ASD. In *Proceedings of the IEEE International Conference in Rehabiliation Robotics* (pp. 1–7).

Golan, O., & Baron-Cohen, S. (2006). Systemizing empathy: Teaching adults with Asperger syndrome or high-functioning autism to recognize complex emotions using interactive multimedia. *Development and Psychopathology, 18,* 591–617.

Gray, C. (2000). *The new social story book.* Future Horizons.

Grynszpan, O., Weiss, P. L., Perez-Diaz, F., & Gal, E. (2013). Innovative technology-based interventions for autism spectrum disorders: A meta-analysis. *Autism, 18*(4), 346–361.

Hall, L. (2009). *Autism spectrum disorders: From theory to practice.* Upper Saddle River, NJ: Pearson Education Inc.

Kearney, A. (2007). *Understanding applied behavior analysis: An introduction for parents teachers, and other professionals.* London: Jessica Kingsley Publishers.

Kelley, F. J. (1983). An empirical methodology for writing user-friendly natural language computer applications. In *Proceedings of the SIGCHI Conference on Human Factors in Computing Systems.*

Kerr, M. M., & Nelson, C. M. (2010). *Strategies for addressing behavior problems in the classroom* (6th ed.). Boston, MA: Pearson.

Khosla, R., Nguyen, K., & Chu, M. T. (2016). Service personalisation of assistive robots for autism care. In *Proceedings of the Annual Conference of the IEEE Industrial Electronics Society,* (pp. 2088–2093).

Lovaas, O. I. (2003). *Teaching individuals with developmental delays: Basic intervention techniques.* Austin, TX: Pro-Ed.

Michaud, F., & Theberge-Turmel, C. (2002). Mobile robotic toys and autism. *Socially Intelligent Agents, 3,* 124–132.

Miller, H. L., & Bugnariu, N. L. (2016). Level of immersion in virtual environments impacts the ability to assess and teach social skills in autism spectrum disorder. *Cyberpsychology, Behavior, and Social Networking, 19*(4), 246–256.

Newbutt, N., Sung, C., Kuo, H. J., & Leahy, M. J. (2016). The potential of virtual reality technologies to support people with an autism condition: A case study of acceptance, presence and negative effects. *Annual Review of CyberTherapy and Telemedicine, 14,* 149–154.

Newbutt, N., Sung, C., Kuo, H. J., & Leahy, M. J. (2017). The acceptance, challenges, and future applications of wearable technology and virtual reality to support people with autism spectrum disorders. *Recent Advances in Technologies for Inclusive Well-Being, 119,* 221–241.

Pennisi, P., Tonacci, A., Tartarisco, G., Billeci, L., Ruta, L., Gangemi, S., & Pioggia, G. (2016). Autism and social robotics: A systematic review. *Autism Research, 9*(2), 165–183.

Petric, F. (2014). Robotic autism spectrum disorder diagnostic protocol: Basis for cognitive and interactive robotic systems.

Pierucci, J., Barber, A., Gilpin, A., Crisler, M., & Klinger, L. (2015). Play assessments and developmental skills in young children with autism spectrum disorders. *Focus on Autism and Other Developmental Disabilities, 30*(1), 35–43.

Rajendran, G. (2013). Virtual environments and autism: A developmental psychopathological approach. *Journal of Computer Assisted learning, 29*(4), 334–347.

Reynhout, G., & Carter, M. (2007). Social story TM efficacy with a child with autism spectrum disorder and moderate intellectual disability. *Focus on Autism and Other Developmental Disabilities, 22*(3), 173–181.

Robins, B., & Kerstin, D. (2005). Robotic assistants in therapy and education of children with autism: Can a small humanoid robot help encourage social interaction skills? *Universal Access in the Information Society, 4*(2), 105–120.

Robins, B., & Kerstin, D. (2010). Developing play scenarios for tactile interaction with a humanoid robot: A case study exploration with children with autism. In *Proceedings of the International Conference on Social Robotics* (pp. 243–252).

Sartorato, F., Leon, P., & Sarko, D. K. (2017). Improving therapeutic outcomes in autism spectrum disorders: Enhancing social communication and sensory processing through the use of interactive robots. *Journal of Psychiatric Research, 90,* 1–11.

Scassellati, B., Admoni, H., Matarić, M., & Admoni, H. (2012). Robots for use in autism research. *Annual Review of Biomedical Engineering, 14,* 275–294.

Scassellati, B., Laura, B., Chien-Ming, H., Marilena, M., Meiying, Q., Nicole, S., ... Frederick, S. (2018). Improving social skills in children with ASD using a long-term, in-home social robot. *Science Robotics, 3*(21), 1–9.

Silvera-Tawil, D., & Roberts-Yates, C. (2018). Socially-assistive robots to enhance learning for secondary students with intellectual disabilities and autism. In *Proceedings of the International Symposium on Robot and Human Interactive Communication*.

Silvera-Tawil, D., Roberts-Yates, C., & Bradford, D. (2018). Talk to me: The role of human-robot interaction in improving verbal communication skills in students with autism or intellectual disability. In *Proceedings of the International Symposium on Robot and Human Interactive Communication*.

Silvera-Tawil, D., Strnadová, I., & Cumming, T. M. (2016). Social stories in robot-assisted therapy for children with ASD. In Y. Kats (Ed.), *Supporting the education of children with autism spectrum disorders* (pp. 225–244). IGI Global.

Simut, R., Pop, C., Saldien, J., Rusu, A., Pintea, S. a., Vanderfaeillie, J., ... Vanderborght, B. (2012). Is the social robot probo an added value for social story intervention for children with autism spectrum disorders? In *Proceedings of the ACM/IEEE International Conference on Human-Robot Interaction* (pp. 235–236).

Tang, H.-H., Jheng, C.-M., Chien, M.-E., Lin, N.-M., & Chen, M. Y. (2013). iCAN: A tablet-based pedagogical system for improving the user experience of children with autism in the learning process. In *Proceedings of the International Conference on Orange Technologies*, (pp. 177–180).

Vanderborght, B., Simut, R., Saldien, J., Pop, C., Rusu, A. S., Pintea, S., ... David, D. O. (2012). Using the social robot probo as a social story telling agent for children with ASD. *Interaction Studies, 13*(3), 348–372.

Vicker, B. (2003). Assessment day: Questions about the communication development of your young child with an autism spectrum disorder. *The Reporter, 8*(2), 18–21.

Warren, Z. E., Zheng, Z., Swanson, A. R., Bekele, E., Zhang, L., Crittendon, J. A., ... Sarkar, N. (2013). Can robotic interaction improve joint attention skills? *Journal of Autism and Developmental Disorders*.

Winkle, K., & Paul, B. (2018). Social robots for engagement in rehabilitative therapies: Design implications from a study with therapists. In *Proceedings of the ACM/IEEE International Conference on Human-Robot Interaction* (pp. 289–297).

Yun, S.-S., Park, S.-K., & Choi, J. (2014). A robotic treatment approach to promote social interaction skills for children with autism spectrum disorders. In *Proceedings of the International Symposium on Robot and Human Interactive Communication*.

Chapter 10
Social Robots and Human Touch in Care: The Perceived Usefulness of Robot Assistance Among Healthcare Professionals

Jaana Parviainen, Tuuli Turja and Lina Van Aerschot

Abstract Touching in care work is inevitable, particularly in cases where clients depend on nurses for many activities of daily living, such as bathing, dressing, lifting and assisting. When new technologies are involved in nurse–client relationships, the significance of human touch needs special attention. Stressing the importance of practitioners' opinions on the usage of robots in care environments, we analyze care workers' attitudes toward robot assistance in the care of older people and reflect on their ideas of the embodied relationship that caregivers and care receivers have with technology. To examine nurses' attitudes toward care robots, we use survey data on professional care workers ($n = 3800$), including random samples of registered and practical nurses working primarily in elderly care. As the theoretical framework for analyzing the empirical data, we apply two different conceptual approaches regarding human touch: nursing ethics and the phenomenological theory of embodiment. The empirical results suggest that the care workers are significantly more approving of robot assistance for lifting heavy materials compared to the moving patients. Generally, the care workers have reservations about the idea of utilizing autonomous robots in tasks that typically involve human touch, such as assisting the elderly in the bathroom.

Keywords Social robots · Care work · Human–robot interaction · Human touch · Care ethics

J. Parviainen (✉)
Faculty of Social Sciences, Research Centre for Knowledge, Science,
Technology and Innovation Studies (TaSTI), Tampere
University, Tampere, Finland
e-mail: jaana.parviainen@tuni.fi

T. Turja
Faculty of Social Sciences, Tampere University, Tampere, Finland
e-mail: tuuli.turja@tuni.fi

L. Van Aerschot
Department of Social Sciences and Philosophy,
University of Jyväskylä, Jyväskylä, Finland
e-mail: lina.vanaerschot@jyu.fi

10.1 Introduction

Some critical voices have brought up concerns about care technologies and have suggested that they may create a risk of dehumanizing and depersonalizing care and objectifying care receivers by jeopardizing their individuality and subjectivity. For example, Barnard and Sandelowski (2001) have suggested that clinical and sterile environments characterized by standardization and strict regulation may fail to uphold and support human-centered care. In these kinds of environments with the highly palpable and audible presence of equipment, people may sometimes become treated as extensions of the machinery. However, many care workers, nurses and caregivers welcome tools, techniques, equipment and robots that can assist them in work tasks, especially in tele-care (Alaiad & Zhou, 2014). Hence, there seems to be some tension between the ideals of "touch-based" care and "technology-driven" care, or in other terms, "humanistic" care and "technocratic" care (Barnard & Sandelowski, 2001).

When new care robots are introduced in nurse–client relationships, the significance of human touch needs special attention. The new generation robots may be equipped with improved sensor technology and artificial intelligence, both increasing the potential for interaction between robots and people. Care ethics are closely connected to professional touching and the physical presence of care workers with clients. The professional standards of nursing work include respectfulness, compassion, partnership, trustworthiness, competence and safety (NMC, 2015). There are concerns that the robotization of care may reduce human contact and increases feelings of objectification (Sharkey & Sharkey, 2012). Medical technologies have often been considered extensions of the nurse's body, but in the context of assisting robotics, the robot can be seen as a technological medium, a co-bot, operating between the care worker and the client. Turkle (2012) and van Wynsberghe (2013) claim that embodied practices in human care, even if technologically assisted, always require a reciprocal interaction between the care receiver and caregiver. If social robots assist in some of the tasks in human care, it is necessary to consider how to arrange mediating interdependencies within care relationships. van Wynsberghe (2013) suggests an approach with a value sensitive design, taking the ethical considerations as the first priority in the design process of care robots.

Stressing the importance of practitioners' opinions on the usage of care robots in care environments, we analyze care workers' attitudes toward robot assistance in care work and in services for older people more specifically. We then reflect on care workers' ideas about the embodied relationship caregivers and care receivers have with technology. To examine nurses' attitudes toward care robots, we use survey data on professional care workers ($n = 3800$) working primarily in the care of older people. The respondents were asked to evaluate how desirable different scenarios of using robot assistance in their work would be.

As the theoretical framework for analyzing the empirical data, we apply two different conceptual approaches regarding human touch. First, we rely on theoretical and practical discussions of touch in nursing ethics and nursing science (Routasalo

& Isola, 1996; Twigg, Wolkowitch, Cohen, & Nettleton, 2011). Based on the traditional distinction between instrumental touch and expressive touch, we consider the role of professional touch in nursing practices. Second, we draw on a philosophical theory of human touch and embodiment to illuminate the human–robot interaction in care work. From the perspective of Merleau-Ponty's (1968) phenomenology, a living being cannot touch without being touched. This implies that touching and being touched are inherently connected among humans and animals. Identifying the significance of touch associated with the use of robots in care for older people, the phenomenological view of touch is a necessary first step toward ethical discussions on social robots and their impression of touch. The phenomenological approach can address senior persons' intimacy, individuality, autonomy and rights to touch and be touched. However, this does not necessarily mean that the touch of social robots could not have significance in human care.

In analyzing the empirical results of our survey data, we will address the meaning of affective touch as regards both functional touch and expressive touch in care work. To concretize the empirical findings, we reflect on two types of social robots that seem relevant and timely for discussing touch in robotizing care work in the future. First, we discuss what kind of affective touch the robotic therapy animals may provide for older patients. We consider how the previous empirical results concerning the use of the Paro seal robot resonate with the views of the respondents regarding affective touch in care work. Second, we reflect on how activities that require more (e.g., bathing, feeding, lifting, dressing) or less (demonstrating light exercises) functional touch are intended to be assisted by care robots. New generation robots are expected to be well suited to lifting and carrying or other tasks, for example, feeding and bathing physically impaired persons. The use of these kinds of robots still remains marginal both in care-giving facilities and home care, but new robots are being developed. For instance, the robot named Robear is intended to overcome its current limitations with added power and functionality. In the future, robots are also expected to interact with people. These typically humanoid robots are often doll-like in appearance and have the functionality of a preprogrammed puppet. In services for older people, these social robots are used for mere entertainment or, when steered by the professionals, for patients' cognitive, emotional or physical activation.

10.2 The Importance of Touching in Care of Older People

In traditional cognitive psychology, touch and haptic sensation refer to a sensory mode in which the body senses pressure, temperature and pain as well as itself through proprioceptive, vestibular and kinesthetic senses (Paterson, 2007, xi). In our social science-driven approach (e.g., Ahmed & Stacey, 2003), touch—and lack of touch—is also seen as a central channel to transmit emotions, affects and moral codes in the society. From the very start, touch plays a crucial role in the early life and parent–child relationships, since touch is an essential channel of communication with caregivers for a child (Field, 1990). A physical and caring touch enhances the attachment between a parent and a child by signifying security through the body: "I am here—you are safe." Of course, depending on the styles of touch, affective touch can generate negative emotions, for the child if the caregiver's touch is rough or abusive.

Merleau-Ponty (1968, 146–149) describes touch as a "double sensation" since touch always occurs between two discrete entities forming the *reversibility of touching*. Merleau-Ponty's idea of the reversibility of touching becomes clear when we consider nurse–patient interaction. When a nurse intentionally touches a patient's hand, the patient feels the nurse's touch but also the nurse is touched by the patient. The reversibility of touching means that there cannot ever be a unidirectional or one-way touch but touching always includes a moment of being touched. The patient can make her own interpretation of the meaning of touching that is not necessarily the same as nurse's intention. So, sometimes a caring touch can be considered patronizing or humiliating, even if that is not the nurse's intention. In this way, the reversibility between touching and being touched forms a highly complex and dynamic structure in nurse–patient interaction.

Recently, empirical studies on touch have shown that adults have an innate ability to decode emotions via touch alone (e.g., Hertenstein, Holmes, Keltner, & McCullough, 2009). Touch can communicate multiple different emotions—anger, fear, joy, love, sympathy disgust, gratitude and sadness—in much more nuanced, sophisticated and precise ways than were expected previously. The phrase "affective touch" describes how touch includes emotional content or conjunction with emotion. The reversibility of affective touch is a crucial issue since touching can transmit and designate the moral values of culture, indicating what kind of behavior is allowed and what is forbidden.

It is relevant to discuss affective touch but also the *lack of touch* in society. Many researchers have stated that some societies are more "touch-phobic cultures" than others (Hertenstein et al., 2009; Kinnunen, 2013). In the touch-phobic cultures, people are not allowed to touch strangers or even their family members or friends; thus, they have no possibility to cultivate their touch skills and develop affective bonding structures with others through touch. Thus, different cultures have different tolerance levels for touch regarding same-sex and opposite-sex touching as well as the quality of the touch, the duration, the intensity and the circumstances. Similarly, there are huge variations regarding individuals and how they enjoy or tolerate touch.

Touch in care work is inevitable, because clients are dependent on nurses for many activities in daily living, such as bathing, eating, lifting, dressing, and other similar types of care activities that are related to the well-being and medical treatment of older, disabled or sick people. Drawing on the discussions of touch in nursing science (e.g., Gleeson & Timmins, 2004; Routasalo, 1999), we differentiate between *instrumental* (physical, functional, necessary, procedural) touch (Routasalo, 1999) and *expressive* (non-necessary, communicative, caring) touch (Belgrave, 2009) in nursing practices. Instrumental touch refers to physical contact between a nurse and a patient when, for instance, the nurse takes a blood test, measures blood pressure or transfers patients between wards or rooms. This kind of physical touch is associated with routine tasks within nursing in the sense that touching has an effect and impact on performing necessary work duties. When expressive touch is used, nurses usually touch the patient's hands, arms or shoulders to say hello and good-bye or show caring, compassion and support to the patients. At its best, expressive touch and the presence of nurses can have a major role in developing care environment that advances patients' recovery. Small gestures can be crucial when professionals face people in vulnerable situations. According to Berg and Hallberg's (2000) empirical findings, caring for people with mental illness demands an intensified presence, and one is not allowed to emotionally glide away, close the door or just disappear (Berg & Hallberg, 2000, p. 329). For patients with depression, the nurse being present by the bedside is beneficial and helps to alleviate the patients' fears (Moyle, 2003).

Instrumental touch is far more common in nursing situations than expressive touch (Routasalo, 1999). According to Gleeson and Timmons (2004), the widespread adoption of touch as a caring intervention is discouraged in the absence of clear guidelines that could develop touch as a nursing work skill. They suggest that many nurses do not touch patients more than is necessary but only to conduct their duties, so most patients do not necessarily receive any affective touch from their professional caregivers when they are most vulnerable.

However, it is important to recall that functional, purposeful and instrumental touch when lifting or dressing the client can still carry affective intentions, such as comforting, reassuring and encouraging the patient or protecting the patient from physical harm (Parviainen & Pirhonen, 2017). Instrumental touch can be affective touch even if it is done for functional purposes. In a similar way, expressive touch can be a strictly formal gesture, for instance, when the nurse routinely shakes the patient's hand to say hello.

Whether nursing touch is functional or expressive, the nurse's touch is always supposed to be a "professional touch." Closely connected to professional ethics, professional touch refers to a special professional and ethical attitude in which the client's body is cared for and attended to mindfully and respectfully but not too personally, emotionally or in an intimate manner. Professional touch is also sharply separated from violence such as sexual abuse and harassment, so it is supposed to be sensitive toward the patient's individual needs and respect her/his personal intimacy (Paterson, 2007). This implies that professional touching is inherently reflective in its nature and that nurses need to consider sensitively the manner in which they touch the patient, considering social and cultural contexts.

Care work involves a great deal of "body work." Body work is an essential part of caring because it involves direct, hands-on activities, handling, assessing and manipulating bodies (Twigg et al., 2011). Professional touch in human care can take different forms. As stated above, all tactile communication is reciprocal in nature: when a nurse touches a client, he/she is also being touched by the client (Belgrave, 2009). Touching a living body, a care worker reflects, usually internally, on how her/his touch is being felt by the other body. All ethically sensitive touch, including professional touch, is a tentative activity as it requires awareness of the patient's intimate space. Touch has very different meanings in a multicultural society where people live together with different systems of touching (van Dongen & Elema,

2001). Touch involves a risk of misinterpretation and misunderstanding, and nurses are usually well aware of the dangers of touch.

Being touched or being seen by others is considered vital for all people but is especially important for the well-being of babies and older people (Routasalo & Isola, 1996). Empirical studies on older people show that those parts of the body that are touched most frequently are the hands, arms, forehead, hair and shoulders; those that are touched less often are the legs, ankles, abdomen, chest and forearm; and those that are touched rarely or not at all are the neck, ears, lips and genitalia (McCann & McKenna, 1993). However, according to Langland and Panicussi (1982), the more unable to communicate elderly people are due to, for example, memory disorders or other cognitive impairments, the more touch-deprived they become. This implies that in human care there is a need for expressive caring touch without any functional purpose. Yet people with communicative or social restrictions often interpret feelings and affects that touching mediates and experience pleasure or displeasure within physical care practices (Bush, 2001). Touching is usually more than just physical contact between bodies; it can include various affective atmospheres such as an icy atmosphere when we feel chilly, an uncanny situation that makes our hair stand on end or a tense interpersonal climate that is felt as oppressive or suffocating (Fuchs, 2013).

Despite the ethics of professional touch, not all touching in care work is pleasurable for care workers or clients. In problematic situations—when a patient is violent, sexually aroused or psychotic—a care worker may need to call on colleagues or safeguards to help. In nurse–client relationships, feelings of disgust, shame, guilt or embarrassment are also common. These negative feelings are not seen to fit into the idea of professional nursing behavior. Some tasks such as removing feces and changing diapers include bodywork and co-presence with patients (Wolkowitch, 2006). These tasks can be considered repulsive even if professionals feel sympathy for the patients.

Touching becomes a more complex phenomenon when new technologies intervene in nurse–client relationships. The use of robotics for lifting patients out of their bed or into the bath, for example, does not necessarily mean limiting the direct touching of patients. New equipment may be used with a minimum of human effort but may still require human presence to support, surveil or encourage the activity.

In Merleau-Ponty's (1968) phenomenology, one of his influential formulations concerns touching inanimate things—touch of artifacts—including natural objects (trees) or human-made artifacts like tea cups and robots. The main difference between human–human interaction and human–artifact interaction is the lack of reversibility: robots and other artifacts do not feel affective touch as humans and animals do. Even if the sensors of robots can be designed to respond to a touching act as if they "feel" touch, the fact is that artifacts do not sense anything.

Drawing on Merleau-Ponty's (1968) formation of touch, Kerruish (2017) discusses tactile sensations that social robots provide to the users. She considers that each tactile perception is embedded in an embodied imagination that includes memories, ideals, cultural norms and values among other things (Merleau-Ponty, 1962/1989). Tactile meanings emerge from this human embodied perception and the messy materialism of the device in which the discrete units of the digital are instantiated. Sensations provoked through touching are never completely precise or predictable.

The difference in touching animate and inanimate beings is fundamental to humans. This becomes clear when we touch something that we expect to be an inanimate object, but only after touching it do we realize that it is alive. Similarly, touching something that is expected to be alive influences our touching style when we notice that the object is an inanimate thing. The latter case is typical, including sometimes embarrassing moments when we come across social robots which appear to be living beings. Nevertheless, touching inanimate objects can involve as much emotional tactile content as touching living beings but reversibility does not exists between humans and artifacts.

10.3   Robotic Technology and Care Work

So far, very little robotic technology is used in care work if we define a robot as a programmable machine, with some degree of autonomy and the capability of performing intended tasks and moving around or otherwise adjusting to its environment (International Federation of Robotics). The most common service robot is still a vacuum cleaner (Hennala et al., 2017) and the robotic pet Paro, the baby seal, is the most common robot used in the care of older people (Van Aerschot, Turja, & Särkikoski, 2017). For the practical tasks of lifting, eating, bathing or moving, some robotic devices are found on the market but they are not yet widely used.

When it comes to using robots in the context of care, it has been shown that clients and patients do not wish robots to replace human contacts with caregivers (Alaiad & Zhou, 2014; Beedholm, Fredriksen, Skovsgaard, Fredriksen, & Lomborg, 2015; Jenkins & Draper, 2015). According to van Wynsberghe (2013) robots should be designed to support and promote the fundamental values of care, for example, patient safety, dignity and well-being. Contemplating care work more concretely, the different tasks can be divided into direct patient care, indirect patient care and other activities, including documentation, administration and planning use of a medication (Ballermann, Shaw, Mayers, Gibney, & Westbrook, 2011). In general, using robotic appliances for indirect activities of care or other, i.e., assistive, activities seem more easily acceptable among people than the idea of using robots in direct patient care which also includes touching, both instrumental and affective (Santoni de Sio & van Wynsberghe, 2016).

The research on care professionals' attitudes and opinions on robotic appliances designed to be used in care-giving shows varying results. On the one hand, care professionals have been found not to welcome robot technology (Katz & Halpern, 2014; Saborowski & Kollak, 2015), but on the other hand, the caregivers attitudes vary according to the kind of care that they are providing and the patients that they work with. According to Mutlu and Forlizzi (2008), nurses' readiness to integrate a delivery robot into their work environment was affected by their job definition, workload and interruptibility. For example, nurses working with cancer patients who demand intensive care and attention often found that the robot was annoying and that it interrupted them in an undesirable way when they were in the middle of trying to do their work. On the contrary, nurses working in a post-partum ward found the delivery robot delightful and it conducted its tasks just fine. Robots are not especially desired for the tasks that require social skills (Alaiad & Zhou, 2014; Jenkins & Draper, 2015), but instead they could be used as tools or equipment for, say, monitoring or measuring (Pfadenhauer, 2015). A qualitative research on using a bathtub robot in a setting of institutional care showed that the employees in managerial positions were more enthusiastic and positive than the staff members about the idea of using the technology, even if it had not been proved to have any economic benefit or to even function properly (Beedholm et al., 2015).

Despite the awareness of the opinion that robots are not wanted to replace human caregivers, the fear of robots diminishing human contacts is genuine. It has been stated that it is very likely that the more technology and robots are introduced in organizing and providing care, the more patients and clients will be left alone (Sharkey & Sharkey, 2012). It has also been shown that the interaction between doctors and patients has decreased since more computers and technologies have been introduced to provide treatment and care (Menon, 2015). However, there are expectations of robots assisting nursing staff in some routine tasks, which would free up working hours for more person-centered tasks (Sparrow & Sparrow, 2006).

10.4   Desirable and Non-desirable Robot Assistance

To analyze care professionals' attitudes toward robot assistance in care tasks, we used two sets of survey data collected from the professional care workers ($N = 3800$). The first sample was randomly selected from the members of The Finnish Union of Practical Nurses, who were currently working with older adults ($n = 2218$). Every other individual in the population was chosen for sampling with an equal likelihood of selection. Participants were aged 17–68 ($M = 45.5$; SD $= 12.1$), and 89.8% were female. The response rate was 11%. The second sample was collected from The Union of Health and Social Care Professionals in Finland. The sample included every nurse and physiotherapist currently working with older adults and homecare services, and every third, randomly selected nurse and physiotherapist working at a health center or a hospital. This sample comprised mostly female (89.0%) nurses ($n = 1701$) and physiotherapists ($n = 81$) aged 19–70 ($M = 47.5$; SD $= 10.4$). The response rate was 9%.

The samples were collected in October–November 2016. Online questionnaires included multiple choice questions about educational and occupational background, experiences with assistive tools in healthcare and attitudes toward robots presented in a variety of care work scenarios and more specifically in services involving older people. Care work consists of a variety of tasks, and physical labor is often a central part of the activities (Wolkowitch, 2002). The questionnaire presented scenarios of care tasks performed or assisted by a robot. The variety of scenarios emphasized tasks that include body work. In assessing the scenarios, the respondents scaled (from 1 to 10) firstly the perceived usability of robotic assistance in care work ($\alpha = 0.93$), and secondly the perceived usefulness of robotic assistance in services for older people. The latter were further categorized into autonomous robot assistance scenarios ($\alpha = 0.97$) and teleoperated robot assistance scenarios ($\alpha = 0.95$). The specific questions are presented in Appendix A.

Measuring the compatibility of personal values with using care robots, we modified three statements (see Appendix A) from the information system acceptance questionnaire validated by Karahanna et al. (2006). The response scale was from 1 (totally agree) to 5 (totally disagree), thus the composite variable ranged from 3 to 15 ($\alpha = 0.929$), with a higher score indicating care robots' compatibility with personal values.

We present our preliminary and descriptive results in percentages, means (M), standard deviations (SD) and differences between means (t). The statistical differences between single assessments of robot-assisted work scenarios are observed by confidence intervals of 95%. Regression tables present standardized betas (β) and the predictive power of the models (R^2).

10.4.1   Results

Most of the 3800 respondents were working in the public sector (78%). Typically, they were practical nurses (56%) or registered nurses (35%), the rest being head nurses, physiotherapists or other care workers (9%). Healthcare technology was fairly familiar to the respondents: safety phone was familiar to 71%, meal automaton to 11% and the Paro seal to 8%, to list a few.

Firstly, the respondents had to evaluate how comfortable they felt with the idea of robot assisting them with moving or lifting patients and heavy materials and also assisting them in threatening situations at work. Secondly, they had to evaluate how useful they perceived robot assistance in elderly care scenarios such as helping a physically impaired resident to move around in the home and in the bathroom.

The respondents were most comfortable with the idea of a robot helping them with physically straining work. Figure 10.1 shows that care workers were significantly more approving of robot assistance for lifting heavy materials compared to lifting patients ($t = -20.77$; $p < 0.001$). Regarding lifting or moving patients, the respondents were more comfortable with the idea of a separate robotic assistant compared to an exoskeleton for a worker to wear ($t = -24.94$; $p < 0.001$). However, transferring patients using an autonomous stretcher was remarkably less welcomed compared to moving patients with any form of other robotic assistance ($t = -8.73$; $p < 0.001$). Summarizing these results, care workers see robots more desirable primarily in other tasks than patient work. In addition, if robots are used in patient work, the care workers prefer situations where a care worker is present.

Care workers saw the potential in robots assisting in threatening situations. This is not surprising as studies have shown that care workers have to endure and be prepared

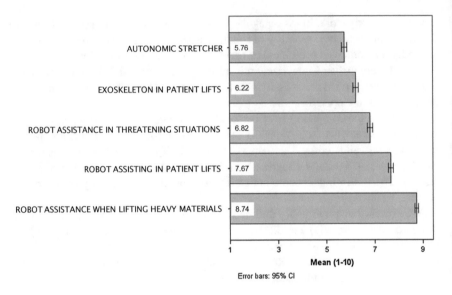

Fig. 10.1 Acceptance of robot assistance at work, means on a scale from 1 to 10

for aggression from patients and those close to them (Kerruish, 2017; Twigg et al., 2011). In care scenarios, touching is usually seen as something that happens on care workers' terms. Here, the respondents suggested that robotic applications could also be suitable for protective use where care workers are the targets of unwanted contact.

When asked specifically about which services for older people could benefit from robotic assistance, the respondents found it easier to see the benefits of teleoperated robots ($M = 5.45$) compared to autonomous robots ($M = 5.16; t = -6.13; p < 0.001$). Figure 10.2 presents the means for some of the scenarios. Of these scenarios, care workers were most willing to see robots in situations where physical contact is not necessary, namely demonstrating light exercises. This kind of entertainment-like coaching by a robot was perceived as more feasible than teleoperated physiotherapy with a therapist ($t = 23.08; p < 0.001$). In addition, most of the respondents did not consider autonomous robots conducting physiotherapy as appropriate. The robotic assistance in bathing, dressing and in the toilet was met with a similar refusal. However, general support in moving around the residence was viewed more positively. A robot which was remotely operated and monitored by care professionals could be used in the homes of older people as an assistant for moving, walking and getting up.

We further analyzed the mechanism of how robot acceptance varies between the least approved (robot assisting in bathing and dressing) and the most approved (robot demonstrating light exercises) examples of care-related tasks (Table 10.1). In the models, the robot's usefulness was mostly explained by the compatibility between care robot use and personal values (e.g., "Using care robots does not fit the way I view the world"). The majority (64.5%) of the respondents did not find robot use in

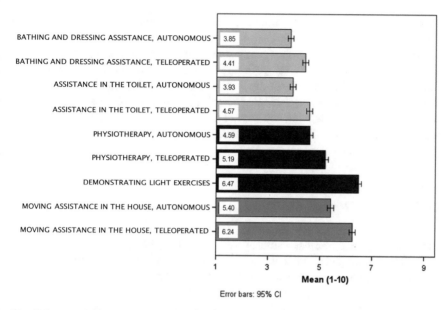

Fig. 10.2 Acceptance of robot assistance in elderly care, means on a scale from 1 to 10

Table 10.1 Perceived usefulness of robot assistance in care-related tasks

	Robot demonstrating light exercises		Robot assisting in bathing/dressing	
	β	p	β	p
(Constant)		<0.001		0.260
Age	0.022	0.173	0.037	<0.05
Male	−0.056	<0.001	0.034	<0.05
Practical nurse	−0.019	0.250	−0.070	<0.001
Familiar with Nao robot	0.035	<0.05	0.025	0.113
Familiar with other care robot[a]	−0.007	0.677	0.028	0.087
Personal values	0.538	<0.001	0.512	<0.001
Adjusted R^2	0.288		0.291	

[a]Physiotherapy, telepresence, therapy animal or patient-lifting robot

care compatible with their personal values ($M = 8.14$; SD $= 3.61$). On average, the more the respondents found care robots compatible with their personal values, the more accepting they were toward robots.

While the acceptance in both the most and the least approved tasks was highly connected with the compatibility with personal values, some differences were found. An autonomous robot assisting in bathing and dressing, as the least approved robot-assisted task, was perceived as more useful among older and male respondents with higher education, compared to younger female practical nurses. Experience with care robots did not quite have the power to explain the acceptance of robots assisting in bathing and dressing.

Among our examples, a robot demonstrating light exercises was perceived as the most useful form of robot assistance in care for older people. Because the robot assistance in demonstrating exercises was more broadly accepted by all of the respondents, fewer significant explanatory factors of perceived usefulness were found. Contrary to the robots assisting in bathing and dressing, the robot demonstrating light exercises was perceived as more useful among female respondents who had experience with the care robot Nao.

10.5   Social Robots Assisting in Human Care

In interpreting the empirical results of our survey data, the care workers emphasized both affective and functional touch as central in care work. In understating how robot solutions could assist in care work regarding both these aspects in the future, two types of social robots seem relevant. The first question is how the affective touch of the robotic therapy animal can meet the needs of care of older people. Secondly, how are activities that require functional touch, such as bathing, feeding, lifting or dressing patients, intended to be assisted by service robots. New generation robots are expected to be useful in lifting and carrying but also in social interaction and, for example, feeding and bathing physically impaired persons. The use of these kinds of social robots still remains marginal both in care-giving facilities and home care, yet new robots are being developed. For instance, the teddy bear-faced patient lifting robot named RIBA (later Robear) was intended to overcome the current operational and technical limitations with added power and functionality. New joints in the robot's base and lower back enabled RIBA-II to crouch down and lift a patient from floor level. This could be a social robot assistant that could remarkably relieve the physically burdening care tasks of lifting and moving patients. However, in 2015, the Riken Center for Human-Interactive Robot Research in Japan was closed, so the Robear project is no longer being carried out at Riken. Unfortunately, Robear did not possess such "embodied intelligence" that it could have replaced human reflectivity of tactile-kinesthetic movements and carry patients safely and autonomously. Moreover, social robots cannot replace humans in providing emotional comfort to people who need another person's co-presence in a vulnerable situation (Sharkey & Sharkey, 2012; Turkle, 2011). Still, new generation robots are being developed with more human-like touch, with better fine motor skills and soft artificial skin (Cabibihan et al., 2016).

Lack of human presence can be seen as one of the reasons why transferring a patient using an autonomous stretcher was notably less welcomed by nurses than lifting patients with robotic assistance. From the physical presence point of view, a seemingly instrumental procedure of moving a patient from one ward to another on a stretcher can actually be a holistic and interactive event. Escorting a patient to an operation room, for example, is not just about the transport since a nurse may also provide attention, comfort and encouragement by being present and able to touch. Even if the automotive navigation system and the sensors of social robots can be programmed to transmit patients in hospitals or care homes, robots do not compensate for an escorting person's presence and being accessible and available (Moyle, 2003).

The empirical results of this study suggest that nurses consider robotic devices beneficial when assisting in physically demanding tasks of lifting patients and, especially, lifting heavy materials. However, care workers see robots as desirable to ease the physical strain of lifting, particularly in a situation where the autonomous system does not block nurses from the patient–robot interaction. In this way, nurse–patient interaction would remain intact and robot assistance would even enhance the interaction by providing more time and opportunities to support the patient emotionally.

Regarding robot assistance in services for older people, scenarios of using a robot for personal care such as assisting in bathing and dressing or going to the toilet were evaluated as the least acceptable. Practical nurses in particular were prone to rejecting robot assistance in these scenarios, which we classified as intimate. This kind of intimate assistance may sometimes be seen as being at the core of care even though at the same time it may also entail negative feelings (e.g., assisting with toileting may provoke aversion). The care workers' opinions on robots assisting in their work and in tasks related to the care of older people reflect the idea that technology should only be used in care-related tasks which are not too intimate, affectionate or personal. The caregiver and care receiver make use of technological devices in ways that suit their needs without losing the possibility for human touch and interaction.

In general, the nurses' opinions on useful and acceptable robot assistance indicated that they resisted the ideas of standardized, technologized care and endorsed the ideas of human dignity and individuality. With regard to this, it is surprising that nurses saw social robots as useful to provide distance and protect them in threatening situations. Recent research findings show that aggression from patients and their visitors in hospitals and other health care organizations has become more common (e.g., Speroni et al., 2014). Nurses certainly need new tools and skills to manage patients' violence, verbal abuse, threats or intimidation (Harwood, 2017), but the kind of "robocops" found in science fiction are hardly good solutions to prevent clients' violent behavior.

For ages, different tools and devices have been used to assist in care tasks or even to conduct them automatically, but nurses expect that care technologies should be designed to enhance the relation between care-giving and care receiving activities. Technologies should allow embodied practices of professional touching when nurses lift, bath, feed or move the care receiver and deliver medications or food or bring sheets to the room. As stressed above, the affective touch involved in care practices transmits complex information about emotions and affects, creating a value-laden milieu (Turkle, 2012). Social robotics is considered valuable as an interpersonal intervention when it can develop a partnership and reciprocity in the nurse–client relationship.

10.6 Discussion

Recent research indicates that modern (western) societies have developed toward touch-deprived cultures (Linden, 2015) or even touch-phobic cultures (Hertenstein et al., 2009). Touching and physical presence are inevitable in care work; however, instrumental touch is used much more commonly than expressive caring touch. There are good reasons why touch has recently become a highly politicized issue; scandals concerning sexual harassment and the #Metoo movement, in particular. Yet touching is also an essential part of human interaction and a way of mediating empathy and comforting feeling. Elderly people in particular may suffer from a lack of touch. For instance in the UK, half a million older people have at least five or six days a week without seeing or speaking to anyone at all (Age UK, 2018). Empirical research has shown that loneliness can increase people's risk of premature death by up to one quarter (Holt-Lunstad, Smith, Baker, Harris, & Stephenson, 2015). Findings also show that, as far life years are concerned, loneliness can be as harmful to people's health as smoking 15 cigarettes a day (Holt-Lunstad, Smith, & Layton, 2017). Protecting vulnerable people from abuse and sexual harassment has increased concern about whether no-touch policies fail to recognize the importance of touch for well-being in their lives (Linden, 2015, 4). Not only elderly people but also children are suffering if their carers are unable to show them affection, intimacy, comforting closeness or to simply give them a hug.[1]

While loneliness, particularly in late life, has been described as an epidemic in many modern countries, the service industry has started to commodify touch in the form of the "touch industry" in Europe, Australia and the US. Professional "cuddlers" operate workshops and services to help the touch-deprived. Similarly, in the industry of social robotics, researchers and designers are developing "huggiebots," humanoid robots programmed to offer hugs to humans. For instance, the Huggable is a new type of robotic companion being developed at the MIT Media Lab for healthcare, education, and social communication applications (Jeong et al., 2015). The Huggable is not designed to replace nurses or other caregivers but rather to enhance human social networks. In Japan, the "tranquility chair" or "anti-loneliness hugging chair" is built in the shape of a large fabric doll with long arms to wrap comfortably around the lonely sitter. Even if huggiebots and hugging chairs are not necessarily designed to replace humans, they are new commodities to relieve angst caused by missing human contact.

The crucial question that remains is why the technical equipment is designed to find solutions to problems that could also be solved by developing social interaction and enhancing the social contact of all people, especially the lonely and isolated. Thus, identifying the significance of touch associated with the use of social robots in care for older people is not necessarily the first step to providing solutions for

[1] While touch should be considered as an important value since it is found to be essential, especially for infants and elderly people, doctors in the UK have been warned by the Medical Defense Union (MDU, 2018) to avoid comforting patients with hugs since physical contact can easily be misinterpreted by patients.

people who suffer from loneliness and lack of touch. This does not mean that some people could not benefit from touching and being touched by a social robot. However, from the ethical point of view, it is important to stress that the reversibility of touch is missing in the touch of an artifact. Still, there is also room for touch intensive social robots when developing societies toward more positive and multi-valued touch cultures.

Drawing on the distinction between instrumental touch and expressive touch, we pinpointed that nurses are not necessarily encouraged to use their affective touch capabilities as prominent work skills. If their touch skills have remained undeveloped, the relevant question is whether their work is more easily replaceable by social robots that can conduct simple instrumental work tasks. In these circumstances, perhaps new resources should be allocated to develop the skills of professional touching to educate nursing students to use their touch. If care professionals can cultivate their work skills, the use of touch and its inherently reflective nature offers great potential for health care organizations. Of course, more research is needed to examine how social robotics will change nurses' capabilities of using touch in human care and to what extent human touch can be replaced by a robot.

Taking seriously the idea that touching and presence are crucial for the well-being of older people, we do not believe the development of social robots should aim at replacing caregivers. We suggest, like many other researchers (Alaid & Zhou, 2014; Jenkins & Draper, 2015; Kristoffersson, Coradeshi, Loufti, & Severinson-Eklund, 2011), that social robots should be designed to improve the quality of care rather than just to save money in the health care sector. When social robots become a part of the network, the distribution of roles and responsibilities as well as the care processes will change (van Wynsberghe, 2013; Verbeek, 2006). van Wynsberghe (2013) states that technologies are the products of our culture and built on societal values and norms. Yet social robots also change the ways human organizations function, and they affect human habits and ways of acting. Also, as already stated, social robots may have a far reaching impact on the touch culture of societies: "social norms, values and morals find their way into technologies both implicitly and explicitly and act to reinforce beliefs or to alter beliefs and practices" (van Wynsberghe, 2013, p. 412). The use of independently functioning social robots, even for just a few tasks, would fundamentally alter relations between caregivers and care receivers and nurses' care practices in care for older people.

Acknowledgements This research is part of the project Robots and the Future of Welfare Services (2015–2020), which is funded by the Academy of Finland's Strategic Research Council (grant number 292980). The authors wish also to thank all nursing personnel who were kind to respond to the questionnaire.

References

Age UK. (2018). *Later life in the United Kingdom.* Retrieved October 16, 2018 from https://www.ageuk.org.uk/globalassets/age-uk/documents/reports-and-publications/later_life_uk_factsheet.pdf.

Ahmed, S., & Stacey, J. (2003). Introduction: Dermographies. In S. Ahmed & J. Stacey (Eds.), *Thinking through the Skin* (pp. 1–18). New York: Routledge.

Alaiad, A., & Zhou, L. N. (2014). The determinants of home healthcare robots' adoption: An empirical investigation. *International Journal of Medical Informatics, 83*(11), 825–840. https://doi.org/10.1016/j.ijmedinf.2014.07.003.

Ballermann, M. A., Shaw, N. T., Mayers, D. C., Gibney, N., & Westbrook, J. (2011). Validation of the work observation method by activity timing (WOMBAT) method of conducting time-motion observations in critical care settings: An observational study. *BMC Medical Informatics and Decision Making, 11*(32), 1–12.

Barnard, A., & Sandelowski, M. (2001). Technology and humane nursing care: (ir)reconcilable or invented difference. *Journal of Advanced Nursing, 34*(3), 367–375.

Beedholm, K., Frederiksen, K., Frederiksen, S., Maria, A., & Lomborg, K. (2015). Attitudes to a robot bathtub in Danish elder care: A hermeneutic interview study. *Nursing & Health Sciences, 17*(3), 280–286.

Belgrave, M. (2009). The effect of expressive and instrumental touch on the behavior states of older adults with late-stage dementia of the Alzheimer's type and on music therapist's perceived rapport. *Journal of Music Therapy, 46*(2), 132–146.

Berg, A., & Hallberg, I. R. (2000). Psychiatric nurses' lived experiences of working with inpatient care on a general team psychiatric ward. *Journal of Psychiatric and Mental Health Nursing, 7,* 323–333.

Bush, E. (2001). The use of human touch to improve the well-being of older adults: A holistic nursing intervention. *Journal of Holistic Nursing, 19*(3), 256–70.

Cabibihan, J.-J., Abu Basha, M. K., & Sadasivuni, K. (2016). Recovery behavior of artificial skin materials, after object contact. In A. Agah, J.-J. Cabibihan, A. M. Howard, M. A. Salichs & H. He (Eds.), *Social Robotics: 8th International Conference,* ICSR 2016 Kansas City, MO, USA, Proceedings.

Field, T. (1990). *Infancy.* Cambridge, MA: Harvard University Press.

Fuchs, T. (2013). The phenomenology of affectivity. In K. W. M. Fulford, M. Davies, R. G. T. Gipps, G. Graham, J. Z. Sadler, G. Stanghellini, & T. Thornton (Eds.), *The oxford handbook of philosophy and psychiatry* (pp. 612–631). Oxford: Oxford University Press.

Gleeson, M., & Timmins, F. (2004). Touch: A fundamental aspect of communication with older people experiencing dementia. *Nursing Older People, 16*(2), 18–21.

Harwood, R. H. (2017). How to deal with violent and aggressive patients in acute medical settings. *Journal of the Royal College of Physicians of Edinburgh, 47,* 176–182. https://doi.org/10.4997/JrCPe.2017.218.

Hennala, L., Koistinen, P., Kyrki, V., Kämäräinen, J.-K., Laitinen, A., Lanne, M., … Van Aerschot, L. (2017). *Robotics in care services: A finnish roadmap.* Retrieved from Tampere University Institutional Repository, http://tampub.uta.fi/handle/10024/101673.

Hertenstein, M. J., Holmes, R., Keltner, D., & McCullough, M. (2009). The communication of emotion via touch. *Emotion, 9*(4), 566–573.

Holt-Lunstad, J., Smith, T. B., Baker, M., Harris, T., & Stephenson, D. (2015). Loneliness and social isolation as risk factors for mortality: A meta-analytic review. *Perspectives on Psychological Science, 10*(2), 227–237.

Holt-Lunstad, J., Smith, T. B., & Layton, J. B. (2017). Social relationships and mortality risk: A meta-analytic review. *PLoS Med, 7*(7).

Jenkins, S., & Draper, H. (2015). Care, monitoring, and companionship: Views on care robots from older people andtheir carers. *International Journal of Social Robotics, 7*(5), 673–683.

Jeong, S., Dos Santos, K., et al. (2015). Designing a socially assistive robot for pediatric care. In *Proceedings of the 14th International Conference on Interaction Design and Children* (pp. 387–390). New York, NY: ACM.

Karahanna, E., Agarwal, R., & Angst, C. M. (2006). Reconceptualizing compatibility beliefs in technology acceptance research. *MIS Quarterly, 30,* 781–804.

Katz, J. E., & Halpern, D. (2014). Attitudes towards robots' suitability for various jobs as affected robot appearance. *Behaviour & Information Technology, 33*(9), 941–953.

Kerruish, E. (2017). Affective touch in social robots. *Transformations, 29,* 116–135.

Kinnunen, T. (2013). *Vahvat yksin, heikot sylityksin: Otteita suomalaisesta kosketuskulttuurista.* (Strong ones manage alone, weak ones on each other's laps: On Finnish touch culture). Helsinki: Kirjapaja.

Kristoffersson, A., Coradeschi, S., Loutfi, A., & Severinson-Eklund, K. (2011). An exploratory study of health professionals' attitudes about telepresence technology. *Journal of Technology in Human Services, 29*(4), 263–283.

Langland, R. M., & Panicucci, C. L. (1982). Effects of touch on communication with elderly confused clients. *Journal of Gerontological Nursing, 8*(3), 152–155.

Linden, D. (2015). *Touch: The science of hand, heart and mind.* New York: Penguin Books.

McCann, K., & McKenna, H. P. (1993). An examination of touch between nurses and elderly patients in a continuing care setting in Northern Ireland. *Journal of Advanced Nursing, 18,* 838–846.

MDU. (2018). *Hugging a patient could land doctors in hot water, MDU advises,* February 12, 2018. Retrieved October 16, 2018 from https://www.themdu.com/press-centre/press-releases/hugging-a-patient-could-land-doctors-in-hot-water-mdu-advises.

Menon, S. P. (2015). Maximizing time with the patient: The creative concept of a physician scribe. *Current Oncology Reports, 17,* 12–59.

Merleau-Ponty, M. (1962/1989). *Phenomenology of perception* (C. Smith, Trans.). London, New York, NY: Routledge.

Merleau-Ponty, M. (1968). *The Visible and the invisible: Followed by working notes* (C. Lefort., Ed. A. Lingis, Trans.). Evinston, IL: Northwestern University Press.

Moyle, W. (2003). Nurse-patient relationship: A dichotomy of expectations. *International Journal of Mental Health Nursing, 12,* 103–109.

Mutlu, B., & Forlizzi, J. (2008). Robots in organizations: The role of workflow, social and environmental factors in human-robot interaction. Human-Computer Interaction Institute. Paper 36. http://repository.cmu.edu/hcii/36.

NMC. *The code, Professional standards of practice and behaviour for nurses and midwives.* Retrieved from https://www.nmc.org.uk/globalassets/sitedocuments/nmc-publications/nmc-code.pdf, 2015 (January 2018).

Parviainen, J., & Pirhonen, J. (2017). Vulnerable bodies in human-robot interaction: Embodiment as ethical issue inrobot care for the elderly. *Transformations, 29,* 104–115.

Paterson, M. (2007). *The senses of touch: Haptics, affects and technologies.* London and New York, NY: Bloomsbury.

Pfadenhauer, M., & Dukat, C. (2015). Robot caregiver or robot—Supported caregiving? *International Journal Social Robotics, 7*(3), 393–406.

Routasalo, P., & Isola, A. (1996). The right to touch and to be touched. *Nursing Ethics, 3*(2), 73–84.

Routasalo, P. (1999). Physical touch in nursing studies: A literature review. *Journal of Advanced Nursing, 30*(4), 843–850.

Saborowski, M., & Kollak, I. (2015). "How do you care for technology?" Care professionals' experiences with assistive technology in care of the elderly. *Technological Forecasting and Social Change, 93,* 133–140.

Santoni de Sio, F., & van Wynsberghe, A. (2016). When should we use care robots? The nature-of activities approach. *Science and Engineering Ethics, 22*(6), 1745–1760.

Sharkey, A., & Sharkey, N. (2012). Granny and the robots: Ethical issues in robot care for the elderly. *Ethics and Information Technology, 14*(1), 27–40.

Sparrow, R., & Sparrow, L. (2006). In the hands of machines? The future of aged care. *Minds and Machines, 16,* 141–161.

Speroni, K. G., Fitch, T., Dawson, E., Dugan, L., & Atherton, M. (2014). Incidence and cost of nurse workplace violence perpetrated by hospital patients or patient visitors. *Journal of Emergency Nursing, 40*(3), 218–228.

Turkle, S. (2011). *Alone together: Why we expect more from technology and less from each other.* New York: Basic Books.

Turkle, S. (2012). *Alone together? We do we expect more from technology and less from each other?.* New York, NY: Basic Books.

Twigg, J., Wolkowitz, C., Cohen, R. L., & Nettleton, S. (2011). Conceptualising body work in health and social care. *Sociology of Health & Illness, 33*(2), 171–188.

Van Aerschot, L., Turja, T., & Särkikoski, T. (2017). Roboteista tehokkuutta ja helpotusta hoito-työhön? Työntekijät empivät, mutta teknologia ei pelota (Robots in care work to ease the burden and increase efficiency? Care professionals are hesitant but not afraid of technology). *Yhteiskun-tapolitiikka, 82*(6), 630–640.

Van Dongen, E., & Elema, R. (2001). The art of touching: The culture of 'body work' in nursing. *Anthropology & Medicine, 8*(2–3), 149–162.

van Wynsberghe, A. (2013). Designing robots for care: Care centered value-sensitive design. *Science and Engineering Ethics, 19*(2), 407–433.

Verbeek, P.-P. (2006). Materializing morality: Design ethics and technological mediation. *Science, Technology and Human Values, 31*(3), 361–80.

Wolkowitz, C. (2006). *Bodies at Work.* London: Sage.

Chapter 11
Attitudes of Professionals Toward the Need for Assistive and Social Robots in the Healthcare Sector

Kimmo J. Vänni and Sirpa E. Salin

Abstract We conducted three studies among healthcare professionals and explored the need for service and social robots in the healthcare sector. The methods consisted of cross-sectional surveys and literature reviews. The survey data were analyzed with cross-tabulations, a logistic regression model, a Pearson correlation test, and a factor analysis. The literature reviews showed that there were only a few papers which discussed the use of service and social robots as tools by healthcare workers. Both professional care workers and healthcare educators perceived that robots were able to increase productivity. The results also showed that robots are able to reduce the mental workload of workers and to increase the diversity of work. Robots were also considered as good devices for activating the patients' motoric and cognitive skills and for making them happy. Even if the attitudes were positive and people were not afraid that robots may take over workplaces, the ecosystem of social robotics is still fragmented and the number of intervention studies among professional care workers is small. Policymakers should create a strategy for promoting service and social robots in the healthcare sector. The strategy should take into account robotics in education and implementation of robots in healthcare facilities.

Keywords Service robots · Health care · Implementation · Attitude · Perception · Workload · Education

11.1 Introduction

Countries have focused on developing the use of services from robots. It is a well-known fact that companies are interested in cutting labor costs and increasing productivity (Boston Consulting Group, 2015). This also applies to healthcare services,

K. J. Vänni (✉) · S. E. Salin
Tampere University of Applied Sciences, Tampere, Finland
e-mail: kimmo.vanni@tuni.fi

S. E. Salin
e-mail: sirpa.salin@tuni.fi

© Springer Nature Switzerland AG 2019
O. Korn (ed.), *Social Robots: Technological, Societal and Ethical Aspects of Human-Robot Interaction*, Human–Computer Interaction Series,
https://doi.org/10.1007/978-3-030-17107-0_11

and the use of social and service robots in healthcare services has recently been one focus area (Vänni & Korpela, 2016).

Robots as co-workers seems to be an emerging topic (Diep, Cabibihan, & Wolbring, 2015; Haddadin et al., 2011; Sauppe & Mutlu, 2015; Vänni & Korpela, 2015; Vänni & Salin, 2017), but the number of studies where employees' attitudes have been assessed is still limited (Vänni & Korpela, 2015). Even though robotics is well researched in industry, there are only a few studies which discuss the use of social robots as co-workers (including assistive social robotics), among employees (Danish Technological Institute, 2015; Haddadin et al., 2011; Sauppe & Mutlu, 2015; Vänni & Korpela, 2015), and especially among professional care workers (Vänni & Korpela, 2016; Vänni & Salin, 2017). We consider that assessing professional care workers' attitudes toward social robots would offer new insights for healthcare organizations such as hospitals, as well as for robot designers and policymakers.

From the philosophical point of view, there are four different beneficiaries of the use of service robots at work: workers, customers, organizations, and a nation. Pressure to develop a new robot application in the healthcare sector is based partly on policy and partly on requirements for cost-effectiveness. There has been pressure to use robots in services due to the increasing healthcare costs in Europe (Munton et al., 2015) and a poor dependency ratio (Muszyńska & Rau, 2012). The trend shows that there is a lack of labor force in the service sector and one solution to this may be to deploy robots. Therefore, the Strategic Research Agenda for Robotics in Europe, SRA2020 (euRobotics aisbl, 2015a) has identified healthcare as a significant sector for the application of robotic technologies, and during the last few years, the European Union has allocated monetary resources for robotics development (European Commission, 2013). A future trend in Europe (Saritas & Keenan, 2004) may be advanced use of robots, and the restructuring of the healthcare services will force nations to find new methods for producing and delivering cost-effective care (Munton et al., 2015).

There are several reasons for considering using social robots at work. First, employees' health and functional capacities with respect to workload may entail assistive methods and tools, and robots may contribute well in this regard. In addition, work itself can be strenuous and can include repetitive movements, e.g., in the case of physiotherapy for stroke patients. Some work may also involve dangerous tasks, and robots are able to cut accident and health risks. More concretely, robots can assist employees whose physical and mental resources as well as skills, knowledge, or motivation do not meet the work demands and for whom workload may thus contribute to stress, depression, and poor performance overall. Employees may already have diagnosed temporary or permanent disorders which prevent them from performing at the normal productivity and quality level. Regarding temporary illnesses or excessive workload, robots can be used to assist employees when needed. From an employer's point of view, cases where employees' work abilities have been reduced permanently are challenging. An employer should make new work arrangements or re-educate employees, but sometimes both an employer and an employee are against any new arrangement. An assistive social robot may offer a solution if the robot is able to do part of the job and reduce the workload (Vänni & Salin, 2017). Traditionally, the service robot development in the healthcare sector has focused on patients and has emphasized assistive technology for elderly care (Kanamori, Suzuki, & Tanaka, 2002; Broadbent, Stafford, & MacDonald, 2009). However, some countries, e.g., Japan, have reported a need for more healthcare workers to ensure high-level health services, and that has prompted companies to develop new robot applications which are able to assist healthcare professionals and increase their productivity (Vänni, 2017).

There are also external reasons for using robots. Working in the healthcare sector is challenging because it is influenced by business and ICT trends but also by trends in public economics. Healthcare providers, which follow business trends and emphasize high productivity and effective processes, may require a high work contribution. The intensification of performance requirements concerning matters such as the number of operations per day and annual turnover and profit may be seen as a hustle in workplaces. Another issue has been possible changes in job descriptions, which means that professional care workers have many other tasks than direct care, such as transporting things (Vänni & Salin, 2017). On the other hand, most healthcare providers are public organizations which are funded by the public sector, and their economies are based on economic conditions and tax revenues. Economic downturns may lead to budget cuts, and service reductions may directly affect layoffs, e.g., of nurses (Alameddine, Baumann, Laporte, & Deber, 2012). That may lead to the case where the number of patients per nurse will increase and the time available for each patient will decrease. In sum, business and ICT trends and conditions of the public economy may create a latent need for social and assistive robots among professional care workers.

11.1.1 Social Robotics

There are various classifications of robotics available (IFR, 2017), but usually, robots are classified into two main categories—industrial robots and service robots—and other service robots can be divided into personal and professional robots (Kumar, Bekey, & Zheng, 2005). It can be stated that personal robots are social or semi-social because they are involved in human–robot interaction (HRI) (Dautenhahn, 2007), but professional service robots can also be non-social, such as manipulators, if they are assisting, for example, in industrial processes without any HRI.

The social robotics (Duffy, Rooney, O'Hare, & O'Donoghue, 1999) domain is quite fresh in the scientific sense. It is heterogeneous and it combines many research fields, such as technical and human behavioral sciences (Budisan, Ignat, Vacariu, & Florea, 2010, Chen et al., 2011).

Researchers defined social robotics about two decades ago (Duffy et al., 1999; Fong, Thorpe, & Baur, 2001; Tapus, Mataric, & Scassellati, 2007), but there is no gold standard or taxonomy concerning which technology and robot applications can be included in the social robotics domain. An example of the difficulty of the classification and the variety of sub-groups of robots is presented by Heerink, Kröse,

Evers, and Wielinga (2010) who studied assistive robots and classified them into two main categories: (1) non-social assistive and (2) social assistive, with sub-groups of (2a) companion robots, and (2b) service robots. In this study, we consider a social robot as a physical entity, which would be able to collaborate with workers and to provide services for them (Vänni & Korpela, 2015). In addition, a new sub-group of social robots seems to be socially and emotionally assistive robotics (Khosla, & Mei-Tai Chu, 2013) which could be categorized as companion robots.

11.1.2 Need for Social Robots

Despite the diversity of robotics definitions, the need for service robots is evident (Andrade et al., 2014; Khosla & Mei-Tai Chu, 2013; Vänni & Korpela, 2016; Vänni & Salin, 2017). The overall interest in human-related robotics has increased, and the special domain of it, social robotics, has become an important research target in the service sector recently. There are many different viewpoints and future plans for using robots in healthcare services, monitoring, and diagnostics (Vänni, 2013). A need for interactive service robots and socially assistive robots which are able to detect users' emotions has also been identified (Tapus et al., 2007; Khosla, & Mei-Tai Chu, 2013; Andrade et al., 2014).

According to the European Commission (EC), there are major healthcare targets where robots and smart ICT are considered to be useful. The Strategic Research Agenda for Robotics in Europe, (SRA2020) (euRobotics aisbl, 2015a) identifies healthcare as a prominent and growing sector for robotics and its applications. The European strategy and the multi-annual roadmap for robotics emphasize robotics systems which are able to exploit the Internet (euRobotics aisbl, 2015b). The reports recommend developing health-monitoring systems which operate over the Internet and systems where customers can customize and adapt robots prior to purchase (euRobotics aisbl, 2015b). This means in practice that robots would be linked to the Internet and data detected by robots would be processed in servers instead of standalone robots' central control units. However, the commercial Internet-based services or conceptual approaches which may support end users, healthcare professionals and robot designers for selecting, modifying and designing social robots have so far been limited.

A trend toward cost-effectiveness in healthcare services will force healthcare providers to rethink how the care should be produced and delivered (Munton et al., 2015). The economic burden of healthcare-related costs as well as costs associated with stroke, loneliness, stress, depression, dementia, and other cognitive disorders are high in the public economy (Okumura & Higuchi, 2011; Sobocki, Angst, Jönsson, & Rehnberg, 2006). Also, a trend from hospital care toward homecare services and patients' possibilities to choose between hospital and home care (Munton et al., 2015) may require new robot-assisted care procedures and robot technologies. Based on specialists' opinions (Taylor, 2015), healthcare providers should be ready to offer homecare services and user-centered technology instead of long-term hospital care

(Kim, Wang, Cai, & Feng, 2008). The voices of the patients and healthcare professionals are crucial (Kollengode, 2015) because they define the need and expectations for social robots.

11.1.3 Examples of Use Cases of Social Robots

Robots are devices that could be used in many fields other than the production industry, including, for example, healthcare services. There are examples of robots being used in rehabilitation of, e.g., autism (Kozima, Michalowski, & Nakagawa, 2008), mental disorders (Rabbitt, Kazdin, & Scassellati, 2015), dementia, or other neurological problems (Ferrari, Robins, & Dautenhahn, 2010). There are studies of how robotics has been exploited among neuro-cognitive patients (Krebs et al., 1999; Takahashi, Der-Yeghiaian, Le, Motiwala, & Cramer, 2008), disabled and injured people (Van der Loos & Reinkensmeyer, 2008) and in medical surgery (Scott, 2015).

There have been many earlier studies regarding the use of service robots by elderly people (Broadbent et al., 2009; Flandorfer, 2012), but there is hardly anything available regarding employees and their health. However, the distinction in health statuses and functional performances between recently retired elderly people and aged workers is not clear, and it is well-known that when employees are getting older their perceived work abilities decrease, whereas their health disorders and need for support may increase (Ilmarinen, Tuomi, & Klockars, 1997). Because the number of focused studies regarding the perception of social robots among employees is limited; some studies regarding the perception of robots among the elderly might be evaluated and found to be useful. For example, it has been reported that sociodemographic factors such as age, gender, and education level are relevant to how well robots are perceived (Alaiad & Zhou, 2014; Flandorfer, 2012). In addition, a user's physical and mental condition and cognitive skills should be taken into account (Scopelliti, Giuliani, & Fornara, 2005). Sekmen and Challa (2013) have reported that a robot's ability to learn is critical for interaction and that might be important regarding a robot's ability to motivate employees. Peine, Rollwagen, and Neven (2014) proposed to consider the older persons as active consumers of technology, which is quite comparable with the older employees as well. Linner et al. (2014) have argued that the integration of the service robot systems into the real world has been difficult because of the gap between the development of robotics systems and the use environment. That should be taken into account when developing social robotics solutions for professional care workers.

According to several studies, healthcare professionals have considered that robots may be useful in nursing practice in different nursing environments. In the home environment, robots can be used for assessing the body functions of patients and to alert emergency services if necessary (Beedholm, Frederiksen, Frederiksen, & Skovsgaard-Lomborg, 2015; Göransson, Pettersson, Larsson, & Lennernäs, 2008; Kristoffersson, Coradeschi, Loutfi, & Severinson-Eklundh, 2011). Falls at home are reported to be major problems, and the ability to manage such emergencies is essential in service robots (Alaiad & Zhou, 2014; Boman & Bartfai, 2015; Pigini, Facal, Blasi, & Andrich, 2012).

Assisting patients' communication with family members and care professionals is one of the key benefits of social robots. From the nurses' points of view, monitoring patients' well-being (Beedholm et al., 2015), location (Boman & Bartfai, 2015; Göransson et al., 2008; Kristoffersson et al., 2011), assessing their medicine use, promoting exercise, and providing assurance (Broadbent et al., 2012) are remarkable advantages of using robots.

According to Cohen-Mansfield and Biddison (2007), nurses would prefer robots that can aid them in the most physically demanding parts of their jobs, such as bathing, toileting, and transferring patients.

Robots often face initial resistance, but nurses have tended to accept them as a possible tool and see them as beneficial to their work, particularly if patients prefer them. Having robots has even strengthened nurses' professional values such as caring for the user's well-being, integrity, and open-mindedness (Beedholm et al., 2015).

Robots are also able to provide increased safety and help with maintaining social contacts for people who live in isolated locations (Kristoffersson et al., 2011). The use of social robots may lead to fewer visits to patients (Beedholm et al., 2015), but robots could help maintain the users' independence (Broadbent et al., 2012) and strengthen the relationship between users and their families (Pigini et al., 2012). Specialized robots have been reported to be useful tools for facilitating patients' rehabilitation (Kristoffersson et al., 2011) and pharmacy operations (Summerfield, Seagull, Vaidya, & Xiao, 2011). Some studies have also reported positive effects on nurses' job satisfaction, safety, working conditions, and stress recognition (Rincon et al., 2012; Zullo et al., 2014).

The usability of technology is crucial (Ge, 2007) and nurses expect robots to be safe (Cohen-Mansfield & Biddison, 2007), funny, exciting, and easy for patients to interact with (Boman & Bartfai, 2015). Nurses consider that even with the help of robots, human-to-human contact is still paramount in nursing (Summerfield et al., 2011). Robots cannot replace nurses or other healthcare professionals, but they can assist with and lighten work tasks.

In sum, a need for interactive service robotics and assistive social robotics is evident (Tapus et al., 2007), but the number of studies regarding experiences of using social robots in workplaces in the healthcare sector is still small.

11.1.4   Challenges of Using Robots

Social robots are a relevant part of future healthcare, but some challenges to make effective use of robots have still been reported. Many famous universities and science centers have been active in researching and developing robotics (Goeldner, Herstatt, & Tietze, 2015). However, the journey of a robot from the research laboratory to end users is expensive and time-consuming. In addition, having a robot does not ensure that employees are able to use its functionalities.

The effective use of robots in healthcare services pre-supposes that the users and patients are familiar with robots. From the employees' point of view, the implementation of a robot as a co-worker (Vänni & Salin, 2017) should be carried out in stages (Ge, 2007; Vänni, 2017). Societies are looking forward to better use of social and service robots, but there is a lack of professionals who are familiar with robots and able to use them as tools in healthcare services. Some workers in the healthcare sector do not know what kinds of robots are available and how robots may assist them (Vänni & Salin, 2017).

Health professionals have addressed some concerns regarding robots. One fear has been robots' unreliability in clinical situations (Boman & Bartfai, 2015). In addition, several studies have highlighted nurses' worries about the privacy of patients (Boman & Bartfai, 2015; Broadbent et al., 2012; Kristoffersson et al., 2011). Professionals have also pointed out that using robots in healthcare may increase unemployment and decrease face-to-face contact (Boman & Bartfai, 2015; Kristoffersson et al., 2011). Also, a robot's physical entity has been commented upon in some studies. Pigini et al., (2012) reported that the size of the robot is important in the home environment but less important in the hospital environment. However, robots' size and features in the hospital environment do not affect their normal activities (Summerfield et al., 2011).

The ecosystem for social robots is still fragmented compared to the industrial robotic ecosystem. There are some robot designers, manufacturers, and traders, but social robot education is not yet a common subject in the curricula of healthcare studies in universities. Courses for healthcare students, which focus on the implementation, usage, and development of social robots from the employees' and the end users' points of view, are scarce.

An increase in productivity (Vänni, 2017; Vänni & Salin, 2017) in the healthcare sector entails that graduate students are familiar with social robots and application areas. In addition, healthcare workers may need vocational courses on implementing and using social robots at work (Diep et al., 2015; Vänni, Cabibihan, & Salin, 2018; Vänni & Salin, 2017).

11.2 Concepts, Approaches, and Factors for Defining Social Robots

Meng and Lee (2006) have argued that the traditional industrial robot engineering approaches for human-related robotics are inappropriate in terms of user-friendliness which means that from end users' point of view robots are difficult to program and use.

There are two main models for studying the acceptance of information technology; the TAM model (Davis, 1989) where perceived usefulness and ease of use have been discussed, and the UTAUT model (Venkatesh, Morris, Davis, & Davis, 2003) which also takes into account a user's age, gender, experience, and voluntariness of use. Even if studies regarding the acceptance of technology and related factors have been discussed in the literature rather well, the number of articles regarding the implementation of robots among healthcare professionals is limited. Only a few studies report on employees' attitudes toward social robots (Vänni & Korpela, 2015; Vänni & Salin, 2017).

Flandorfer (2012) reported that sociodemographic factors such as users' age, gender, and education have significant impacts on a robot's acceptance, and users' previous experience with technology may mitigate stress in the adoption process. It has also been suggested that moderating factors such as users' physical and mental conditions and cognitive skills should be taken into account (Scopelliti et al., 2005). In addition, in some cases, religious and cultural backgrounds may be significant factors in the adoption process (Arras & Cerqui, 2005; MacDorman, Vasudevan, & Ho, 2009). Meng and Lee (2006) presented a human-centered approach for designing autonomous assistive robots and argued that industrial robot engineering approaches are inappropriate to take into account users' need and expectations such as ease of use robots. Saborovski and Kollak (2014) argued that patients' needs have been taken into account at some level in assistive technology design, but the healthcare professionals' experiences have often been ignored. Compagna and Kohlbacher (2015) researched participatory technology development and reported that healthcare workers are seen as incapable of assessing innovative robot technology. Alaiad and Zhou (2014) studied the determinants of healthcare robots' adoption and reported that sociotechnical factors may play an important role in this. Andrade et al. (2014) concluded that the cost of robots is still prohibitive and limits the wide use of robots. Chibani et al. (2013) reported on the recent challenges and future trends in ubiquitous robotics and argued that the integration of web services and ambient intelligence technologies will offer better options than standalone robots.

11.3   Study Design

The aim of the study was to explore the need for robots among healthcare workers. A relevant research question is whether healthcare employees perceive social robots as co-workers and/or tools. This study is based on three cross-sectional surveys on healthcare professionals' attitudes toward social robots, conducted between 2016 and 2018. The core of the study was a survey among nurses ($N = 220$) in 2017. Studies among health and well-being technology professionals ($N = 33$) in 2016 (Vänni & Korpela, 2016) and healthcare educators ($N = 21$) in 2018 (Vänni et al. 2018) supported the results of the main study.

11.4 Study Among Healthcare Workers

11.4.1 Materials and Methods

The main sub-study was carried out in 2017 (Vänni & Salin, 2017), and it was a part of the PALROB project, which focuses on developing an open web-based platform for service robotics innovations in the healthcare sector (Vänni, Savolainen, Salin, & Haho, 2017). The study was based on a literature review and a cross-sectional survey questionnaire conducted among healthcare workers (registered and practical nurses, head nurses, physiotherapists, managers, and directors) in Finnish hospitals and housing services. The number of respondents was 224 (206 women and 15 men), and they represented six organizations (1 hospital and five housing service organizations). All the participating organizations gave their consent to the survey. The response rate varied from organization to organization but was on average about 30%. The mean age of the respondents was 38.7 (Md 38.0, SD 11.7). To ensure that the respondents understood what a service or a social robot was, we provided them with three web links with which they were able to explore various robots. We analyzed 148 scientific articles in a literature review and evaluated articles where the associations between robots and healthcare workers were studied. Our hypotheses were that workers who stated that their work was physically strenuous as well as older workers (age ≥ 50) may need robots.

11.4.2 Survey Data, Variable Design, and Analyses

Data on a need for service robots was based on the question: "How much do you need a service robot in healthcare work?" We also asked the respondent to assess the need to develop direct and indirect nursing care and the benefits and possibilities of service robots for patients and healthcare workers. The questionnaire also included questions regarding workers' perceived work ability and physical and mental workload as well as an open-ended question asking the respondents to tell us which tasks robots could be used for in their organizations. The response options were on a five-point Likert scale: (1) "Not at all," (2) "Little" (3) "To some extent," (4) "Much," and (5) "Very much" except in perceived work ability where a ten-point scale was used. The quantitative survey results were analyzed with SPSS 25 software (Statistical Package for the Social Sciences).

Workers' need for service robots at work was selected as a dependent variable. The main independent variable was a robot's ability to help at work. The following questions were asked: "How much might a service robot (a) increase the quality of work, (b) save time, (c) increase meaningfulness at work, and (d) lighten workload?" We analyzed the respondents' age, perceived mental and physical workload, and perceived work ability variables. In addition, we analyzed variables related to patient and material logistics and the robot's role in assisting patients and workers. We also constructed sum variables of single items. For example, the sum variable "Robot as an assistant for work" consisted of the following items: The robot may lighten work; the robot may increase meaningfulness; the robot may save time at work; and the robot may increase work quality. In all, five sum variables were designed. Regarding factor analysis, we selected 13 factors that may associate with each other and impacted on a need for robots.

First, the data was analyzed with a logistic regression model. The variables were classified and dichotomized to assess the odds ratios (OR). A chi-square (χ^2) test was performed and p-values were assessed with 95% confidence intervals (95% CI). Second, data was analyzed with factor analysis. We also checked the validity and reliability of the factors. Third, correlations between single variables as well as sum variables were analyzed. Fourth, we analyzed the selected articles from a literature review and responses to an open-ended question "For which tasks could robots be recruited in your work or organization?" with a content analysis method. Responses were coded according to the type of tasks mentioned.

11.4.3   Results Among Healthcare Workers

Table 11.1 presents the background characteristics of the respondents. The majority of the respondents worked in housing services. Altogether, 88 (39%) of the respondents claimed that they might need service robots at work. The need for service robots at work was almost equal among nurses in a hospital and housing services.

Table 11.2 shows that robots' abilities to lighten work, increase meaningfulness at work, save time and increase the quality of work had significant associations with a need for robots. In addition, perceived high mental workload had a significant association with a need for robots, whereas high physical workload did not. We also tested whether being aged 50 or over would have an association with a need for robots but that factor showed a low OR and was non-significant.

Table 11.3 presents the mean values and standard deviations of selected variables in factor analysis. Likert scale was from 1 to 5, and the mean values of most of the variables are close to 3, which mean that a service robot would be useful to some extent. The highest mean values were reported regarding a robot's abilities to activate a patient's cognitive and motoric skills. Also, robots' ability to motivate patients and

Table 11.1 Characteristics of respondents ($n = 224$)

Organization type	Participants			Need for service robots			
	n	%	Age (Mn)	Yes	%	No	%
Hospital	58	25.9	38.6	24	41.4	34	58.6
Housing services	166	74.1	38.8	64	38.5	102	61.5
Sum	224	100		88		136	

Table 11.2 Associations between selected variables and a need for a service robot using a logistic regression model ($n = 221$)

Variables	OR	95% CI	χ^2	p
Robot may increase meaningfulness	26.1	11.5–59.4	84.3	<0.001
Robot may lighten my work	18.9	8.4–42.4	68.4	<0.001
Robot may save time at work	18.5	8.4–40.3	70.4	<0.001
Robot may increase quality	17.3	8.6–34.7	77.9	<0.001
High mental workload	3.0	1.7–5.3	15.4	<0.001
Age 50 or over	1.3	0.7–2.5	0.8	0.38
High physical workload	1.2	0.7–2.1	0.6	0.46

Table 11.3 Descriptive statistics of selected variables regarding a robot's function ($n = 221$)

Variables	Mn	SD	χ^2	P
Activate a patient's cognitive skills	2.96	1.16	52.7	0.00
Activate a patient's motoric skills	2.92	1.12	54.3	0.00
Connect a patient to relatives	2.82	1.38	8.3	0.08
Make patients happy	2.79	1.06	75.1	0.00
Support work tasks	2.70	1.04	73.6	0.00
Motivate a patient	2.66	1.15	55.7	0.00
Save time concerning routine work	2.65	1.12	50.8	0.00
Increase meaningfulness of work	2.52	1.05	90.2	0.00
Increase the quality of work	2.49	1.13	55.1	0.00
To be a discussion companion	2.25	1.26	66.5	0.00
Assist patients in eating	2.19	1.14	75.3	0.00
Assist patients in toilet visits	2.16	1.12	83.0	0.00
Assist patients in bathing	2.02	1.12	108.9	0.00

make them happy was high. Correspondingly, the mean value of a robot's ability to assist patients physically was lower compared to a robot's ability to assist mentally.

Table 11.4 presents a rotated factor matrix for tested variables. There were two relevant factors which supported the use of service robots in hospital and housing services. One was related to healthcare work and the other to activating patients. The correlations between the variables were high ($p < 0.01$) and the Kaiser-Meyer-Olkin Measure was 0.93. The Chi-Square value for the test was 331.4 ($p < 0.01$) and Cronbach's Alpha was 0.94.

Table 11.5 presents the Pearson correlations between employees' work ability, workload, and the need for a social robot. Even if work ability often correlated with workload (Vänni et al., 2018), in this case, there was no correlation. Physical and mental workloads had a significant association. A need for a robot was associated

Table 11.4 Rotated factor matrix for tested variables

	Factor	
	1	2
Make patients happy	0.385	*0.716*
Activate a patient's cognitive skills		*0.814*
Activate a patient's motoric skills	0.385	*0.786*
Motivate a patient	0.379	*0.831*
Support work tasks	*0.824*	
Increase meaningfulness of work	*0.767*	0.409
Save time concerning routine work	*0.856*	0.309
Increase the quality of work	*0.833*	0.388
Assist patients in toilet visits	0.640	0.384
Assist patients in bathing	0.642	0.426
To be a discussion companion		*0.663*
Assist patients in eating	0.533	0.573
Connect a patient to relatives	0.301	0.533

Extraction Method: Maximum Likelihood. Rotation Method: Varimax with Kaiser Normalization
Rotation converged in three iterations
Bold indicate the relevant variables

Table 11.5 Correlations between workloads and a need for a robot

		Work ability	Need for a robot	Mental workload	Physical workload
Work ability	R	1	−0.087	−0.129	−0.138
	Sig.		0.259	0.088	0.071
	N	225	172	175	172
Need for a robot	R	−0.087	1	0.106	0.192**
	Sig.	0.259		0.118	0.005
	N	172	220	219	215
Mental workload	R	−0.129	0.106	1	0.316**
	Sig.	0.088	0.118		0.000
	N	175	219	223	219
Physical workload	R	−0.138	0.192**	0.316**	1
	Sig.	0.071	0.005	0.000	
	N	172	215	219	219

**Correlation is significant at the 0.01 level (2-tailed)

with a physical workload. The mean of perceived work ability score of the respondent was 8.0 (SD 1.4). The mean of a need for a robot was 2.2 (SD 1.1) on a Likert Scale 1–5.

Table 11.6 Correlations between work-related variables and need for a robot

		Need for a robot	Robot may increase work quality	Robot may save time at work	Robot may increase meaningful-ness at work	Robot may lighten work tasks
Need for a robot	R	1	0.767**	0.725**	0.750**	0.722**
	Sig.		0.000	0.000	0.000	0.000
	N	220	215	218	219	216
Robot may increase work quality	R	0.767**	1	0.817**	0.808**	0.736**
	Sig.	0.000		0.000	0.000	0.000
	N	215	219	219	219	217
Robot may save time at work	R	0.725**	0.817**	1	0.754**	.765**
	Sig.	0.000	0.000		0.000	0.000
	N	218	219	222	221	218
Robot may increase meaningful-ness at work	R	0.750**	0.808**	0.754**	1	0.746**
	Sig.	0.000	0.000	0.000		0.000
	N	219	219	221	223	220
Robot may lighten work tasks	R	0.722**	0.736**	0.765**	0.746**	1
	Sig.	0.000	0.000	0.000	0.000	
	N	216	217	218	220	220

**Correlation is significant at the 0.01 level (2-tailed)

Table 11.6 presents the Pearson correlation between work-related variables and explains the need for a robot in more detail. All the variables had a strong and statistically significant association. High associations were found between a robot's ability to increase meaningfulness at work, a robot's ability to save time, and a robot's ability to increase the quality of work.

Table 11.7 presents the associations between five patient-related and work-related sum variables which were constructed from single items. The variables had significant associations.

Table 11.8 presents the possibilities and current needs for developing patients' transfers, monitoring, guidance, and lifting. It seems that current needs were ranked highly but also to use robots was seen as highly possible.

11.4.3.1 Results of an Open-Ended Question

In all, 97 respondents of 224 (43%) replied to the open-ended question and wrote 210 suggestions on robot use (Table 11.9). Of these statements, 120 (57%) concerned indirect nursing care, which is work where a patient is not present. In this category,

Table 11.7 Correlations between various sum variables and a need for a robot

		Robots as an assistant for nurses	Robot as an assistant for patients	Robot as an activator and a motivator for patients	Robot as an assistant for developing work	Robot as a tool in material logistics
Robots as an assistant for nurses	R	1	0.565**	0.506**	0.612**	0.610**
	Sig.		0.000	0.000	0.000	0.000
	N	225	225	225	225	225
Robot as an assistant for patients	R	0.565**	1	0.684**	0.652**	0.546**
	Sig.	0.000		0.000	0.000	0.000
	N	225	225	225	225	225
Robot as an activator and a motivator for patients	R	0.506**	0.684**	1	0.605**	0.452**
	Sig.	0.000	0.000		0.000	0.000
	N	225	225	225	225	225
Robot as an assistant for work	R	0.612**	0.652**	0.605**	1	0.629**
	Sig.	0.000	0.000	0.000		0.000
	N	225	225	225	225	225
Robot as a tool in material logistics	R	0.610**	0.546**	0.452**	0.629**	1
	Sig.	0.000	0.000	0.000	0.000	
	N	225	225	225	225	225

**Correlation is significant at the 0.01 level (2-tailed)

Table 11.8 Needs and possibilities to exploit robots

	Need to develop transfers inside a hospital	Possibilities to exploit robots in transfers inside a hospital	Need to develop monitoring	Possibilities to exploit robots in monitoring	Need to develop guiding of customers	Possibilities to exploit robots in guiding	Need to develop lifting	Possibilities to exploit robots in lifting
Mean	2.91	2.80	3.19	3.30	2.74	3.15	3.13	3.20
N	223	221	222	219	221	219	223	221
SD	0.97	1.10	1.04	1.15	1.14	1.17	0.91	1.07

All the variables employ Likert scale 1–5

Table 11.9 Nursing staff's suggestions on the use of robots in nursing and/or in their own organization (210 statements)

Category	Percentage
Indirect nursing care	57
Laundry service	
Tasks related to food distribution	
Cleaning	
Shelving	
Maintaining and transporting	
Turning down patient beds	
Distributing medicine	
Speech recognition and record-keeping	
Direct nursing care	37
Monitoring and raising alarms	
Companionship	
Guiding and advising	
Transferring and lifting patients	
Giving the patient reminders	
Motivation and activation	
No need/no possibilities	6
No use	
No information about robots' capabilities	

logistics-related tasks were seen as the most important area of robot use. According to the respondents, filling shelves, cleaning (especially floors), and managing food and laundry services with robots would allow healthcare professionals to focus more on direct nursing care. To aid record-keeping, respondents wished for speech recognition software that could add text directly to the patient's medical record.

In direct nursing care, in which a patient is present, 78 statements (37%) made clear that the presence of a robot could assist in tasks which related to patient safety, such as monitoring, alarm-raising, and giving reminders. Respondents believed that robots could be used for motivating and activating patients by stimulating activities. Lifting and transferring patients, in particular, were seen as tasks that could benefit from a robot's assistance. Robots could also guide patients through twisting hospital corridors.

Only 12 (6%) statements reported that robots had no potential use in their work. Respondents felt that nursing was based on human interaction, which could not be replaced by a machine. Some respondents wrote that they did not know enough about the possibilities of robots to suggest use cases for them.

11.5 Study Among Well-Being and Healthcare Technology Professionals

11.5.1 Materials and Methods

Another sub-study was aimed at exploring the attitudes of healthcare and well-being professionals toward social robots, and it was carried out in 2016 (Vänni & Korpela, 2016). The study investigated the attitudes toward social and assistive robots and was sent to ten leading professionals who represented assistive technology units in ten of the largest cities and hospital districts in Finland. Those ten hospital districts covered about 90% of the Finnish population and cases, which were potential for assistive or social robots used in the hospital environment. Seven professionals of ten hospital districts replied, covering about 70% of potential assistive robots cases in Finland. Also, a pilot survey of attitudes toward social robots was conducted among 33 health and well-being technology professionals (Median age 38, SD 7.7). Of these, 11 had engineering, 15 nursing, and seven computer programming backgrounds.

11.5.2 Survey Data, Variable Design, and Analyses

The following questions were asked to assess the need for designing and selecting social robots: (a) would an assessment of the need for social robots be useful, (b) what kind of computer-based approaches are hospital districts using for assessing the need for robots for patients if any, (c) what are the current procedures for selecting robots for patients, and (d) do hospitals have advisory services for patients for choosing a robot. In a pilot survey, 33 health and well-being technology professionals were asked (a) to evaluate the relevance of an approach to assess a need for a social robot, and (b) to evaluate the relevance of the suggested factors in the model. Replies from hospital districts were analyzed with the content analysis method and data from health and well-being technology professionals were analyzed with descriptive statistics. A chi-square (χ^2) test was performed and p-values were assessed with 95% confidence intervals (95% CI).

11.5.3 Results Among Well-Being and Healthcare Technology Professionals

11.5.3.1 Design of Approach in a Study Regarding Professionals

Based on our previous research, literature review and the International Classification of Functioning, Disability, and Health (ICF) (WHO, 2002), we have created a list of five aspects and 34 variables (Table 11.10), which may help healthcare workers to assess a need for robots and suggest a type of robot (social and/or assistive) for patients. Aspect A concerns a user's demography and profile. Aspect B concerns a user's social relationships. Aspect C takes into account a user's overall health, aspect D takes into account a user's functional capacity, and aspect E discusses a user's skills and learning capacity. ICF-based approach can be also used for assessing a need for robots as tools for healthcare professionals. In that case, aspects D (functional capacity) and E (skills and learning) will be applied.

The basic idea behind an approach was to help a care worker to choose a robot that might assist a patient and thus lighten a care worker's workload, especially if it was possible to find a robot which assisted patients autonomously and was able to do part of the professional's job. For example, patients who have good health and functional capacities may have poor social connections and a companion robot would be useful.

Furthermore, persons who have limitations in health, functionalities, and social connections might need assistive social robots, which are able to assist both emotionally and physically. However, we suggest healthcare workers take into account that they would not recommend robots for patients whose mental health status or willingness or ability to use robots was poor.

Our philosophy is that a robot should be selected or designed based on the end users' (patients') needs and expectations, but also healthcare workers' needs and expectations should be taken into account. Therefore, we have designed a flowchart (Fig. 11.1), which may help healthcare professionals, together with robot designers, to consider the details which should be taken into account when selecting or designing a social robot. We found that the number of theoretical models for assessing a need for social robots is small, and therefore a flowchart may be needed.

Table 11.11 presents the attitudes of well-being technology professionals toward the relevance of variables in the assessment model. They considered that aspects and variables were relevant overall, but some adjustments would be needed. They considered that issues regarding users' health, functional capacities, and social networks were important. Major differences between the respondents' attitudes occurred regarding the relevance of assessing users' former occupations and the level of religiousness. Most respondents argued that religion and former occupations do not have anything to do with the perception of robots.

The professionals suggested improvements to an approach for selecting a robot. Their main criticism concerned patients' abilities to understand questions associated with technical issues. In addition, they considered that some questions were too general and not able to reveal a focused need for robots. They commented that an approach should assess directly what kind of robot would be needed and whether patients would be willing to use robots if robots were able to entertain them. In addition, the professionals stated that assessment factors should be more explicit and terms should be explained. The professionals also stated that some examples of how to use robots would clarify the benefits of robots to patients and encourage them to think of uses for robots. Finally, the healthcare professionals reported that both they and patients needed more information and knowledge about social and assistive robots before using them at work. Even though criticism was presented,

Table 11.10 Topics under evaluation

No	Topic	Method[a]	Option	ICF code
A. User's profile				
1	Gender		Male/female	n/a
2	Age		Current age	n/a
3	Interest areas		Many options	d9204
4	Former occupation		Many options	d859
5	Level of religiousness	Likert	Low to high	d9301
6	Perceived need for assistive robot	Likert	Low to high	e1158
7	Frequency of using robots	Likert	Days and hours	e1158
B. Social relationships				
8	Frequency of meeting family members	Likert	Low to High	e310
9	Frequency of meeting a caregiver	Likert	Low to High	e340
10	Frequency of meeting friends	Likert	Low to high	e320
11	Level of loneliness	Likert	Low to high	d9100
12	Level of fear	Likert	Low to high	b198
13	Extent of social connections	Likert	Passive to active	d9100
C. Overall health				
14	Perceived physical health	Likert	Poor to excellent	b7300
15	Perceived mental health	Likert	Poor to excellent	b122
16	Level of functional capacity	Likert	Poor to excellent	b7402
17	Level of body strength	Likert	Poor to excellent	b7306
18	Level of cognitive capacity	Likert	Poor to excellent	b117
19	Need for rehabilitation	Likert	A little to a lot	e5800
D. Functional capacity				
	Level of moving			
20	Legs	Likert	Poor to excellent	b7303
21	Hands	Likert	Poor to excellent	b7300
22	Head/neck	Likert	Poor to excellent	b7300
23	Back	Likert	Poor to excellent	b7305
24	Level of hearing	Likert	Poor to excellent	b230
25	Level of seeing	Likert	Poor to excellent	b210
26	Ability to communicate	Likert	Poor to excellent	d330
E. Skills and learning				
27	Experience of technology overall	Likert	Low to high	e1250
28	Experience of robots	Likert	Low to high	e1258
29	Experience of smart phones	Likert	Low to high	e1250
30	Experience of computers	Likert	Low to high	e1251
31	Experience of Internet and applications	Likert	Low to high	e1251
32	Programming skills	Likert	Low to high	d1551
33	Attitude toward robotics	Likert	Neg. to pos.	e498
34	Willingness to learn new things	Likert	Low to high	d198

[a]Likert scale is from 1 to 5

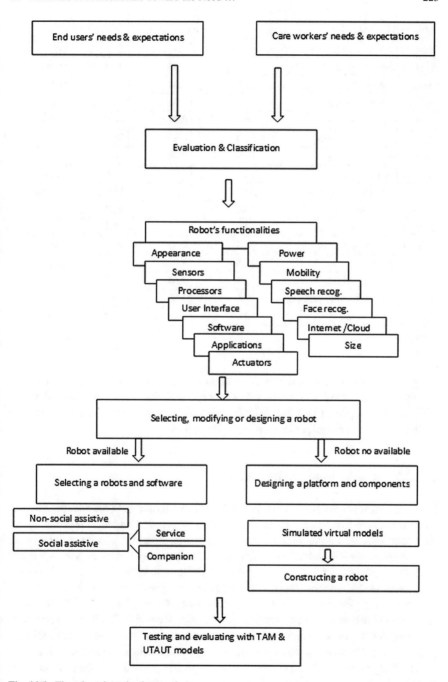

Fig. 11.1 Flowchart for selecting a robot

Table 11.11 Evaluation of some factors from well-being professionals' viewpoints

Relevance of assessing	Mean[a]	SD	χ^2	P
Functionality of senses	4.1	1.05	22.9	0.00
Former experience of technical devices	4.0	0.81	23.2	0.00
Physical and mental health	4.0	1.02	19.6	0.00
Ability to communicate	4.0	0.94	17.5	0.00
Functionality of body parts	3.9	0.83	22.3	0.00
Attitude toward robotics	3.9	0.93	16.5	0.00
Current need for a robot	3.9	1.05	14.1	0.01
End user's feeling of loneliness	3.7	0.98	15.9	0.00
Need for rehabilitation	3.6	1.00	14.7	0.00
Prevalence of chronic illnesses	3.6	1.27	9.9	0.04
Hobbies and interest areas	3.5	0.91	20.2	0.00
Social networks	3.5	1.03	15.9	0.00
End user's feeling of fear	3.3	1.05	13.5	0.01
Former occupations and work career	2.8	1.00	12.6	0.01
Level of religiousness	2.0	1.10	19.0	0.00

[a]Scale from 1 to 5

the professionals considered robots positive overall. Many of them reported that assessments of a need for a social robot aroused patients' interest toward robots. In addition, they reported that the variables were relevant and the questions were able to give an important insight into the need for robots.

11.5.3.2 Inquiry from Ten Big Cities

Leading professionals from seven hospital districts replied to open-ended questions and commented that there was no computer- or web-based method for defining and selecting assistive and/or social robots for patients. According to them, healthcare professionals are still unfamiliar with robots and the number of cases where patients or staff members ask for assistive robots is still limited. However, the professionals were interested in whether there would be a computer-based method that could help both them and patients to select a robot. For example, one healthcare professional stated that "We don't have any method for that but we would definitely be interested if someone developed it". Another professional commented, "I suppose that healthcare professionals are not ready to select an assistive robot for the patients because of lack of knowledge and tools". The comments were positive overall and the professionals understood that the new era of assistive and social robots will soon begin. However, they also commented that much introduction and training will be needed if robots are to become embedded in the healthcare sector.

11.6 Study Among Healthcare Educators

11.6.1 Materials and Methods

The third sub-study (Vänni et al., 2018) was aimed at exploring the attitudes and perceptions of the heads of education and directors of research toward a future need for social robotics education in Finnish Universities of Applied Sciences (FUAS), especially in healthcare faculties. There are 23 FUASs in Finland and 21 professionals (14 women and seven men) from 16 FUASs participated in the survey. All the participants gave their consent to the survey. The study was based on an explorative cross-sectional survey questionnaire conducted in 2016. Before the survey, the respondents participated in a two-hour workshop where basic information about social robots was presented.

11.6.2 Survey Data, Variable Design, and Analyses

Data on attitudes toward social robots was obtained from multiple questions. The respondents were asked, for example, to assess the readiness level of the social and healthcare sector in using robots at work and to predict when the social robots would be needed. Also, the need for robotics education and its significance for healthcare students were assessed with six separate questions, e.g., "Does your university offer robotics courses for social and healthcare students?" and "How important would it be that your university offered robotics courses for social and healthcare students?" The response options were on a five-point Likert scale (1 = poor/minor/not at all, 5 = excellent/major/a lot). The survey data was analyzed with SPSS 25 software (Statistical Package for the Social Sciences).

The study was explorative and it emphasized cross-tabulations and Pearson Correlation test. Additionally, the χ^2—test was used.

11.6.3 Results Among Healthcare Educators

Table 11.12 presents the current state and the need for robot education for social and healthcare students. The responses were re-classified and dichotomized into two categories "Yes" (Likert options from 3 'some extent' to 5 'a lot') and "No" (Likert options from 1 'not at all' to 2 'a little'). The respondents perceived that it would be important that universities be able to offer social robotics courses for social and healthcare students. They stated that robotics education should be a regular part of the curriculum in healthcare studies. The respondents assumed that robotics training during their studies might familiarize students with the use of robots at work and may educate students to face real patients better at work. Even if the robotics education

Table 11.12 Current state and the need for robotics training in FUASs ($n = 21$)

		Total ($n = 21$)		Statistics[a]	
		n	%	χ^2	P
Importance of offering robot courses to healthcare students					
	Yes	20	95	14.0	0.007
	No	1	5		
Robotics education prepares the students to face real patients					
	Yes	18	86	15.9	0.003
	No	3	14		
Robotics courses should be part of the healthcare curriculum					
	Yes	19	90	21.1	0.000
	No	2	10		
University offers robotics courses for healthcare students					
	Yes	6	29	14.5	0.006
	No	15	71		
Studies should familiarize students with using robots at work					
	Yes	20	95	22.6	0.000
	No	1	5		
Which robotics topics would be important in the curriculum?					
		n	%		
Programming		2	10		
Implementation		11	52		
User interfaces		4	19		
Linking robots to other systems		9	43		
Robot usage		17	81		
Information systems		4	19		

[a]χ^2 and P—values based on five-point Likert scale

was found to be essential, only a few universities offered robotics courses for social and healthcare students. The most relevant topics for education purposes were the implementation and usage of robots in the real environment. The importance of linking robots to other technical systems was found to be moderate, but the courses regarding programming were found to be poor. The results showed that the heads of education and directors of research emphasized the need for some technical training which would prepare the students to use robots at work.

Table 11.13 presents the means and standard deviations of the used variables. It shows that the respondents were not worried about the robots that would substitute for nurses and take over healthcare professionals' workplaces. Instead, robots were considered as team members and workmates. The results showed that robots would be

Table 11.13 Means and standard deviations of the used variables

	Mn	SD
Need for robots in the healthcare sector	4.20	0.93
Robotics education is essential for universities	4.05	0.86
Robots would be useful in the healthcare sector	3.71	0.78
The level of familiarizing students to face robots at work	3.52	0.68
A robot may increase productivity	3.52	0.93
Robots will be common in the near future	3.52	0.75
A robot as a work mate	3.43	0.87
A robot as a team member	3.43	0.93
Robotics training may educate students to face real patients	3.24	0.89
A robot as a substitute	2.33	0.66
Readiness of the healthcare sector to use robots	2.14	0.48
The present state of robotics training for healthcare students	1.95	0.80
Robots will take over workplaces	1.86	0.65

The range was from 1 (poor, disagree) to 5 (excellent, agree) $n = 21$

common and needed in the healthcare sector and that they may increase productivity. The respondents reported that robotics education would be essential for students even though the present state of education was quite poor. The respondents also claimed that training with social robots might help students face real patients and prepare them to use social robots at work.

Table 11.14 shows some correlation coefficients between the selected variables. The correlations showed two clusters. One relates to the usage of robots and their usefulness at work. For example, respondents who perceived that robots are useful, also stated that robots may increase productivity ($r = 0.628$, $P < 0.01$), robots will be common in the near future ($r = 0.608$, $P < 0.01$) and studies should familiarize students to face robots at work ($r = 0.577$, $P < 0.01$). Another cluster relates to social robotics education. The importance of robotics training correlates with the importance of having robotics training as a part of the curriculum ($r = 0.702$, $P < 0.01$). Familiarizing students during studies to face social robots at work correlates with an increase of productivity ($r = 0.494$, $P < 0.05$), the usefulness of robots ($r = 0.577$, $P < 0.01$), and all the variables that concern robotics education.

11.7   Discussion

The results of the studies were tangential and showed that healthcare workers and educators as well as well-being technology developers stated a future need for social robots in the healthcare sector. The results also showed that the respondents were not worried that robots would take over workplaces, or the robots would hamper

Table 11.14 Pearson correlation coefficients between some of the variables ($n = 21$)

		Usefulness of robots	Robots will be common	Robot education is essential	Education prepares students to face patients	State of robot education	Familiarizing students with robots	Robot education in the curriculum
Robots increase productivity	R	0.628**	0.233	0.217	0.326	0.370	0.494*	0.226
	Sig.	0.002	0.310	0.346	0.149	0.099	0.023	0.326
Robots will be common	R	0.608**	1	0.345	0.329	0.375	0.710**	0.363
	Sig.	0.003		0.125	0.146	0.094	0.000	0.106
Education about robots education is essential	R	0.316	0.345	1	0.570**	0.291	0.551**	0.702**
	Sig.	0.163	0.125		0.007	0.201	0.010	0.000
Familiarizing students with robots	R	0.577**	0.710**	0.551**	0.528*	0.688**	1	0.678**
	Sig.	0.006	0.000	0.010	0.014	0.001		0.001
Robot education in the curriculum	R	0.240	0.363	0.702**	0.589**	0.494*	0.678**	1
	Sig.	0.294	0.106	0.000	0.005	0.023	0.001	

*Correlation is significant at the 0.05 level. **Correlation is significant at the 0.01 level

healthcare workers' well-being at work. For example, the study among healthcare workers reported that robots were able to increase productivity and meaningfulness at work, and the study among healthcare educators confirmed that robots may increase productivity. Also, a study of well-being and healthcare technology professionals showed that robots may assist workers, and professionals from hospital districts looked forward to methods and tools for assessing the need for robots. It is remarkable that the attitudes of both healthcare workers and healthcare educators were positive toward the use of social robots. A relevant question may be why the number of social and service robots is limited in hospitals and care homes even though robots are considered to be effective. One explanation may be that only a very limited number of healthcare workers have experience with robots, and the number of real cases where robots have been used is small.

Overall, there seemed to be a lack of information on how common social robots would be in a healthcare sector. A common denominator of all three studies was that healthcare professionals might need an introduction to social robots and vocational courses should be organized. Another common factor was that the ecosystem of social robots was found to be fragmented and actors were not yet defined clearly (Vänni et al., 2018). This means that from healthcare organizations' point of view, it is challenging to find designers and robot companies that can provide cost-effective robots which can be used in real cases. Also, the conceptual approaches for defining, designing, and selecting robots from both healthcare professionals' and patients' points of view are missing. A challenge seems to be that there are only a few channels for assessing healthcare professionals' needs and expectations (Vänni & Salin, 2017). There are few possibilities to find a robot for professional use for nurses is challenging. Many robotics companies focus on technical issues, and focus on marketing and selling is still poor (Newman, 2013). The users do not know how to choose from the available options and decide if there is a solution which will match their expectations.

Another challenge seems to be that there are only a few robots (Wu et al., 2013) or robot-like actuators available even though the technology is quite mature (Tobe, 2012). A common strategy seems to be to design more or less closed, single-purpose systems and to wait for better markets for multi-purpose robots (Tobe, 2012). On the other hand, there are laboratory versions of robots that are much more capable but they are neither robust nor simple or cost-effective enough for markets (Tobe, 2012). There are many prototypes or beta versions (Tobe, 2012), but the research has focused on technical development, not on customer-oriented service models. Some service platforms have been introduced (Bartneck & Forlizzi, 2004; Lee & Forlizzi, 2009; Vänni & Korpela, 2016) but more applications, concepts, and approaches are needed, especially from the healthcare professionals' point of view.

In the main sub-study, we assumed that healthcare workers whose work is strenuous might favor service robots at work. In addition, we hypothesized that workers' being aged 50 or over may be a relevant factor for showing a need for service robots but our hypothesis did not match the results of the study. The OR showed that physical workload and age were non-significant factors in assessing a need for service robots, even though Cohen-Mansfield and Biddison (2007) reported that robots are needed for physical tasks. However, a Pearson correlation coefficient presented an association between physical workload and a need for robots. Correspondingly, odds ratios showed that workers' perceived mental workload had an association with a need for service robots, even though that was not found from Pearson correlations. The differences between the results were due to differences between the statistical analysis methods used, the sample sizes, the nature of ordinal variables, and the dichotomy procedure. We suggest taking into account that both physical and mental workloads may be a trigger for a need for a robot, but we emphasize that mental workload seems to be a more sensitive and robust indicator than physical workload.

The reason for an association between a need for a robot and mental workload may be that healthcare work is mentally strenuous because it entails various routine tasks and there is a lack of time. Poor time resources may be related to business trends which emphasize high human productivity and cost-effectiveness. Studies have shown that robots are needed to carry out some daily basic tasks but also to assist nurses to some extent in patient-related actions. The results of factor analysis indicated two relevant factors. One was related to healthcare professionals' work and the other to patients' motivation and skills. A work-related factor supports the hypothesis that service robots are needed to increase meaningfulness at work and to cut time-consuming routine work.

We also found that service robots are needed for motivating patients and activating their cognitive and motoric skills. Patients' good motivation and cognitive skills are related to nurses' workload, which would be less strenuous if patients were independent and motivated and eager to learn to use robots. Another emerging issue was the robot's function as a discussion companion and a communication aid (Kristoffersson et al., 2011; Pigini et al., 2012) which are traditional functions of social robots (Dautenhahn, 2007; Duffy et al., 1999; Heerink et al., 2010; Fong et al., 2001; Tapus et al., 2007). However, a robot as a discussion companion entails that a robot is able to discuss various topics and content designers are able to create new discussion topics. A favorable solution would be to connect a robot to the Internet and its popular topics such as sports, news, and entertainment. That has been tested in some cases and the results have been promising (Jokinen & Wilcock, 2017). The Pearson correlation tests showed that both single-work-related variables and sum variables strongly associated. The high correlations between factors may show that variable options are alike or respondents' opinions have polarized. Another explanation may be that the respondents really perceived that robots are able to improve their performance and advance job descriptions.

The results of the literature review showed that there are some studies which have focused on the benefits of robots in the healthcare sector overall, but the number of studies regarding robots as co-workers is still small (Danish Technological Institute, 2015; Haddadin et al., 2011; Sauppe & Mutlu, 2015; Vänni & Korpela, 2015).

The responses to the open-ended question emphasized that social and service robots are needed for indirect nursing care where nurse–patient interaction is not prevalent. The responses regarding direct nursing care emphasized patients' safety and lifting and transferring patients, which are common arguments for robot usage in healthcare facilities.

The strength of the study was that it consisted of three sub-studies and it employed both quantitative and qualitative methods. The study material was examined using various analyses, and sub-studies reported on the relevant factors, which were tangential and supported the use of robots. In addition, survey questionnaires were implemented among various organizations and among different participants who represented all the levels of healthcare organizations. Overall, the number and quality of respondents were adequate for assessing the need for robots in the healthcare sector.

The limitation of the study was that it was based on respondents' subjective attitudes and preconceptions but not on objective field tests of robots. The number of respondents who had previous experience of social robots was limited and only a few participants had seen a social robot in action. Some respondents argued that it was difficult to answer the questions because they lacked information on robots' possible functionalities. As far as we know, the participants and their organizations had not used service robots but some of them had seen demos such as a demo of a Zora robot (Bots, 2016). Even if we offered an option (web links) to familiarize themselves with service robotics, we were not able to confirm that the respondents visited those pages before replying to the questionnaire. However, it might be challenging to organize a field study of service robots on a large scale because many robots would be needed and the introduction of robots should be organized before the implementation of the study (Ge, 2007; Vänni, 2017). The limited number of eligible social robots and use cases in healthcare organizations reflects healthcare education in universities. It is challenging to persuade the board of a university to invest in social robots and curriculum work if there is no clear vision of what kind of social robots are needed in the field, and when.

Another limitation from healthcare organization management's point of view is that there is a lack of applied frameworks and methods such as Cost-Effectiveness Analysis (CEA) (Robinson, 1993a) or Cost-Benefit Analysis (CBA) (Robinson, 1993b) for evaluating the economic benefits of social robots. Healthcare organizations may need evaluation models to convince policymakers that the purchase of social robots can cut productivity loss and improve the quality of work. It would be challenging to evaluate the effect of social robots in the long-term concerning, e.g., customer relationships or the image of the hospital. It is even more demanding to monetarize such values. Even if CEA and CBA are common, the use of them for reporting the economic benefits of social robots in the healthcare sector is still limited.

In conclusion, this study showed that service robots are needed as co-workers or substitutes. A fear that social robots will take over workplaces is irrelevant. In fact, social robots were considered to decrease workload and develop a diversity of work overall. The most important fact was that social robots were considered to promote patients' cognitive and motoric skills and provide happiness and motivation. Promoting social robots in the near future requires a clear strategy which takes into account monetary issues, basic and voluntary education, the implementation of robots in workplaces (Ge, 2007), and their use in real cases. Some studies have reported that healthcare workers have been seen as incapable of assessing innovative technology (e.g., Compagna & Kohlbacher, 2015). Therefore, we emphasize the role of professional care workers and conclude that their experience and insights should be taken into account in robots' design and use processes.

References

Alaiad, A., & Zhou, L. (2014). The determinants of home healthcare robots adoption: An empirical investigation. *International Journal of Medical Informatics, 83*(11), 825–840.

Alameddine, M., Baumann, A., Laporte, A., & Deber, R. (2012). A narrative review on the effect of economic downturns on the nursing labour market: Implications for policy and planning. *Human Resources for Health, 10,* 23.

Andrade, A., Pereira, A., Walter, S., Almeida, R., Loureiro, R., Compagna, D., & Kyberd, P. (2014). Bridging the gap between robotic technology and health care. *Biomedical Signal Processing and Control, 10,* 65–78. https://doi.org/10.1016/j.bspc.2013.12.009.

Arras, K., & Cerqui, D. (2005). Do we want to share our lives and bodies with robots? *Tech. Rep.,* 0605-001, Swiss Federal Institute of Technology, Lausanne, Switzerland.

Bartneck, C., & Forlizzi, J. (2004). A design-centred framework for social human-robot interaction. In *Proceedings of the Ro-Man 2004* (S. 591–594).

Beedholm, K., Frederiksen, K., Frederiksen, A.-M., & Skovsgaard-Lomborg, K. (2015). Attitudes to a robot bathtub in Danish elder care: A hermeneutic interview study. *Nursing & Health Sciences, 17*(3), 280–286.

Boman, I.-L., & Bartfai, A. (2015). The first step in using a robot in brain injury rehabilitation: Patients' and health-care professionals' perspective. *Disability & Rehabilitation: Assistive Technology, 10*(5), 365–370.

Boston Consulting Group. (2015). Takeoff in robotics will power the next productivity surge in manufacturing. https://globenewswire.com/news-release/2015/02/13/924190/0/en/Takeoff-in-Robotics-Will-Power-the-Next-Productivity-Surge-in-Manufacturing.html.

Bots, Z. (2016). Zora the first social robot already widely used in Healthcare. http://www.roboticstomorrow.com/article/2016/04/zora-the-first-social-robot-already-widely-used-in-healthcare/7927.

Broadbent, E., Stafford, R., & MacDonald, B. (2009). Acceptance of health care robots for the older populations: Review and future directions. *International Journal of Social Robotics, 1,* 319–330.

Broadbent, E., Tamagawa, R., Patience, A., Knock, B., Kerse, N., Day, K., & MacDonald, B. A. (2012). Attitudes towards health-care robots in a retirement village. *Australasian Journal on Ageing, 31*(2), 115–120.

Budisan, O., Ignat, I., Vacariu, L., & Florea, C. (2010). Social interaction in systems of humans and mobile robots. *Solid State Phenomena, 166–167,* 89–94. https://doi.org/10.4028/www.scientific.net/SSP.166-167.89.

Chibani, A., Amirat, Y., Mohammed, S., Matson, E., Hagita, N., & Barreto, M. (2013). Ubiquitous robotics: Recent challenges and future trends. *Robotics and autonomous systems, 61*(11), 1162–1172. https://doi.org/10.1016/j.robot.2013.04.003.

Chen, Y-Y., Wang, J-F., Lin, P-C., Shih, P-Y., Tsai, H-C., & Kwan, D-Y. (2011). Human-robot interaction based on cloud computing infrastructure for senior companion. In: *TENCON 2011 IEEE Region 10 Conference* (pp. 1431–1434).

Cohen-Mansfield, J., & Biddison, J. (2007). The scope and future trends of gerontechnology: Consumers' opinions and literature survey. *Journal of Technology in Human Services, 25*(3), 1–19.

Compagna, D., & Kohlbacher, F. (2015). The limits of participatory technology development: The case of service robots in care facilities for older people. *Technological Forecasting and Social Change, 93,* 19–3. https://doi.org/10.1016/j.techfore.2014.07.012.

Danish Technological Institute. (2015). Robot co-worker for assembly. http://www.dti.dk/services/robot-co-worker-for-assembly/32733.

Dautenhahn, K. (2007). Socially intelligent robots: Dimensions of human–robot interaction. *Philosophical Transactions of the Royal Society of London. Series B, Biological sciences, 362*(1480), 679–704.

Davis, F. (1989). Perceived usefulness, perceived ease of use, and user acceptance of information technology. *MIS Quarterly, 13*(3), 319–340.

Diep, L., Cabibihan, J. J., & Wolbring, G. (2015). Social robots: Views of special education teachers. In *Proceedings of the 3rd Workshop on ICTs for Improving Patients Rehabilitation Research Techniques* (S. 160–163).

Duffy, B., Rooney, C., O'Hare, G., & O'Donoghue, R. (1999). What is a Social Robot? *10th Irish Conference on Artificial Intelligence & Cognitive Science*, University College Cork, Ireland. http://www.csi.ucd.ie/csprism/publications/pub1999/AICS99Duf.pdf.

euRobotics aisbl. (2015a). Strategic research agenda for robotics in Europe 2014–2020, applications: Societal challenges (pp. 59–64). http://www.eu-robotics.net/cms/upload/PPP/SRA2020_SPARC.pdf.

euRobotics aisbl. (2015b). Robotics 2020 multi-annual roadmap for robotics in Europe. Call 1 ICT23—Horizon 2020. Initial Release B 15/01/2014. http://www.eu-robotics.net/cms/upload/PDF/Multi-Annual_Roadmap_2020_Call_1_Initial_Release.pdf. Retrieved January 16, 2015.

European Commission. (2013). EU-funded research into robotics for ageing well. https://ec.europa.eu/digital-single-market/node/376.

Ferrari, E., Robins, B., & Dautenhahn, K. (2010). Does it work? A framework to evaluate the effectiveness of a robotic toy for children with special needs. In *19th International Symposium in Robot and Human Interactive Communication*, RO-MAN.

Flandorfer, P. (2012). Population ageing and socially assistive robots for elderly persons: The importance of sociodemographic factors for user acceptance. *International Journal of Population Research*. ID 829835.

Fong, T., Thorpe, C., & Baur, C. (2001). Collaboration, dialogue, and human-robot interaction. In *10th International Symposium of Robotics Research*, Lorne, Victoria, Australia.

Ge, S. S. (2007). Social robotics: Integrating advances in engineering and computer science. In: *Proceedings of Electrical Engineering/Electronics, Computer, Telecommunications and Information Technology International Conference* (S. 9–12).

Goeldner, M., Herstatt, C., & Tietze, F. (2015). The emergence of care robotics—A patent and publication analysis. *Technological Forecasting and Social Change, 92,* 115–131. https://doi.org/10.1016/j.techfore.2014.09.005.

Göransson, O., Pettersson, K., Larsson, P., & Lennernäs, B. (2008). Personals attitudes towards robot assisted health care—a pilot study in 111 respondents. *Studies in Health Technology & Informatics, 137,* 56–60.

Haddadin, S., Suppa, M., Fuchs, S., Bodenmüller, T., Albu-Schäffer, A., & Hirzinger, G. (2011). Towards the robotic co-worker. In *Robotics Research, The 14th International Symposium ISRR* (S. 261–282). Springer Tracts in Advanced Robotics, Springer Berlin Heidelberg.

Heerink, M., Kröse, B., Evers, B., & Wielinga, B. (2010). Assessing acceptance of assistive social agent technology by older adults: The almere model. *International Journal of Social Robotics, 2*(4), 361–375.

IFR. (2017). The international federation of robotics. https://ifr.org/img/office/Service_Robots_2016_Chapter_1_2.pdf.

Ilmarinen, J., Tuomi, K., & Klockars, M. (1997). Changes in the work ability of active employees over an 11-year period. *Scandinavian Journal of Work, Environment & Health, 23*(1), 49–57.

Jokinen, K., & Wilcock, G. (2017). Expectations and first experience with a social robot. In *Proceedings of the 5th International Conference on Human Agent Interaction* (S. 511–515).

Kanamori, M., Suzuki, M., & Tanaka, M. (2002). Maintenance and improvement of quality of life among elderly patients using a pet-type robot. Nihon Ronen Igakkai Zasshi. *Japanese Journal of Geriatrics, 39*(2), 214–218.

Khosla, R., & Mei-Tai Chu., M-T. (2013). Embodying care in Matilda: An affective communication robot for emotional wellbeing of older people in Australian residential care facilities. *ACM Transactions on Management Information Systems., 4*(18), 33.

Kim, J., Wang, Z., Cai, W., & Feng, D. (2008). Multimedia for Future Health-Smart Medical Home. In *Biomedical Information Technology*. Burlington: Academic Press, 23. (S. 497–512). https://doi.org/10.1016/b978-012373583-6.50027-x.

Kollengode, A. (2015). Voice of the customer (Patient) for six sigma processes in healthcare. Process Excellence Network. http://www.processexcellencenetwork.com/lean/columns/voice-of-the-customer-patient-for-six-sigma-proces/.

Kozima, H., Michalowski, M., & Nakagawa, C. (2008). Keepon—A playful robot for research, therapy, and entertainment. *International Journal of Social Robotics, 1*(1), 3–18.

Krebs, H. I., Hogan, N., Volpe, B. T., Aisen, M. L., Edelstein, L., & Diels, C. (1999). Overview of clinical trials with MIT-MANUS: A robot-aided neuro-rehabilitation facility. *Technology and Health Care, 7,* 419–423.

Kristoffersson, A., Coradeschi, S., Loutfi, A., & Severinson-Eklundh, K. (2011). An exploratory study of health professionals' attitudes about robotic telepresence technology. *Journal of Technology in Human Services, 29*(4), 263–283.

Kumar, V., Bekey, G., & Zheng, Y. (2005). Assessment of international research and development in robotics: Industrial, personal, and service robots, Chapter 5 (S. 55–62). http://www.wtec.org/robotics/report/05-Industrial.pdf.

Lee, M., & Forlizzi, J. (2009). Designing adaptive robotic services. In Proceedings of IASDR'09.

Linner, T., Pan, W., Georgoulas, C., Georgescu, B., Güttler, J., & Bock, T. (2014). Co-adaptation of robot systems, processes and in-house environments for professional care assistance in an ageing society. *Procedia Engineering, 85,*328–338. https://doi.org/10.1016/j.proeng.2014.10.558.

MacDorman, K., Vasudevan, S., & Ho, C. (2009). Does Japan really have robot mania? Comparing attitudes by implicit and explicit measures. *AI & Society, 23*(4), 485–510.

Meng, Q., & Lee, M. (2006). Design issues for assistive robotics for the elderly. *Advanced Engineering Informatics, 20*(2), 171–186. https://doi.org/10.1016/j.aei.2005.10.003.

Munton, T., Alison, M., Marrero, I., Llewellyn, A., Gibson, K., & Gomersall, A. (2015). Getting out of hospital? The evidence for shifting acute inpatient and day case services from hospitals into the community. London, UK.: The Health Foundation.http://www.health.org.uk/publications/getting-out-of-hospital.

Muszyńska, M., & Rau, R. (2012). The old-age healthy dependency ratio in Europe. *Journal of Population Ageing, 5*(3), 151–162.

Newman, P. (2013). Patchy robotics industry growth doesn't fit with aging population growth. Pergali, Editorial. http://pergali.com/patchy-robotics-industry-growth-doesnt-fit-with-aging-population-growth/.

Okumura, Y., & Higuchi, T. (2011). Cost of depression among adults in Japan. *The Primary Care Companion for CNS Disorders, 13*(3). https://doi.org/10.4088/pcc.10m01082.

Peine, A., Rollwagen, I., & Neven, L. (2014). The rise of the "innosumer"—Rethinking older technology users. *Technological Forecasting and Social Change, 82,* 199–214. https://doi.org/10.1016/j.techfore.2013.06.013.

Pigini, L., Facal, D., Blasi, L., & Andrich, R. (2012). Service robots in elderly care at home: Users' needs and perceptions as a basis for concept development. *Technology & Disability, 24*(4), 303–311.

Rabbitt, S., Kazdin, A., & Scassellati, B. (2015). Integrating socially assistive robotics into mental healthcare interventions: Applications and recommendations for expanded use. *Clinical Psychology Review, 35,* 35–46. https://doi.org/10.1016/j.cpr.2014.07.001.

Rincon, F., Vibbert, M., Childs, V., Fry, R., Caliguri, D., Urtecho, J., … Jallo, J. (2012). Implementation of a model of robotic tele-presence (RTP) in the neuro-ICU: Effect on critical care nursing team satisfaction. *Neurocritical Care, 17*(1), 97–101.

Robinson, R. (1993a). Economic evaluation and health care, cost-effectiveness analysis. *BMJ, 307,* 793–795.

Robinson, R. (1993b). Economic evaluation and health care, cost-benefit analysis. *BMJ, 307,* 924–926.

Saborovski, M., & Kollak, I. (2014). How do you care for technology?—Care professionals' experiences with assistive technology in care of the elderly. *Technological Forecasting and Social Change, 93,* 133–140. https://doi.org/10.1016/j.techfore.2014.05.006.

Saritas, O., & Keenan, M. (2004). Broken promises and/or techno dreams? The future of health and social services in Europe. *Foresight, 6*(5), 281–229.

Sauppe, A., & Mutlu, B. (2015). The social impact of a robot co-worker in industrial settings. http://pages.cs.wisc.edu/~bilge/pubs/2015/CHI15-Sauppe.pdf.

Scopelliti, M., Giuliani, M., & Fornara, F. (2005). Robots in a domestic setting: A psychological approach. *Universal Access in the Information Society, 4*(2), 146–155.

Scott, C. (2015). Is da vinci robotic surgery a revolution or a ripoff? *HealthlineNews.* http://www.healthline.com/health-news/is-da-vinci-robotic-surgery-revolution-or-ripoff-021215.

Sekmen, A., & Challa, P. (2013). Assessment of adaptive human–robot interactions. *Knowledge-Based Systems, 42,* 49–59. https://doi.org/10.1016/j.knosys.2013.01.003.

Sobocki, P., Angst, J., Jönsson, B., & Rehnberg, C. (2006). Cost of depression in Europe. *The Journal of Mental Health Policy and Economics, 9*(2), 87–98.

Summerfield, M., Seagull, F., Vaidya, N., & Xiao, Y. (2011). Use of pharmacy delivery robots in intensive care units. *American Journal of Health-System Pharmacy, 68*(1), 77–83.

Takahashi, C. D., Der-Yeghiaian, L., Le, V., Motiwala, R. R., & Cramer, S. C. (2008). Robot-based hand motor therapy after stroke. *Brain, 131,* 425–437.

Tapus, A., Mataric,' M., & Scassellati, B. (2007). The grand challenges in socially assistive robotics. Robotics Autom Mag IEEE, 14(1), 35–42.

Taylor, R. (2015). Hospitals are very bad places for the elderly, says head of the NHS as he calls for expansion of community care. Daily Mail. Associated Newspapers Ltd. http://www.dailymail.co.uk/news/article-2265692/Hospitals-bad-places-elderly-says-head-NHS-compares-treatment-national-scandals-asylum-care.html.

Tobe, F. (2012). Where are the elder care robots? IEEE Spectrum 2012. http://spectrum.ieee.org/automaton/robotics/home-robots/where-are-the-eldercare-robots.

Van der Loos, H. F. M., & Reinkensmeyer, D. J. (2008). Rehabilitation and health care robotics, handbook of robotics. New York, NY: Springer. (S. 1223–1251).

Vänni, K. (2013). Social robotics as a tool for promoting occupational health. In COST event: The future concept and reality of Social Robotics, Brussels, Belgium (2013).

Vänni, K. (2017). Robot applications in communication. In Smart technology solutions support the elderly to continue living in their own homes. Reports of the Ministry of the Environment 7/2017, (S. 44–51). Ministry of the Environment in Finland.

Vänni, K., & Korpela, A. (2015). Role of social robotics in supporting employees and advancing productivity. In *Social Robotics.* ICSR 2015. Lecture Notes in Computer Science, vol. 9388. Cham: Springer.

Vänni, K., & Korpela, A. (2016). An effort to develop a web-based approach to assess the need for robots among the elderly. In *Social Robotics.* ICSR 2016. Lecture Notes in Computer Science, vol. 9979. Cham: Springer.

Vänni, K., & Salin, S. (2017). A need for service robots among health care professionals in hospitals and housing services. In *Social Robotics.* ICSR 2017. Lecture Notes in Computer Science, vol. 9979. Cham: Springer.

Vänni, K., Savolainen, J., Salin, S., & Haho, P. (2017). Innovation platform for service robotics. https://www.researchgate.net/project/Innovation-platform-for-service-robotics.

Vänni, K., Cabibihan, J.-J., & Salin, S. (2018). Attitudes of heads of education and directors of research towards the need for social robotics education in Universities. In *Social Robotics.* ICSR 2018. Lecture Notes in Computer Science, vol. 9979. Cham: Springer.

Venkatesh, V., Morris, M., Davis, G., & Davis, F. (2003). User acceptance of information technology: Toward a unified view. *Management Information Systems Quarterly, 27*(3), 425–478.

WHO. (2002). Towards a common language for functioning, disability and health ICF. The international classification of functioning, disability and health. World Health Organization, Geneva.

Wu, Y.-H., Wrobel, J., Cristancho-Lacroix, V., Kamali, L., Chetouani, M., Duhaut, D., … Rigaud, A.-S. (2013). Designing an assistive robot for older adults: The ROBADOM project. *IRBM, 34*(2), 119–123. https://doi.org/10.1016/j.irbm.2013.01.003.

Zullo, M., McCarroll, M., Mendise, T., Ferris, E., Roulette, G., Zolton, J., … Gruenigen, V. (2014). Safety culture in the gynecology robotics operating room. *Journal of Minimally Invasive Gynecology, 21*(5), 893–900.

Chapter 12
Evaluating the Sense of Safety and Security in Human–Robot Interaction with Older People

Neziha Akalin, Annica Kristoffersson and Amy Loutfi

Abstract For many applications where interaction between robots and older people takes place, safety and security are key dimensions to consider. 'Safety' refers to a perceived threat of physical harm, whereas 'security' is a broad term which refers to many aspects related to health, well-being, and aging. This chapter presents a quantitative evaluation tool of the sense of safety and security for robots in elder care. By investigating the literature on measurement of safety and security in human–robot interaction, we propose new evaluation tools specially tailored to assess interaction between robots and older people.

Keywords Sense of safety and security · Quantitative evaluation tool · Social robots · Elder care

12.1 Introduction

The focus of social robotics research is on designing, developing, and evaluating robots that interact with humans in social environments. The rapid growth of the aging population in Europe (European Commission, 2014) and worldwide attracts researchers' attention as they work to develop assistive technologies for improving elder care. In the effort to support older people in their domestic environments, to preserve their independence and to relieve the burden of caregivers, social robots

N. Akalin (✉) · A. Kristoffersson · A. Loutfi
School of Science and Technology, Örebro University,
Örebro, Sweden
e-mail: neziha.akalin@oru.se

A. Loutfi
e-mail: amy.loutfi@oru.se

A. Kristoffersson
School of Innovation, Design and Engineering, Mälardalen University,
Västerås, Sweden
e-mail: annica.kristoffersson@mdh.se

© Springer Nature Switzerland AG 2019
O. Korn (ed.), *Social Robots: Technological, Societal and Ethical Aspects of Human-Robot Interaction*, Human–Computer Interaction Series,
https://doi.org/10.1007/978-3-030-17107-0_12

have great potential. Many studies have examined the usage of robots in elder care settings (Bedaf, Gelderblom, & De Witte, 2015), including design issues (Broadbent, Jayawardena, Kerse, Stafford, & MacDonald, 2011; Wu, Fassert, & Rigaud, 2012), acceptance (De Graaf, & Allouch, 2013; Heerink, Kröse, Evers, & Wielinga, 2010), older people's experience, and their attitudes toward robots (Vandemeulebroucke, de Casterle, & Gastmans, 2018). For example, Fischinger et al. (2016) found that older people (aged over 70) had a positive reaction to the socially assistive care robot developed in the Hobbit project in terms of perceived usability, acceptance, and affordability.

To use social robots in homes and care facilities, understanding the underlying reasons for the acceptance or rejection of this technology is crucial. The acceptance of social robots among older people includes several dimensions, and the one studied in this chapter focuses on how social robots impact 'sense of safety and security'. In human–robot interaction (HRI) literature, we see a variety of terms being used that relate to the sense of safety and security. These include perceived safety (Bartneck, Kulić, Croft, & Zoghbi, 2009), psychological safety (Lasota, Rossano, & Shah, 2014), and mental safety (Nonaka, Inoue, Arai, & Mae, 2004). However, a broader term, 'sense of safety and security' (Fonad, Wahlin, Heikkila, & Emami, 2006), is used in gerontology literature.

For effective HRI, understanding the target user and designing the interaction based on the user needs are key challenges. In this chapter, we exclusively focus on older people's sense of safety and security in the context of using social robots in elder care. Boström, Ernsth Bravell, Lundgren, and Björklund (2013) pointed out that *secure relationships, sense of control, and perceived health were significantly related to the subjects' sense of security.*, p. 1. The sense of security differs depending on context but is not only affected by access to an emergency response alarm system. Rather, the sense of security is more associated with the existence or establishment of secure relationships. Thus, the social robots' ability to foster secure relationships with the older people is an important aspect of effective HRI. The main challenge is understanding the underlying factors affecting an older person's sense of safety and security, and how to model these factors for better HRI. Addressing this challenge requires the consultation of gerontology studies. With the intention of narrowing the gap between the terminology used in HRI and gerontology literature, this chapter presents the notion of *the sense of safety and security* for HRI, and introduces an approach for measuring it based on developed tools and evaluations in two user studies.

In the remainder of this chapter, the sense of safety and security is explained, with its components, in Sect. 12.2. The tool developed was tested in a video-based study which is described in Sect. 12.3. A revised version of the tool that was tested with older people is presented in Sect. 12.4. Section 12.5 presents our proposed initial model of sense of safety and security. The chapter concludes in Sect. 12.6 by summarizing the open problems and future directions. This chapter is an extended version of the paper presented at the International Conference on Social Robotics (ICSR) 2017 (Akalin, Kiselev, Kristoffersson, & Loutfi, 2017) in which the initial results of the video-based study were published.

12.2   The Sense of Safety and Security

In order to measure the sense of safety security in older person–robot interaction, it is important to clearly define the term 'sense of safety and security'. In this section, we first provide definitions of safety and security in general in Sect. 12.2.1. Thereafter, we discuss outcomes of gerontology research on older people's sense of safety and security in Sect. 12.2.2. The various terms relating to safety and security in HRI literature are discussed in Sect. 12.2.3.

12.2.1   Conceptual Definitions of Sense of Safety and Security

The terms 'safety' and 'security' have extensive usage in different senses and contexts (Boholm, Möller, & Hansson, 2016; Burns, McDermid, & Dobson, 1992). Therefore, it is difficult to define and generalize these concepts. They have many similarities; both are associated with harm (Burns et al., 1992), and often they are used as synonyms (Boholm et al., 2016). Although both terms deal with risks (Eames & Moffett, 1999), the origin of risk creates the difference between them. Safety refers to hazards that the system may cause and how the system can harm its environment. Security refers to the threats from the system's environment that can negatively affect the system (Kornecki & Liu, 2013). Another approach to differentiating between safety and security is the intentionality of the harm, i.e., safety refers to unintentional harm (e.g., accidents, occupational injuries, and food poisoning) and security refers to intentional harm (e.g., military occupation and computer viruses) Boholm et al. (2016).

The term 'safety' has one subjective and one objective dimension. The subjective dimension includes physical, social, and psychological aspects. The objective dimension is assessed by behavioral and environmental parameters (Maurice et al., 1998). The dimensions can affect each other positively or negatively. In a report by the World Health Organization (Maurice et al., 1998), 'safety' is defined as *a state in which hazards and conditions leading to physical, psychological or material harm are controlled in order to preserve the health and well-being of individuals and the community.*, p. 6. Another study (Torstensson Levander, 2007) discusses the social phenomena of safety, security, and risk, where the definition of security is given as having no fear.

12.2.2   Older People's Sense of Safety and Security

In terms of human perceptions and feelings, the concepts of safety and security are quite difficult to understand and measure. Safety is an essential property of daily life. The importance of safety, including security, was stated in Maslow's Needs Theory as being one of the fundamental needs of human beings (Maslow, 1943).

It is troublesome to find a general definition for 'sense of security' since there is no general consensus about the definition of 'security'. Other difficulties are the translation between different languages and finding studies that focus on how to promote a sense of security rather than relating it to different risk perspectives (Boström et al., 2013). Older people's quality of life and well-being is related to their sense of safety and security (Boström et al., 2013; Fonad et al., 2006). Security is a multidimensional concept that is associated with safety, confidence, and trust. One of the dimensions of security includes interaction within the surrounding environment where the sense of knowledge and control (having knowledge about what is required to cope with and manage situations) are the elements of this interaction (Boström et al., 2013). Petersson and Blomqvist (2011) explored the Swedish concept 'trygghet' by using the story dialog method which involves telling case stories from everyday life to solve problems or reflect on an incident. Their results showed that older people's sense of security is affected by external factors such as being part of a community, having trust in others, being familiar with things and situations, and using various kinds of aids. It is separated into internal (e.g., feeling or state) and external (e.g., safe environment) 'trygghet'. The authors also noted that knowing that it is possible to get in contact with someone by using the phone or an alarm enhances older people's sense of security. There is a close connection between trust and sense of security.

Another study explored the factors that are related to sense of security among older people who were receiving care in nursing homes in Sweden (Boström et al., 2013). The paper stated that sense of security is related to sense of control, secure relationships, and perceived health. For older people, having control over service routines is important. In the study, Boström et al. (2013) confirmed that there is a correlation between sense of security and sense of control and knowledge. The factors promoting sense of security for older people are the sense of control, knowledge, and having good social relationships. One of the factors negatively affecting the sense of security is living alone. The sense of security highly depends on the personality of the individual, every older person can perceive it differently. Boström, Ernsth Bravell, Björklund, and Sandberg (2016) reported a case study with in-depth interviews and observations to understand the sense of security of older people when moving into and living in a nursing home. Based on their findings, the factors that positively affect the sense of security for older people are: having secure relationships with healthcare service staff, having control over daily routines, and being informed about them. A perceived lack of influence over the daily life and lack of information about routines are associated with a lower sense of security.

Many older people prefer to live in their own home and to continue with their familiar habits which can help them feel more secure (Fonad et al., 2006). In their study, Fonad et al. (2006) investigated the sense of safety and security of older people after moving to a retirement home. They reported that important factors for feeling safe and secure were: continuation of daily routines, familiar habits and practices, and having trust in staff.

12.2.3 Sense of Safety and Security in HRI

There are a considerable number of contributions dealing with the physical safety of robots in the literature (Bicchi, Peshkin, & Colgate, 2008; Haddadin, Albu-Schäffer,

& Hirzinger, 2010; Wyrobek, Berger, der Loos & Salisbury, 2008). Nevertheless, there is only a small amount of research in HRI that considers the sense of safety. The terms that convey a similar meaning as sense of safety are 'perceived safety' (Bartneck et al., 2009), 'psychological safety' (Kamide et al., 2012; Lasota, Rossano, & Shah, 2014; Lasota, Fong, & Shah, 2017), and 'mental safety' (Nonaka et al., 2004). In 2009, Bartneck et al. proposed a series of questionnaires to measure the key concepts in HRI, including perceived safety. They defined 'perceived safety' as follows: *Perceived safety describes the user's perception of the level of danger when interacting with a robot, and the user's level of comfort during the interaction*, p. 76. Their Godspeed V: Perceived safety questionnaire is commonly used in different HRI scenarios. For example, Lichtenthäler, Lorenzy, and Kirsch (2012) used the Godspeed V questionnaire to compare two different navigation algorithms to investigate the effect of legibility on the perceived safety in a path crossing scenario.

In Lasota, Fong, and Shah (2017), the authors presented a survey of methods for providing safety during HRI. In their survey, they categorized the studies in literature as: safety through control, motion planning, prediction, and consideration of psychological factors of which the last category is our focus of interest. They associated psychological safety with robot attributes including the robot's motion, appearance, embodiment, gaze, speech, posture, and social conduct. In another paper, Lasota et al. (2014) defined 'psychological safety' as: *Ensuring that human–robot interaction does not cause excessive stress and discomfort for extended periods of time*, p. 339. Kamide et al. (2012) also used the term psychological safety and presented a new scale for measuring safety quantitatively. They showed movies of 11 humanoids and asked the participants open-ended questions about safety. Kamide et al. (2012) categorized and analyzed the results and came up with six factors for measuring the psychological safety of humanoids. The factors were as follows: performance, acceptance, harmlessness, toughness, humanness, and agency, where the former four factors are more important than the latter two.

Nonaka et al. (2004) defined the term 'mental safety' as *humans do not feel fear of or surprised at robots* and physical safety as *robots do not injure humans*, p. 2770, when evaluating the sense of security. They conducted experiments using virtual robots of varying shape, size, and motions. While the emotions fear and surprise are related to sense of security, disgust and unpleasantness are related to comfort (Nonaka et al., 2004). Through questionnaires evaluating surprise, fear, disgust, and unpleasantness, they observed that robots' human-like behaviors made the humans feel more comfortable. Weiss and Bartneck (2015) presented a meta-analysis of the usage of the Godspeed Questionnaire Series in HRI research which covers studies reported upon between 2009 and October 2014. They reported that the perceived safety questionnaire had been used in 37 different studies. Another study using the term 'sense of security' is Zhang, Zhang, Qi, and Zhang (2016) which proposes a fall detection application for older people by using the Nao robot. The authors claim that the application will increase the sense of security of older people but they do not give a definition thereof.

Although these studies give valuable insights, the *sense of safety and security* has not been fully addressed in HRI research. We believe that the sense of safety and security is a key requirement for HRI and should be addressed in a deeper sense including all its aspects and implications for using robots in elder care. In this research, we attempt to identify the factors influencing the sense of safety and security in HRI, especially for older person–robot interaction. For effective HRI, fostering secure relationships between older people and social robots is crucial. To address this, besides consulting the gerontology literature, we conducted two user studies. We aim not only to advance the state of the art with respect to the understanding of acceptance of social robots, but also to provide insights for robot designers.

12.3   Designing a Tool for Evaluating Sense of Safety and Security in Social HRI: A Video-Based Study

In this section, we describe the first steps in the design of a tool for evaluating the sense of safety and security in social HRI. In order to have control over the scenario in which the tool was tested, the first version of the tool was developed for and tested in a video-based study. In the remainder of this section, we describe the robot used in our study, the experimental design, the participants, the evaluation tool, and experimental results.

12.3.1   The Robot

The robot used in our studies was Pepper, a humanoid robot with 20 degrees of freedom (DOF), a height of 1.2 m, and a weight of 28 kg. There are two DOFs in the head (pan and tilt), two DOFs in the hips, one DOF in the knee, and three DOFs in the base. Each arm has six DOFs: two DOFs in the shoulder, two DOFs in the elbow, one DOF in the wrist, and one DOF in the hand. The robot is equipped with three multidirectional wheels, four directional microphones, six touch sensors, several infrared sensors, laser sensors and sonar sensors, and two loudspeakers. There are three cameras: two RGB cameras (forehead and mouth) and one 3D camera located behind the eyes (Pandey & Gelin, 2018).

12.3.2   Experimental Design

To create more accurate real-world scenarios of HRI, nonverbal gestures were included in the video-recorded scenarios. Gestures accompanied by speech help to convey meaning and semantic information in the social interaction McNeill(1992). The robot gestures involved head gestures such as head nodding and head shaking as well as arm gestures that follow the terminology used by McNeill (1992): (1) deictics, (2) beats, (3) iconics, and (4) metaphorics. Deictic gestures are gestures pointing at objects or abstract space in the environment. Beat gestures are simple up-and-down or back-and-forth hand movements keeping the rhythm of speech and indicating notable points. Iconic gestures are gestures depicting images or actions. Metaphorics represent abstract concepts or objects such as moving one hand toward the shoulder to refer to the past.

To gain a better understanding of how these gestures affect the sense of safety and security, we conducted a between-subjects video-based study using the Pepper humanoid robot. The study considered four different scenarios comprising daily life activities in which older people may be engaged. The participants' sense of safety and security was measured through questionnaires administered after having watched four videos featuring Pepper using *one* of the following variations of nonverbal gestures: only arm gestures (configuration 1), only head gestures (configuration 2), head and arm gestures (configuration 3), and no gestures (configuration 4). Detailed explanations about the configurations are given in Sect. 12.3.2.1.

The scenarios used in this study were recorded as short videos which were between 20 and 30 s long. We selected four scenarios from Cortellessa et al. (2008) which presented a video-based evaluation to compare older people's perception of socially assistive domestic robots in two different cultural backgrounds: Italian and Swedish user groups. Two of the selected scenarios were proactive (i.e., the robot was the initiator), and two of them were on-demand scenarios (i.e., the user was the initiator). Detailed explanations of the scenarios are given in Sect. 12.3.2.2.

12.3.2.1 Robot Configurations

The selected configurations of nonverbal gestures used in this video-based study are briefly described below.

Configuration 1: The robot uses only arm and hand gestures (six DOFs in each arm) including deictic, beat, metaphoric, and iconic gestures.

Configuration 2: The robot uses only head gestures (two DOFs in the head which are pan and tilt) including head nodding and shaking. The head nodding gesture is used for agreement and the head shaking gesture for disagreement.

Configuration 3: The robot uses both head and arm gestures (14 DOFs, i.e., two arms and head DOFs) in the same scenario.

Configuration 4: No gestures; the Pepper robot does not use any of its DOFs, i.e., Pepper performs no nonverbal gestures while speaking.

12.3.2.2 Videos

The selected scenarios used in this video-based study, which are taken from Cortellessa et al. (2008), are briefly described below. Scenario numbers 1 and 2 represent proactive situations; scenario numbers 3 and 4 are on-demand interactions.

(a) Scenario 1 - Environmental safety   (b) Scenario 2 - Reminding analysis

Fig. 12.1 Snapshots from the proactive scenarios **a** Scenario 1—Environmental safety, the robot uses both arm and head gestures (configuration 3), **b** Scenario 2—Reminding analysis, the robot uses only head gestures (configuration 2)

Scenario 1: Environmental safety. *The actor is sitting on the sofa, watching TV. Meanwhile, in the kitchen, the sauce on the stove is overcooking. The robot moves toward the actor and says: "The pot is burning. You should turn it off." The actor immediately goes to the kitchen and turns the stove off* (Fig. 12.1a).

Scenario 2: Reminding analysis. *The actor is in the kitchen. He is about to have breakfast. When he puts the pot on the stove, the robot says: "You cannot have breakfast now. You have an appointment for a medical analysis." The actor answers: "You're right. I have forgotten all about it!"* (Fig. 12.1b).

Scenario 3: Finding objects. *The actor is sitting on the sofa and picks up a magazine to read. Suddenly, he realizes that his glasses are not on the table in front of him. The actor calls the robot and asks: "Where are my glasses?" The robot answers: "Just a minute, I am checking" and then the robot answers: "The glasses are in the kitchen." The actor goes to the kitchen and gets the glasses* (Fig. 12.2a).

Scenario 4: Reminding about medication. *The actor is sleeping on the sofa, and suddenly wakes up. He does not realize what time it is, and thus he asks the robot. The robot answers: "It is four o'clock." The actor does not remember whether or not he took his medicine after lunch, and asks the robot. The robot answers: "Yes, you took it"* (Fig. 12.2b).

Figures 12.1 and 12.2 show snapshots from each scenario and robot configuration.

12.3.3 Participants

In Akalin et al. (2017), we reported the initial results from the video-based study including 100 participants. In this chapter, we include 24 additional participants in the analysis, i.e., in total 124 participants. There were 58 males and 66 females, whose ages ranged from 14 to 65 years ($\mu_{age} = 35.44$, $\sigma_{age} = 10.55$) who took part

(a) Scenario 3 - Finding objects   (b) Scenario 4 - Reminding about medication

Fig. 12.2 Snapshots from the on-demand scenarios **a** Scenario 3—Finding objects, the robot uses only arm gestures (configuration 1), **b** Scenario 4—Reminding about medication, the robot uses no gestures (configuration 4)

in the online survey. The advertising for the study was carried out in social media and through mailing lists. The numbers of participants for configurations 1–4 were as follows: 46, 28, 25, 25 participants. In our online survey, participants could select any of the configurations (1–4). The most selected configuration was the first one (configuration 1). As reported in Sect. 12.3.2, each participant watched four videos featuring Pepper using the same configuration. The participants' familiarity with robots was as follows: 31% of them had seen a real robot before but were not familiar with robots, 16% had already interacted with robots, 23% worked with robots, and 30% had previously seen a robot on TV/Internet.

12.3.4 Questionnaires

Using the literature study presented in Sect. 12.2, we developed a first version of a questionnaire for evaluating the sense of safety and security. The questionnaire could be divided into four different areas: sense of safety, sense of security, acceptance, and emotions.

The questionnaire batch used for data collection in the video-based study consisted of one socio-demographics form plus four separate sections. In the socio-demographics form, the participants were asked about gender, age, educational level, country of residence, and their familiarity with robots. A brief explanation of the other sections is given below:

Section 1: Sense of safety. Six questions were designed on a five-point semantic differential scale to assess sense of safety with regard to the four videos that were presented. The questionnaire included Section V from Bartneck et al.'s (2009) Godspeed questionnaire series (anxious–relaxed, agitated–calm, quiescent–surprised) and the following items: threatening–safe, uncomfortable–comfortable and predictable–unpredictable.

Section 2: Sense of security. Six questions were designed in a five-point semantic differential scale to assess sense of security in response to the four videos that were presented. We designed a questionnaire for the sense of security including the items: insecure–secure, unfamiliar–familiar, fear–ease, unreliable–reliable, unnatural–natural, lack of control–in control. The questionnaire items were decided based on the gerontology studies summarized in Sect. 12.2.2.

Section 3: Acceptance. There were seven questions rated on a five-point Likert scale, ranging from 'Strongly disagree' to 'Strongly agree' to assess the acceptability of the robot. Four of them were asked after each video, and three of them were asked at the end of the survey. These questions were taken from Fischinger et al. (2016) and the trust construct of the Almere model (Heerink et al., 2010).

Section 4: Emotions. Self-Assessment Manikin (SAM) by Bradley and Lang (1994) was used to evaluate the participants' emotions. Manikin is a scale for assessing emotions in the valence-arousal emotion space (Russell, 1980) which ranges from unpleasant to pleasant on the valence scale and calm to excited on the arousal scale. In this study, a nine-point semantic scale was used. At the end of the survey, an additional text field was provided allowing for free comments. The questions are given in Table 12.1.

Table 12.1 The questions asked in the online survey

	Questionnaire items	Scale
Scenarios	The scenario seems realistic, one can encounter this scenario in the daily life	1 = Strongly disagree 5 = Strongly agree
	I liked the scenario	
	I can use the robot for a long period in my home	
	I can imagine having a robot taking care of me	
Sense of safety	Threatening–safe, anxious–relaxed, agitated–calm, quiescent–surprised, uncomfortable–comfortable, unpredictable–predictable	1–5
Sense of security	Insecure–secure, unfamiliar–familiar, fear–ease, unreliable–reliable, unnatural–natural, lack of control–in control	1–5
Emotions	Unpleasant–pleasant, calm–excited	1–9
Acceptance	I think the robot would be helpful in my home	1 = Strongly disagree 5 = Strongly agree
	I would trust the robot if it gave me advice	
	I would follow the advice the robot gave me	

12.3.5   Procedure

The data was collected using an online survey in which the participants watched four videos featuring a robot and an older person. The videos and questionnaires were in English.[1] The participants were provided with a short introduction about the survey and filled out the socio-demographics form. Then, the video sessions started. The participants were asked to watch the four videos one by one. At the end of each video, a short questionnaire was used to assess the participants' views on the scenario and the robot. After having seen all the videos, the participants filled out four questionnaires to assess their sense of safety, sense of security, emotions, and acceptance. No explanation of the real aim of the study was given in order to avoid influencing and biasing the participants.

12.3.6   Experimental Results

To compare the effects of nonverbal gestures on the sense of safety, sense of security, and acceptance, a one-way analysis of variance (ANOVA) was applied. The results seem to indicate that the participants did not notice the gestures of the robot (sense of security $F(3,120) = 0.49$, $p \prec 0.69$, sense of safety $F(3,120) = 0.50$, $p \prec 0.68$, and acceptance $F(3,120) = 0.96$, $p \prec 0.41$). Throughout the online survey, we did not mention anything about the gestures. We expected that the gestures would affect the participants, but the results show that there is no statistically significant difference between the four configurations for any of the measures and only a few participants commented on the gestures in the free text field. One reason for the lack of a statistically significant difference between the four configurations could be that the participants were concentrating on the scenarios and the idea of using robots in homes. Another explanation might be the choice of conducting a video-based study online with participants recruited through mailing lists and social media. Some participants may have watched the videos on a smartphone screen and never noticed the robot's gestures. In addition, we also investigated whether the participants' response to the questions within the scenario section varied as a result of having seen on-demand or proactive scenarios. The results of the one-way ANOVA tests show that the participants liked the on-demand scenarios more than the proactive scenarios, $F(1,494) = 4.89$, $p \prec 0.05$ ($\mu_{proactive} = 3.76$, $\sigma_{proactive} = 0.93$ and $\mu_{on-demand} = 3.95$, $\sigma_{on-demand} = 0.89$).

In the remainder of this section, we present the results from measuring sense of safety, sense of security, as well as the combined results for sense of safety and sense of security, acceptance, and emotions.

---

[1]The videos and questionnaires used during the video-based study: https://bit.ly/2DkiOHo.

12.3.6.1 Sense of Safety

The sense of safety questionnaire consisted of six items (threatening–safe, anxious–relaxed, agitated–calm, quiescent–surprised, uncomfortable–comfortable, unpredictable–predictable). We conducted a reliability analysis to check the internal consistencies within the items in the questionnaire. Cronbach's α was used for measuring the internal consistencies within the items of a test or scale (Cronbach, 1951). The α coefficient ranges from 0 to 1 showing the overall assessment of a measure's reliability, and a value over 0.7 is considered as acceptable.

The Cronbach's α for all participants and configurations including all items was 0.79. Comments provided in the free text field showed that many participants were confused about the quiescent–surprised scale so we excluded this item and recalculated the Cronbach's α. The final Cronbach's α value was 0.87. This indicates that the questionnaire items had a good internal consistency. The calculated Cronbach's α values for each configuration and in total are given in Table 12.2a, where the new Cronbach's α shows the values after the quiescent–surprised item is excluded.

The semantic differential scale descriptive statistics (mean and standard deviation) for each configuration and in total can be seen in Table 12.2b. The results show that the participants' experience was on the positive side for the bipolar adjectives used in the sense of safety questionnaire. This can also be seen in Table 12.3, which shows how the number of responses varied from the negative side to the positive side.

Table 12.2 Sense of safety statistics for each configuration and in total. Five-point semantic scale

Configuration	Cronbach's α	New Cronbach's α	Configuration	μ	σ
c1	0.57	0.78	c1	3.83	0.95
c2	0.84	0.85	c2	3.67	0.96
c3	0.86	0.93	c3	3.84	1.22
c4	0.81	0.86	c4	3.70	1.02
Total	0.79	0.87	Total	3.77	1.02

(a) Cronbach's α values.   (b) Descriptive statistics.

Table 12.3 Number of responses for the sense of safety questionnaire

	1	2	3	4	5	
Threatening	1	4	20	52	47	Safe
Anxious	2	7	29	53	33	Relaxed
Agitated	2	4	20	51	47	Calm
Quiescent	12	28	48	28	8	Surprised
Uncomfortable	4	11	24	46	39	Comfortable
Unpredictable	2	5	40	52	25	Predictable

The numbers in columns 1–5 show the response distribution

Table 12.4 Sense of security statistics for each configuration and in total. Five-point semantic scale

Configuration	Cronbach's α
c1	0.83
c2	0.88
c3	0.93
c4	0.83
Total	0.87

(a) Cronbach's α values.

Configuration	μ	σ
c1	3.62	0.97
c2	3.72	1.00
c3	3.72	1.24
c4	3.49	1.10
Total	3.64	1.07

(b) Descriptive statistics.

Table 12.5 Number of responses for the sense of security questionnaire

	1	2	3	4	5	
Insecure	3	3	26	50	42	Secure
Unfamiliar	6	23	33	38	24	Familiar
Fear	1	5	33	46	39	Ease
Unreliable	0	13	34	44	33	Reliable
Unnatural	12	27	40	35	10	Natural
Lack of control	3	16	29	46	30	In control

The numbers in columns 1–5 show the response distribution

12.3.6.2 Sense of Security

The Cronbach's α value for all participants and all configurations was 0.87, which shows that the questionnaire has good internal consistency. The internal consistencies for each configuration and in total are given in Table 12.4a. The semantic differential scale descriptive statistics of sense of security (mean and standard deviation) for each configuration can be seen in Table 12.4. The results show that the participants' experience was on the positive side also for the bipolar adjectives used in the sense of security questionnaire. This can also be seen in Table 12.5, which shows how the number of responses varied from the negative side to the positive side.

12.3.6.3 Sense of Safety and Security

We also calculated the Cronbach's α for sense of safety and security by using all the items (excluding quiescent–surprised, as mentioned before) in both questionnaires (the sense of safety and the sense of security). The Cronbach's α value for all participants and all configurations was 0.91, which again shows that combining the questionnaires resulted in a good internal consistency.

12.3.6.4 Emotions

In the valence-arousal emotion space, each emotion can be placed on a 2D graph with a horizontal axis (valence) and a vertical axis (arousal) where valence ranges from unpleasant to pleasant, and arousal ranges from calm to excited (Russell, 1980). The emotion section of the questionnaire rated emotional states using SAM to place emotions onto Russell's 2D graph. The graph has four quadrants: high valence-high arousal (HVHA), low valence-high arousal (LVHA), low valence-low arousal (LVLA), and high valence-low arousal (HVLA). The first quadrant, HVHA, includes emotions such as happy and excited, the second quadrant, LVHA, includes emotions such as angry and frustrated, the third quadrant, LVLA, includes emotions such as sad and bored, and the fourth quadrant, HVLA, includes emotions such as calm, pleased, and relaxed. Based on the SAM ratings, the average of the rated emotional states was in the HVLA quadrant. As shown in Fig. 12.3, the majority of the participants felt high valence and low arousal about the robot videos.

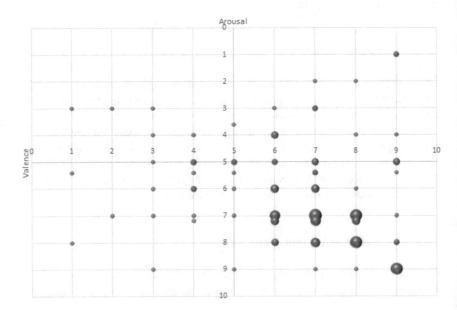

Fig. 12.3 SAM results on the 2D valence-arousal space

Table 12.6 Acceptance Likert scale descriptive statistics for each configuration and in total

Question	c1	c2	c3	c4	Total
I can imagine using the robot for a long period in my home	$\mu = 3.74$	$\mu = 3.42$	$\mu = 3.28$	$\mu = 3.44$	$\mu = 3.52$
	$\sigma = 0.95$	$\sigma = 1.23$	$\sigma = 1.33$	$\sigma = 1.29$	$\sigma = 1.17$
I can imagine having a robot taking care of me	$\mu = 3.87$	$\mu = 3.68$	$\mu = 3.44$	$\mu = 3.72$	$\mu = 3.71$
	$\sigma = 0.86$	$\sigma = 1.12$	$\sigma = 1.26$	$\sigma = 1.20$	$\sigma = 1.08$
I think the robot would be helpful in my home	$\mu = 3.83$	$\mu = 3.64$	$\mu = 3.48$	$\mu = 3.76$	$\mu = 3.70$
	$\sigma = 0.93$	$\sigma = 1.25$	$\sigma = 1.26$	$\sigma = 1.23$	$\sigma = 1.13$

1 = Strongly disagree, 5 = Strongly agree

Table 12.7 Number of responses for acceptance questionnaire

	1	2	3	4	5
I can imagine using the robot for a long period in my home	10	18	15	60	21
I can imagine having a robot taking care of me	4	18	17	56	29
I think the robot would be helpful in my home	6	16	19	51	32

The numbers in columns 1–5 show the response distribution. 1 = Strongly disagree and 5 = Strongly agree

12.3.6.5 Acceptance

The descriptive statistics from the acceptance questionnaire are provided in Tables 12.6, and 12.7 shows the distribution of responses. Interestingly, even though the participants did not notice the gestures according to the one-way ANOVA, the mean had the lowest values for every question in the c3 configuration (head and arm gestures). This may imply that the robot should have enough gestures, but this should be further investigated.

12.4 Revising and Validating the Tool for Evaluating Sense of Safety and Security in Social HRI: Older People Interacting with Pepper

In the first study presented in Sect. 12.3, we developed a questionnaire for measuring sense of safety and sense of security. As shown in Sect. 12.3.6.3, the Cronbach's α value was high for the combined safety and security questionnaire. In this second study, we used a revised and smaller version of the combined questionnaire. The questionnaire was tested with older people at the Senior Festival in Örebro, Sweden. The robot used in the study was, once again, the Pepper robot. At the Senior Festival, we had a stand where older people had a chance to come and interact with the robot. Since the participants could freely interact with the robot, the items unnatural–

Table 12.8 The revised questions used at the Senior Festival

Questionnaire items	Scale
While interacting with the robot, I felt:	
Insecure–secure	1–5 (Semantic scale)
Anxious–relaxed	
Uncomfortable–comfortable	
Lack of control–in control	
I think the robot is:	
Unsafe–safe	
Unfamiliar–familiar,	1–5 (Semantic scale)
Unreliable–reliable,	
Scary–calming	
Using a robot would make my life easier	
I think the robot would be helpful in my home	1 = Strongly disagree
	5 = Strongly agree
Familiarity with robots	This is the first time I have seen a real robot.
	I've interacted with a robot.

natural and unpredictable–predictable were not suitable for the interaction and were therefore excluded from the questionnaire. We replaced the adjective *agitated* with *scary* to make it clearer, thus we excluded fear–ease since scary and fear are quite similar words. The exclusion of the item quiescent–surprised has already been discussed (confusion of the item). In this way, four items were removed and eight items remained (see Table 12.8).

The aim of the study was to understand older people's sense of safety and security regarding social robots. Approximately 80 older people visited our stand and interacted with the robot. During the interaction with Pepper, they could, for example, speak with the robot, dance with the robot, or watch the robot dancing. After the interaction, we asked them to fill out our questionnaire. Even though they enjoyed interacting with the robot and found it very cute and interesting, many interactants declined to fill out the questionnaire. In total, 44 older people ($\mu_{age} = 70.08$, $\sigma_{age} = 8.30$) filled out the questionnaire but only 36 of them (25 females and 11 males) answered all the questions.

For the majority of the participants, this was the first time they had interacted with a robot. Eight of them had interacted previously with a robot while participating in HRI experiments conducted at Örebro University.

Figure 12.4 shows two photographs from the Senior Festival. The questionnaire administered was in Swedish and had 13 questions including the demographics questions (age and gender). The questionnaire and questionnaire results are provided in Tables 12.8 and 12.9. We conducted a reliability analysis to check the internal consistency of the items of our revised questionnaire. The Cronbach's α for the participants who filled out the entire questionnaire was 0.89. Hence, the revised questionnaire has a good internal consistency and can be used in HRI scenarios.

Fig. 12.4 Two photographs from the Senior Festival

Table 12.9 Senior festival questionnaire descriptive statistics

Item	μ	σ
While I interacted with the robot I felt		
Insecure–secure	3.67	0.79
Anxious–relaxed	3.53	0.88
Uncomfortable–comfortable	3.61	1.23
lack of control–in control	3.25	1.30
I think the robot is		
Unsafe–safe	3.72	1.09
Unfamiliar–familiar	3.03	1.21
Unreliable–reliable	3.81	0.92
Scary–calming	3.56	0.97
Using a robot would make my life easier	3.22	1.02
I think the robot would be helpful in my home	3.14	0.99

[a]Descriptive statistics are presented on different scales (see Table 12.8)

12.5   The Initial Model of Sense of Safety and Security

In order to measure the sense of safety and security, we propose a model that considers different properties of the user profile as well as different robot properties. This model includes two parts: (1) human-related factors; and (2) robot-related factors. The factors are determined based on the literature provided in Sect. 12.2 and the insights obtained while conducting the user studies.

Only the human-related factors were taken into consideration in the questionnaires used in the studies discussed in Sects. 12.3 and 12.4. Therefore, we have designed an initial model of sense of safety and security which is based on only human-related factors. In the remainder of this section, we discuss human-related factors in Sect. 12.5.1 while the proposed model of sense of safety and security based on human-related factors is provided in Sect. 12.5.2. Finally, the robot-related factors are elaborated upon in Sect. 12.5.3. More user studies are needed to cover all the factors.

12.5.1   Human-Related Factors

Feeling comfortable during the interaction has been reported to be one of the key issues in HRI scenarios (Dautenhahn et al., 2006). The human comfort was taken into account in different HRI scenarios, such as by addressing the effects of using a simple handheld device to measure the participant's comfort level with changing robot behaviors (Koay, Dautenhahn, Woods, & Walters, 2006), and comfortable distance in a scenario involving a robot following a human (Koay et al., 2006). Lauckner, Kobiela, and Manzey (2014) attempted to determine the threshold of comfort for frontal and lateral distances in human–mechanoid interaction in a hallway scenario. It was also argued that human comfort can change with different scenarios, robots, tasks, application areas as well as users (Dautenhahn, 2007).

A lack of sense of safety and security affects people's feeling and is associated with emotional responses such as fear, stress, anxiety, surprised, and anger (Nonaka, Inoue, Arai, & Mae, 2004; Zheng et al., 2016). Our semantic differential scale covered these items as: fear–ease, anxious–relaxed, quiescent–surprise, scary–calm. In Sect. 12.5.2, we included these items in our model (see Fig. 12.5).

Having experience of and familiarity with robots facilitates a more natural HRI according to, e.g., De Graaf and Allouch (2013). To eliminate the novelty effect of the robot and to gain familiarity with it, long-term studies are required. Throughout the long-term interaction, novelty effects wear off and fade away over time, causing human expectations and behaviors to change (Sung, Christensen, & Grinter, 2009), whereas the short-term interaction with a robot may result in a trade-off between the robot's verbal behavior and the desired positive impact on learning gains with a robot tutor (Kennedy, Baxter, Senft, & Belpaeme, 2016). Previous research in HRI has demonstrated that a long-term interaction between a user and a robot can impact the attitude and behavior of the person and hence the user experience (Leite, Martinho, & Paiva, 2013).

Being in control is one of the most important factors affecting the sense of security among older people (Boström et al., 2013, 2016). We use the term 'sense of control' as "the user feels that he/she is in charge of the system". The sense of control is also known as 'sense of agency' which is defined as a *feeling of control over actions and their consequences* (Moore, 2016). In human–computer interaction, sense of agency has been taken into account in one of Shneiderman (1992)'s golden rules of interface design. The seventh rule (support internal locus of control) is based on the idea that operators desire the sense of being in charge of an interface and that the interface responds to the operators actions. In HRI, the user's sense of control during the interaction with a robot was found to be linearly related to the expected level of autonomy (Chanseau, Dautenhahn, Koay, & Salem, 2016).

Trust has been remarked as being an important factor affecting older people's sense of safety and security (Petersson & Blomqvist, 2011) as well as in HRI (Charalambous, Fletcher, & Webb, 2016; Salem, Lakatos, Amirabdollahian, & Dautenhahn, 2015). Trust is indicated as being one of the factors responsible for the increased acceptance of robots and as contributing to the establishment of effective relationships between humans and robots, as well as to humans' willingness to cooperate with a robot (Salem et al., 2015).

In Sect. 12.5.2, we described our development of an initial model which categorizes the items in the combined sense of safety and security questionnaire into comfort, emotional responses, experience, sense of control, and trust (see Fig. 12.5).

12.5.2 The Model of Sense of Safety and Security

In this section, we first present the initial model (Fig. 12.5) that was designed intuitively. The model was updated after an exploratory factor analysis (EFA) of the video-based study, and validated in a confirmatory factory analysis (CFA) of the older people's study. The Kaiser–Meyer–Olkin (KMO) test indicates the suitability of the data for factor analysis (FA). The result of the KMO test suggests that both the video-based study and the older people study was suitable for FA; 0.91 and 0.86, respectively.

To discover possible underlying factors and to check if these factors match with our initial model, we conducted an EFA by using the R psych package. Ten items (threatening–safe, anxious–relaxed, agitated–calm, uncomfortable–comfortable, unpredictable–predictable, insecure–secure, unfamiliar–familiar, unreliable–reliable, unnatural–natural, lack of control–in control) of the sense of safety and sense of security questionnaires used in the video-based study were subjected to an EFA with varimax rotation and extraction using minres (ordinary least squares) with five factors. We followed Jolliffe's criterion and retained all factors with eigenvalues greater than 0.7. The results of the EFA are presented in Table 12.10 and Fig. 12.6. This model showed a good fit (Goodness of Fit Index (GFI) = 0.937 and Comparative Fit Index (CFI) = 0.974). The video-based study included the fear–ease item, and the older people study included the scary–calming item. In order to ensure consistency

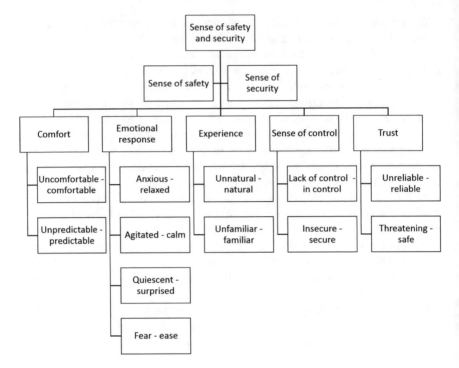

Fig. 12.5 Initial model of sense of safety and security based on human-related factors

Table 12.10 Exploratory factor analysis factor loadings concerning the items of the sense of safety and sense of security

Questionnaire items	Factors				
	F1	F2	F3	F4	F5
Threatening–safe	0.59	0.14	0.31	0.18	0.20
Anxious–relaxed	0.83	0.33	0.15	0.14	0.20
Agitated–calm	0.64	0.27	0.33	0.18	0.14
Uncomfortable–comfortable	0.61	0.15	0.32	0.23	0.37
Unpredictable–predictable	0.27	0.30	0.46	0.19	0.17
Insecure–secure	0.47	0.26	0.75	0.16	0.31
Unfamiliar–familiar	0.25	0.83	0.27	0.11	0.14
Unreliable–reliable	0.33	0.27	0.33	0.14	0.68
Unnatural–natural	0.29	0.49	0.10	0.23	0.37
Lack of control–in control	0.24	0.17	0.17	0.93	0.14

F1 = comfort, F2 = experience, F3 = sense of security, F4 = trust, and F5 = sense of control

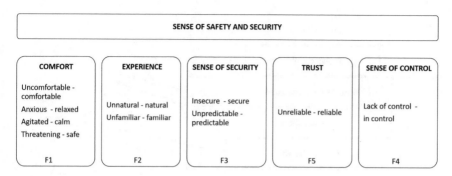

Fig. 12.6 Five factors model, item loadings to factors

between them, we excluded the item fear–ease when conducting the EFA. As already discussed, the quiescent–surprised item was excluded due to confusion. The exclusion of the item was further motivated by the KMO test which also showed that the value of this item for Measure of Sampling Adequacy was 0.45, which is lower than the minimum acceptable value, 0.5.

In our initial model, we had the categories: comfort, emotional responses, experience, trust, and sense of control. After conducting the EFA, we updated our model's factors and items. For example, the items under emotional response in the initial model clustered with the item uncomfortable-comfortable. However, we kept the naming comfort for that category (shown in the Fig. 12.8). The factor three (F3 in Fig. 12.7) had two items: unpredictable–predictable and insecure–secure. We named that factor as sense of security (see the item loadings and namings in Fig. 12.6).

We verified the factor structure obtained using the video-based data on the older people data by conducting CFA. For CFA, we used the R lavaan package. CFA for the model shown in Fig. 12.8 on the older people data again showed a good fit (GFI = 0.882 and CFI = 0.944).

12.5.3 Robot-Related Factors

We categorized robot-related factors that affect the sense of safety and security as gestures, functional properties of the robot, physical properties of the robot, and social properties of the robot. Gestures can be verbal, nonverbal, and gaze; functional properties are ease of use, autonomy, and performance; physical properties are anthropomorphism, embodiment, and size of the robot; and social properties are personality of the robot and friendliness.

The gestures of a robot will make the interaction more natural and comprehensible, i.e., similar to human–human interaction. The manner in which a robot reacts during an interaction with a human may affect the human's perception, well-being, the sense of support and security, and willingness to interact. A responsive robot (robot

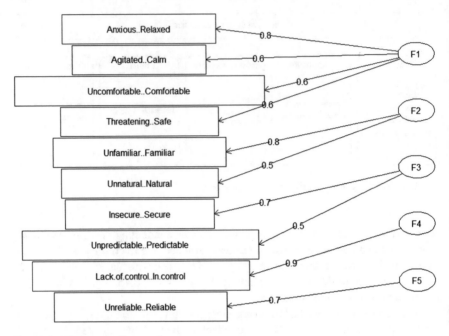

Fig. 12.7 EFA item loadings on factors

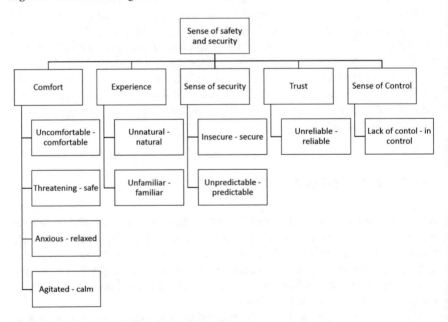

Fig. 12.8 Updated model of sense of safety and security

Physical properties	Functional properties	Gestures	Social properties
• Anthropomorphism • Embodiment • Size	• Ease of use • Autonomy • Performance	• Verbal cues • Nonverbal cues • Gaze	• Personality • Friendliness

Fig. 12.9 Robot-related factors of sense of safety and security

with gestures) may facilitate an increased sense of security and also increase the willingness to use the robot in stressful settings (Birnbaum et al., 2016). Despite the studies investigating gestures and gaze, it has not yet become apparent how the robot should behave to achieve natural communication resulting in a safe and secure relationship between a robot and an older person (Muto, Takasugi, Yamamoto, & Miyake, 2009).

Using assistive technologies in elder care might contribute to promoting a sense of safety and security. However, using these technologies can be challenging for older people. Therefore, ease of use is one of the concerns that should be addressed (Yusif, Soar, & Hafeez-Baig, 2016). Another concern is the performance of the robot, since higher performance of robots is perceived as safer (Kamide, Kawabe, Shigemi, & Arai, 2013). The autonomy of the robot is another issue that should be considered.

Humans attribute human-like features to the agents (e.g., robots) because it allows the agents to become more familiar, understandable, and predictable (Epley, Waytz, & Cacioppo, 2007). Anthropomorphism is important for helping humans to become familiar with robots. On the other hand, if a robot resembles a human greatly but still remains an unnatural copy, the person's response to this agent would shift from empathy to revulsion or fear. This emotional response is called 'uncanny valley' (Mori, 1970). Wu et al. (2012) compared different robots and their results showed that small humanoid robots with some traits between human/animal and machine were appreciated more by their older participants. Figure 12.9 summarizes the robot-related factors.

It is worth mentioning that the environment also affects the sense of safety and security, but it is not in the scope of this study. There are studies examining the effect of the environment on sense of safety and security.

12.6   Discussion and Conclusion

Modern assistive technology can contribute to improving the quality of life of the older population, and to promoting older people's sense of safety and security (Yusif et al., 2016). Social robots have the potential to be used in elder care to preserve independence and to relieve the burden of caregivers. To use social robots in homes and care facilities, understanding the underlying reasons for the acceptance or rejection of this technology is crucial. While physical safety and security have received particular attention in the robotics literature, little research has focused on examining the feeling of safety and security as a determining factor for acceptance of robots in elder care. Feeling safe and secure can increase older people's quality of life and well-being (Fonad et al., 2006). Social robots' ability to foster secure relationships with older people is an important aspect of effective HRI. We believe that the sense of safety and security is a key requirement for HRI which should be addressed in a deeper sense, considering all its aspects and implications for using robots in elder care. In this research, we attempted to identify the factors influencing the sense of safety and security in HRI, especially for older person–robot interaction. To address this, besides consulting the gerontology literature, we conducted two user studies.

Measuring the sense of safety and security is a challenging task since there is no general consensus about the definition of the term and there are many factors affecting an individuals' sense of safety and security. The factors affecting the sense of safety and security can be quite personal; every individual can perceive it differently based on their personal attributes such as background and personality. In the context of HRI, there are many factors affecting the sense of safety and security including subjective preferences as well as a robot's properties. We adopted a categorization of the components of sense of safety and security: human-related factors and robot-related factors. We provided a thorough review of the meaning of safety and security in HRI and gerontology. The review resulted in an evaluation tool and a model of sense of safety and security. The presented model is based on only human-related factors, since we did not vary robot-related factors in our user studies.

The model was the foundation for a tool that can be used for evaluating the sense of safety and security in social HRI. The tool was applied in a between-subject video-based study in which the Pepper humanoid robot interacted with a person. This study had limitations since the experience was not derived from direct interaction with a robot; however, this was a deliberate choice in order to control the experimental setup. In the online video-based study, participants watched four different scenarios showing different daily life activities that older people may engage in. The scenarios were of a proactive and on-demand nature.

The Cronbach's α value for the sense of safety and security was 0.91, which indicates that the questionnaire items had a good internal consistency.

Using the results of the online video-based study, we revised the questionnaire based on the participants' recommendations and the statistical analysis; the items that were negatively correlated with other factors were removed. We conducted a second controlled study in which older people could interact freely with Pepper. After the interaction, they filled out the revised questionnaire assessing the sense of safety and security. We further analyzed the user studies by applying EFA and CFA and updating our intuitive model.

Our experimental results from the two studies, which comprised 160 participants in total, suggest that the questionnaires assessing sense of safety, security, and the combination thereof, are suitable for use within the social HRI domain. Moreover, the model suggested can be used for evaluating the sense of safety and security in older person–robot interaction. However, the model is not complete yet. Robot-related factors require a much larger study with other robots and/or variations in functionality, allowing for gestures, functional, physical, and social properties of the robot to be evaluated and compared. Thus, a possible future direction of the study could be to implement an experimental design that reveals which characteristics of robot systems affect the perceived sense of safety and security. Such insights could be used for updating the initial model of sense of safety and security that was outlined in this chapter.

Acknowledgements This work has received funding from the European Union's Horizon 2020 research and innovation program under the Marie Skłodowska-Curie grant agreement No 721619 for the SOCRATES project.

References

Akalin, N., Kiselev, A., Kristoffersson, A., & Loutfi, A. (2017). An evaluation tool of the effect of robots in eldercare on the sense of safety and security. In *9th International Conference on Social Robotics (ICSR 2017)*, (pp. 628–637). Springer, Cham.

Bartneck, C., Kulić, D., Croft, E., & Zoghbi, S. (2009). Measurement instruments for the anthropomorphism, animacy, likeability, perceived intelligence, and perceived safety of robots. *International Journal of Social Robotics*, *1*(1), 71–81.

Bedaf, S., Gelderblom, G. J., & De Witte, L. (2015). Overview and categorization of robots supporting independent living of elderly people: What activities do they support and how far have they developed. *Assistive Technology*, *27*(2), 88–100.

Bicchi, A., Peshkin, M. A., & Colgate, J. E. (2008). Safety for physical human-robot interaction. In B. Siciliano & O. Khatib (Eds.), *Springer Handbook of Robotics* (pp. 1335–1348). Berlin, Heidelberg: Springer.

Birnbaum, G. E., Mizrahi, M., Hoffman, G., Reis, H. T., Finkel, E. J., & Sass, O. (2016). What robots can teach us about intimacy: The reassuring effects of robot responsiveness to human disclosure. *Computers in Human Behavior*, *63*, 416–423.

Boholm, M., Möller, N., & Hansson, S. O. (2016). The concepts of risk, safety, and security: Applications in everyday language. *Risk Analysis*, *36*(2), 320–338.

Boström, M., Ernsth Bravell, M., Björklund, A., & Sandberg, J. (2016). How older people perceive and experience sense of security when moving into and living in a nursing home: A case study. *European Journal of Social Work*, *20*(5), 1–14.

Boström, M., Ernsth Bravell, M., Lundgren, D., & Björklund, A. (2013). Promoting sense of security in old-age care. *Health*, *5*(6B), 56–63.

Bradley, M. M., & Lang, P. J. (1994). Measuring emotion: The self-assessment manikin and the semantic differential. *Journal of Behavior Therapy and Experimental Psychiatry*, *25*(1), 49–59.

Broadbent, E., Jayawardena, C., Kerse, N., Stafford, R. Q., & MacDonald, B. A. (2011). Human-robot interaction research to improve quality of life in elder care–an approach and issues. In *12th AAAI Conference on Human-Robot Interaction in Elder Care, held in conjunction with 25th AAAI Conference on Artificial Intelligence*.

Burns, A., McDermid, J., & Dobson, J. (1992). On the meaning of safety and security. *The Computer Journal*, *35*(1), 3–15.

Chanseau, A., Dautenhahn, K., Koay, K. L., & Salem, M. (2016). Who is in charge? Sense of control and robot anxiety in human-robot interaction. In *2016 25th IEEE International Symposium on Robot and Human Interactive Communication (RO-MAN)* (pp. 743–748). IEEE.

Charalambous, G., Fletcher, S., & Webb, P. (2016). The development of a scale to evaluate trust in industrial human-robot collaboration. *International Journal of Social Robotics*, *8*(2), 193–209.

Cortellessa, G., Scopelliti, M., Tiberio, L., Svedberg, G. K., Loutfi, A., & Pecora, F. (2008). A cross-cultural evaluation of domestic assistive robots. In *AAAI Fall Symposium: AI in Eldercare: New Solutions to Old Problems* (pp. 24–31).

Cronbach, L. J. (1951). Coefficient alpha and the internal structure of tests. *Psychometrika, 16*(3), 297–334.

Dautenhahn, K. (2007). Socially intelligent robots: Dimensions of human-robot interaction. *Philosophical Transactions of the Royal Society of London B: Biological Sciences, 362*(1480), 679–704.

Dautenhahn, K., Walters, M., Woods, S., Koay, K. L., Nehaniv, C. L., Sisbot, A., ... Siméon, T. (2006). How may I serve you?: A robot companion approaching a seated person in a helping context. In *1st ACM SIGCHI/SIGART conference on Human-robot interaction (HRI 2006)* (pp. 172–179). ACM.

De Graaf, M. M., & Allouch, S. B. (2013). Exploring influencing variables for the acceptance of social robots. *Robotics and Autonomous Systems, 61*(12), 1476–1486.

Eames, D., & Moffett, J. (1999). The integration of safety and security requirements. Computer Safety: Reliability and Security (SAFECOMP 1999), 468–480.

Epley, N., Waytz, A., & Cacioppo, J. T. (2007). On seeing human: A three-factor theory of anthropomorphism. *Psychological Review, 114*(4), 864–886.

European Commission. (2014). The 2015 ageing report: Underlying assumptions and projection methodologies. In *Joint Report prepared by the European Commission (DG ECFIN) and the Economic Policy Committee (AWG), Directorate-General for Economic and Financial Affairs*.

Fischinger, D., Einramhof, P., Papoutsakis, K., Wohlkinger, W., Mayer, P., Panek, P., ... Vincze, M. (2016). Hobbit, a care robot supporting independent living at home: First prototype and lessons learned. *Robotics and Autonomous Systems, 75*(A), 60–78.

Fonad, E., Wahlin, T.-B. R., Heikkila, K., & Emami, A. (2006). Moving to and living in a retirement home: Focusing on elderly people's sense of safety and security. *Journal of Housing for the Elderly, 20*(3), 45–60.

Haddadin, S., Albu-Schäffer, A., & Hirzinger, G. (2010). Safe physical human-robot interaction: Measurements, analysis and new insights. In M. Kaneko & Y. Nakamura (Eds.), *Robotics Research* (Vol. 66, pp. 395–407). Berlin, Heidelberg: Springer.

Heerink, M., Kröse, B., Evers, V., & Wielinga, B. (2010). Assessing acceptance of assistive social agent technology by older adults: The Almere model. *International Journal of Social Robotics, 2*(4), 361–375.

Jolliffe, I. T. (1972). Discarding variables in a principal component analysis. i: Artificial data. *Journal of the Royal Statistical Society. Series C (Applied Statistics)*, 160–173.

Kamide, H., Kawabe, K., Shigemi, S., & Arai, T. (2013). Social comparison between the self and a humanoid. In *5th International Conference on Social Robotics (ICSR 2013)*, (pp. 190–198). Springer International Publishing.

Kamide, H., Mae, Y., Kawabe, K., Shigemi, S., Hirose, M., & Arai, T. (2012). New measurement of psychological safety for humanoids. In *7th ACM/IEEE Conference on Human-Robot Interaction* (pp. 49–56). ACM.

Kennedy, J., Baxter, P., Senft, E., & Belpaeme, T. (2016). Social robot tutoring for child second language learning. In *11th ACM/IEEE International Conference on Human Robot Interaction* (pp. 231–238). IEEE Press.

Koay, K. L., Dautenhahn, K., Woods, S., & Walters, M. L. (2006). Empirical results from using a comfort level device in human-robot interaction studies. In *1st ACM SIGCHI/SIGART conference on Human-robot interaction (HRI 2006)* (pp. 194–201). ACM.

Koay, K. L., Zivkovic, Z., Krose, B., Dautenhahn, K., Walters, M. L., Otero, N., & Alissandrakis, A. (2006). Methodological issues of annotating vision sensor data using subjects' own judgement of comfort in a robot human following experiment. In *2006 15th IEEE International Symposium on Robot and Human Interactive Communication (RO-MAN)* (pp. 66–73). IEEE.

Kornecki, A. J., & Liu, M. (2013). Fault tree analysis for safety/security verification in aviation software. *Electronics, 2*(1), 41–56.

Lasota, P. A., Fong, T., & Shah, J. A. (2017). A survey of methods for safe human-robot interaction. *Foundations and Trends® in Robotics, 5*(4), 261–349.

Lasota, P. A., Rossano, G. F., & Shah, J. A. (2014). Toward safe close-proximity human-robot interaction with standard industrial robots. In *2014 IEEE International Conference on Automation Science and Engineering (CASE)* (pp. 339–344). IEEE.

Lauckner, M., Kobiela, F., & Manzey, D. (2014). 'Hey robot, please step back!'-exploration of a spatial threshold of comfort for human-mechanoid spatial interaction in a hallway scenario. In *2014 23rd IEEE International Symposium on Robot and Human Interactive Communication (RO-MAN)* (pp. 780–787).

Leite, I., Martinho, C., & Paiva, A. (2013). Social robots for long-term interaction: A survey. *International Journal of Social Robotics, 5*(2), 291–308.

Lichtenthäler, C., Lorenzy, T., & Kirsch, A. (2012). Influence of legibility on perceived safety in a virtual human-robot path crossing task. In *2012 21st IEEE International Symposium on Robot and Human Interactive Communication (RO-MAN)* (pp. 676–681). IEEE.

Maslow, A. (1943). A theory of human motivation. *Psychological Review, 50*(4), 370–396.

Maurice, P., Lavoie, M., Levaque Charron, R., Chapdelaine, A., Bélanger Bonneau, H., Svanström, L., ... Romer, C. (1998). *Safety and safety promotion: Conceptual and operational aspects* (pp. 1–20). Québec: WHO.

McNeill, D. (1992). *Hand and mind: What gestures reveal about thought.* University of Chicago Press.

Moore, J. W. (2016). What is the sense of agency and why does it matter? *Frontiers in psychology, 7.* Article ID: 1272.

Mori, M. (1970). The uncanny valley. *Energy, 7*(4), 33–35.

Muto, Y., Takasugi, S., Yamamoto, T., & Miyake, Y. (2009). Timing control of utterance and gesture in interaction between human and humanoid robot. In *2009 18th IEEE International Symposium on Robot and Human Interactive Communication (RO-MAN)* (pp. 1022–1028). IEEE.

Nonaka, S., Inoue, K., Arai, T., & Mae, Y. (2004). Evaluation of human sense of security for coexisting robots using virtual reality. 1st report: evaluation of pick and place motion of humanoid robots. In *IEEE International Conference on Robotics and Automation (ICRA 2004)* (Vol. 3, pp. 2770–2775). IEEE.

Pandey, K. A., & Gelin, R. (2018). A mass-produced sociable humanoid robot: Pepper: The first machine of its kind. *IEEE ROBOTICS & AUTOMATION MAGAZINE, 25*(3), 40–48.

Petersson, P., & Blomqvist, K. (2011). Sense of security-searching for its meaning by using stories: a participatory action research study in health and social care in sweden. *International Journal of Older People Nursing, 6*(1), 25–32.

Russell, J. (1980). A circumplex model of affect. *Journal of Personality and Social Psychology, 39*(6), 1161–1178.

Salem, M., Lakatos, G., Amirabdollahian, F., & Dautenhahn, K. (2015). Towards safe and trustworthy social robots: Ethical challenges and practical issues. In *ICSR* (pp. 584–593).

Shneiderman, B. (1992). *Designing the User Interface: Strategies for Effective Human-Computer Interaction* (2nd ed.).

Sung, J., Christensen, H. I., & Grinter, R. E. (2009). Robots in the wild: Understanding long-term use. In *4th ACM/IEEE international Conference on Human-Robot Interaction (HRI 2009)*, (pp. 45–52). ACM.

Torstensson Levander, M. (2007). *Trygghet, säkerhet, oro eller risk?* Sveriges kommuner och landsting.

Vandemeulebroucke, T., de Casterle, B. D., & Gastmans, C. (2018). How do older adults experience and perceive socially assistive robots in aged care: a systematic review of qualitative evidence. *Aging & Mental Health, 22*(2), 149–167.

Weiss, A. & Bartneck, C. (2015). Meta analysis of the usage of the Godspeed Questionnaire Series. In *2015 24th IEEE International Symposium on Robot and Human Interactive Communication (RO-MAN)* (pp. 381–388). IEEE.

Wu, Y.-H., Fassert, C., & Rigaud, A.-S. (2012). Designing robots for the elderly: Appearance issue and beyond. *Archives of Gerontology and Geriatrics, 54*(1), 121–126.

Wyrobek, K. A., Berger, E. H., der Loos, H. F. M. V., & Salisbury, J. K. (2008). Towards a personal robotics development platform: Rationale and design of an intrinsically safe personal robot. In *2008 IEEE International Conference on Robotics and Automation (ICRA)* (pp. 2165–2170).

Yusif, S., Soar, J., & Hafeez-Baig, A. (2016). Older people, assistive technologies, and the barriers to adoption: A systematic review. *International Journal of Medical Informatics, 94*, 112–116.

Zhang, T., Zhang, W., Qi, L., & Zhang, L. (2016). Falling detection of lonely elderly people based on NAO humanoid robot. In *2016 IEEE International Conference on Information and Automation (ICIA)* (pp. 31–36).

Zheng, Z., Gu, S., Lei, Y., Lu, S., Wang, W., Li, Y., & Wang, F. (2016). Safety needs mediate stressful events induced mental disorders. *Neural Plasticity, 2016*. Article ID: 8058093, 6 pages.

Chapter 13
AMIGO—A Socially Assistive Robot for Coaching Multimodal Training of Persons with Dementia

Lucas Paletta, Sandra Schüssler, Julia Zuschnegg, Josef Steiner, Sandra Pansy-Resch, Lara Lammer, Dimitrios Prodromou, Sebastian Brunsch, Gerald Lodron and Maria Fellner

Abstract In the context of assistive robotics in health care, we introduce the AMIGO system with its innovative "Coach" framework that uses social robots for the entertaining motivation of persons with dementia. The overarching objective is to empower persons with dementia to perform daily stimulating training activities within the concept of an integrated multimodal intervention. The "Coach" frame is complemented by a "Companion" frame that involves the client in a long-term relationship with the robot which will care by asking about the client's health status, remind about

L. Paletta (✉) · G. Lodron · M. Fellner
JOANNEUM RESEARCH Forschungsgesellschaft mbH, Graz, Austria
e-mail: lucas.paletta@joanneum.at

G. Lodron
e-mail: gerald.lodron@joanneum.at

M. Fellner
e-mail: maria.fellner@joanneum.at

S. Schüssler · J. Zuschnegg
Medical University of Graz, Graz, Austria
e-mail: sandra.schuessler@medunigraz.at

J. Zuschnegg
e-mail: julia.zuschnegg@medunigraz.at

J. Steiner · S. Pansy-Resch
Sozialverein Deutschlandsberg, Deutschlandsberg, Austria
e-mail: josef_steiner@gmx.at

S. Pansy-Resch
e-mail: s.pansy-resch@sozialverein-deutschlandsberg.at

L. Lammer · D. Prodromou · S. Brunsch
Humanizing Technologies GmbH, Vienna, Austria
e-mail: lara.lammer@humanizing.com

D. Prodromou
e-mail: dimitrios.prodromou@humanizing.com

S. Brunsch
e-mail: sebastian.brunsch@humanizing.com

© Springer Nature Switzerland AG 2019
O. Korn (ed.), *Social Robots: Technological, Societal and Ethical Aspects of Human-Robot Interaction*, Human–Computer Interaction Series,
https://doi.org/10.1007/978-3-030-17107-0_13

important events or tasks, involve the client in dialog, invite the client to engage in multimodal training, and provide entertainment such as reading the news from all over the world. A research objective is to adjust Pepper's dialog and motivation style based on emotional feedback sensed in interaction. The system will motivate the client to perform personalized exercises and to maintain and extend social bonds and will stimulate cognitive processes and physical activities. Sensors for eye tracking and motion analysis technologies will offer affordances for entertaining, sensorimotor sequences and for data capture and analysis of cognition and locomotion-specific behavioral parameters. Easily usable interfaces enable planning and autonomous daily practice to formal as well as to the informal caregiver in a weekly rhythm so that people with dementia can stay at home longer and the progress of dementia is slowed down. The AMIGO system is motivated from the viewpoint of health care, neuropsychology, and ICT systems. The first implementation of the prototype system and first results of a mixed-method study are presented in detail.

Keywords Dementia · Home care · Socially assistive robot (SAR) · Motivation · Cognitive training · Physical training

13.1 Introduction

Worldwide, dementia rates are increasing and consequently burden global healthcare resources to a serious degree (ADI, 2018; Alzheimer's Association, 2018). In contrast, there are a decreasing number of available caregivers to provide (nursing) care (Robert Koch Institute, 2015; Rösler, Schmidt, Merda, & Melzer, 2018; Rosseter, 2017).

(Nursing) care of people with dementia usually takes place at home, especially in the early stage of the disease (OECD, 2015). Due to the progression of dementia and the growing (nursing) care needs, because of care dependency (e.g., in mobility, learning ability) and (nursing) care problems (e.g., falls, malnutrition), the necessity of professional care and possibly a nursing home transition becomes increasingly likely (ADI, 2013; Braunseis, Deutsch, Frese, & Sandholzer, 2012; OECD, 2015). One of the most important aims in the (nursing) care of people with dementia is to promote their independence, taking into account their current stage of dementia and their individual abilities. Such (nursing) care can counteract an excessively rapid increase in care dependency (Schüssler, 2015). It is in this context that new technologies, such as socially assistive robots, may constitute a supportive tool for caregivers because they have the potential to promote the independence and well-being of older people (Smarr et al., 2012; WHO, 2007).

The market in robots for the assistance of older care-dependent persons is seeing strong worldwide growth and will continue to increase (IFR, 2018). Robots are increasingly applied, with growing acceptance in the health care of older people (IFR, 2018; Spero, 2017). Socially assistive robotics (SAR) can be defined as representing an intersection of assistive robots and socially interactive robots (Feil-Seifer

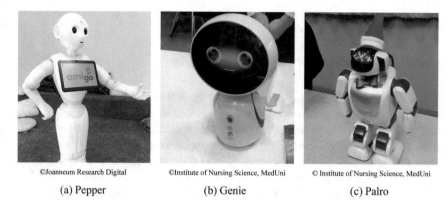

(a) Pepper   (b) Genie   (c) Palro

Fig. 13.1 Representative examples of socially assistive robots with a potential for use in long-term health care: **a** Pepper from SoftBank, **b** Genie from Service Robotics Ltd. (https://serviceroboticsltd. co.uk/genie/), **c** Palro from Fujisoft (https://palro.jp/en/case)

& Matari, 2005). The goal of SAR is to create a close and effective interaction with a human user for the purpose of giving assistance, for instance in daily activities ranging from cognitive to physical tasks, or for encouraging emotional expression, conversation, and gestures (Feil-Seifer & Matari, 2005). Figure 13.1 shows representative examples of SAR for health care.

Most of the socially assistive robots have so far been tested only in laboratory settings. Consequently, there is a scarcity of knowledge about their use in real care situations from the point of view of older care-dependent people, especially regarding their use by persons with dementia, and particularly considering the application in domestic environments. Concretely, the testing of robots in laboratory settings cannot capture the complexity and high variability of everyday real-life situations in nursing care (Bioethics Commission, 2018).

The Austrian project AMIGO is working toward refining an innovative technology on the basis of an already established socially assistive robot platform (Softbank Robotics, 2015) with the goal of developing an **emotional and motivating coach and companion**. Motivation is a key human factor in any activity involving behavioral change or learning actions and is a particular challenge for those affected by dementia (Forstmeier & Maercker, 2015). AMIGO is developing the design of a socially assistive robot in order to enable it to analyze the motivation of persons with dementia. This, in turn, provides additional entertainment which, together with personalized communication and interaction, recommendations and playful training, promotes the empowerment, activation and autonomy of elderly people with dementia.

13.2 Related Work

13.2.1 Motivation and Dementia

A deficit in motivation, some degree apathy, as well as depression are among the most frequent non-cognitive symptoms in the case of mild cognitive impairment (MCI) as well as of Morbus Alzheimer (Robert, Mulin, Malle, & David, 2010; Starkstein, Jorge, & Mizrahi, 2006). Measurements from research studies demonstrate that persons with the Alzheimer type of dementia have lower capacities for motivation processes (Forstmeier & Maercker, 2015).

Current models of motivation identify and discriminate two phases, (1) goal-setting and (2) goal pursuit. The latter requires self-regulatory capacities for decision-making, regulation of activation, and regulation of motivation. The capacity for impulse control and goal pursuit can be correlated with a reduced risk for Alzheimer's (Wilson et al., 2007). Forstmeier and Maercker (2015) conclude from their research that cognitive and physical training should be complemented by motivation-supporting training strategies such as goal-setting and self-motivation (Forstmeier & Maercker, 2011). In addition, motivation-oriented interventions support the reduction of neuropsychiatric symptoms such as apathy and depression.

AMIGO aims at the implementation of motivation-oriented strategies into an overall technological coach concept, and from this will contribute innovatively to AAL technologies.

13.2.2 Socially Assistive Robots for Dementia in Health Care

Japan recognized the challenges of demographic change and the shortage of nurses as early as the 1980s (Wallenfels, 2016). This led to the assumption that robotic technologies might be able to support caregivers (Wallenfels, 2016) and people with dementia (Sugihara, Fujinami, Phaal, & Ikawa, 2013). Research into socially assistive robots is a relatively young field (Bekey et al., 2006), and many robots are inapplicable in current nursing practice because most of them are still in development and have been tested only in laboratory settings or within institutional health care, like nursing homes. Many studies in the field of robotics generally focus on elderly people without dementia. These studies show that people want robots that help them promote their independence in order to avoid a premature increase of care dependency (Bedarf, Draper, Gelderblom, Sorell, & De Witte, 2016).

Particular risk factors for care dependency are problems with mobility, self-care, and social interaction. Such activities are, however, a prerequisite for an independent life at home (Bedaf et al., 2013). These risk factors also apply to persons with dementia, irrespective of their cognitive deficits. In comparison with persons with other chronic conditions, persons with dementia develop a risk for care dependency earlier in their course of illness (ADI, 2013). This demonstrates the need for early interventions with people with dementia to promote independence. According to Bedaf et al. (2013), future robots need to be able to provide support in various activities of daily living (ADL) and not just with regard to one specific activity only, because people are often care-dependent in several activities of daily living at the same time. A key issue for older people is the promotion of mobility and cognition (Montero-Odasso et al., 2015). The first results of a study by Mann, McDonald, Kuo, Li, and Broadbent (2015) show that older people (without dementia) prefer healthcare instructions given by a robot to instructions provided by means of a tablet PC alone. The participants showed, for instance, an increase in communication and positive emotions. The robot was also perceived as more trustworthy and as a greater fun factor. This motivated the participants to engage in further interactions with the robot. Another study showed the use of a tablet PC in people with severe dementia. This intervention also revealed positive results with respect to, for example, communication, motivation, and behavior (Nordheim, Hamm, Kuhlmey, & Suhr, 2015). However, there was no comparison with a robot, and it is therefore unclear whether a robot might not have produced better results. Bedarf, Gelderblom, and de Witte (2015) identified 107 potential socially assistive robots for the care of older people in their literature review, with six in the concept phase, 95 in the development phase, and only six robots being already commercially available. This illustrates the imperative necessity to test robots that are already commercially available, like Pepper, in nursing practice. In general, the literature shows that older people are indeed open to using robots that help them promote their independence, but that such robots are still in need of development (Bedarf et al., 2016). A systematic review by Ienca et al. (2017) focusing on intelligent assistive technologies (including socially assistive robots) for people with dementia identified only 17 out of 539 studies that included socially assistive robots. The authors stated that it was of critical importance to urgently test the clinical effectiveness of available technologies and that it would be necessary to increase the development of technologies focusing on emotional support, besides the standard cognitive and physical assistance.

A literature review by Mordoch, Osterreicher, Guse, Roger, and Thompson (2013) aimed to identify socially assistive robots for use with persons with dementia and to describe their effectiveness. Most of the research was conducted on Paro (a robotic seal baby), AIBO (a robotic dog) and NeCoRo (a robotic cat). The results show that social robots are potentially useful for therapeutic interventions, but that much development work is still needed to obtain evidence-based knowledge with regard to this specific target group. It was also emphasized that interdisciplinary collaboration with various disciplines, including researchers from the field of robotics, was needed to improve the quality of care for the target group of persons with dementia.

A scoping review by Adbi, Al-Hindawi, Ng, and Vizcaychipi (2018) has also shown that the robot Paro (the robotic seal baby) is tested most often with people with dementia. Only one of the 61 included studies tested a social robot in the home care setting. All other studies were performed mainly in nursing homes or day care centers. Buhtz et al. (2018) stated in their scoping review that future studies testing the effectiveness of robotic systems need to especially focus on the domestic setting.

In Austria, some research efforts have been made to develop socially assistive robots for use in nursing practice. One example is the early prototype Henry, which was tested in a pilot project in a nursing home in Vienna. Henry provides information, entertainment, and security information to residents. He needs a lot of further development work (Wallenfels, 2016). Another project in Austria was focused on the socially assistive robot Hobbit. This was also an early prototype, tested on seven people without dementia living at home. The authors concluded that robots like Hobbit have the potential to support older people living at home, but that a lot of further development work has to be done before the robots reach market readiness (Pripfl et al., 2016). In general, it is highly recommended and necessary to include persons with dementia and cognitive impairment in the design iteration cycles (Boman, Lundberg, Starkhammar, & Nygård, 2014; Mao, Chang, Yao, Chen, & Huang, 2015; Span, Hettinga, Vernooij-Dassen, Eefsting, & Smits, 2013; Wu et al., 2014) because their feedback, requirements, and recommendations are very relevant for the appropriate and user-friendly development of novel technologies (Boman et al., 2014). Furthermore, people with dementia are indeed able to learn to make use of robot technologies (Lauriks et al., 2007; Wu et al., 2014).

Currently, several EU-funded research projects have been investigating the benefit of applying SAR. RADIO (H2020, Robots in assisted living environments: Unobtrusive, efficient, reliable and modular solutions for independent aging, 2015–2018) is developing an integrated "smart home" assistance robot system including an unobtrusive monitoring system. GrowMeUp (H2020, 2015–2018) is developing a low-cost service robot that learns specific requirements and habits from elderly people and aims at compensating for the impaired capacities of elderly people. MARIO (H2020, Managing active and healthy aging with use of caring service robots; 2015–2018) is addressing isolation, loneliness, and dementia by means of an innovative diversity of functionalities. ENRICHME (H2020, Enabling Robot and assisted living environment for Independent Care and Health Monitoring of the Elderly, 2015–2018) is developing a mobile service robot for long-term monitoring and interaction with elderly people suffering from a progressive decrease in cognitive capacities. VictoryAtHome (AAL-JP5, 2013–2016) has developed an unobtrusive robot for smart homes. Hobbit (FP7, HOBBIT, The Mutual Care Robot, 2011–2015) specified interaction between robots and end users for mutual care optimization.

In numerous projects, the needs of people with dementia (in terms of technologies) have so far mostly been considered from the caregiver's perspective. However, there are isolated studies showing that people with dementia or cognitive impairment want technologies that provide cognitive support, promote communication and social interaction, provide security, and promote ADL (e.g., mobility) (Lauriks et al., 2007; Pino, Boulay, Jouen, & Rigaud, 2015; Wang, Sudhama, Begum, Huq, & Mihailidis, 2016). In order to develop future socially assistive robots for the individual needs of persons with dementia, it is of vital importance to take into account their personal views, in addition to the views of others (e.g., relatives, nursing staff).

13.3 The Coaching System AMIGO

13.3.1 Robot-Based Coaching for Playful Training

Daily units of training in the modalities of cognition and motion are pivotal to sufficiently stimulating the mental processes of persons with dementia in the frame of various decision-making situations, and through this to enable a long-term effect on life at home with functioning self-care and autonomy.

Training units in terms of preparations for everyday challenges can only be effective if they are performed as motivated, repeated exercises with relevance to everyday life and situations.

The focus of the "Coach" approach of AMIGO is on the motivation performance and implemented by means of an assisting social robot for the appropriate engagement of persons with dementia, with the purpose of improving and intensifying the playful multimodal training using the theratainment app for the targeted stimulation with satisfactory frequency. Figure 13.2 demonstrates relevant aspects of the overall AMIGO system concept including the "Coach" framework and its application. Figure 13.2a schematically depicts the full AMIGO system, including (in pink color) the "Coach"-based interactions between Pepper and the tablet-based training (pink), while the "Companion" refers to further interactions that support social bonding via various information and entertainment services. Figure 13.2b demonstrates a typical scenario in which the client is using the multimodal training suite while Pepper coaches with motivating comments. Figure 13.2c refers to the recommendation provided by persons with dementia to use the tablet-based training while sitting on a chair near the robot. Figure 13.2d depicts a home-based scenario from the first field trials with Pepper motivating the person with dementia to interact (Companion mode).

13.3.2 Playful Multimodal Training

Recently, serious games have been developed particularly in the context of "dementia-oriented serious games" that focus on cognitive and physical aspects of intervention (McCallum & Boletsis, 2013) but less on social or emotional consequences.

Selected sample games demonstrate the diversity of the games. Lumosity games were applied in several neuroscientific research projects for the application of health care in dementia (Finn & McDonald, 2011). Cogmed offers various exercises for the improvement of working memory (auditive and visual) which is substantially relevant for learning and cognitive performance; however, the impact of Cogmed is rather disputed (Shipstead, Hicks, & Engle, 2012). Cognifit offered as a fitness training for the brain has been studied and found to provide substantial progress (Korczyn, Peretz, & Aharonson, 2007; Shatil, Metzer, Horvitz, & Miller, 2010).

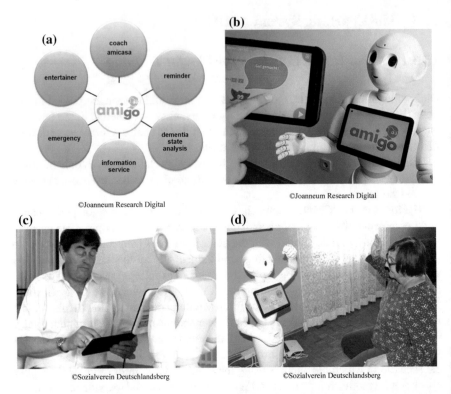

Fig. 13.2 Overall AMIGO system concept including the "Coach" framework and its application.
a The "Coach" refers to interactions between Pepper and the tablet-based training (pink) while the
"Companion" refers to further interactions that support social bonding via various information and
entertainment services. **b** While using the multimodal training suite, Pepper coaches with motivating
comments. **c** It is recommended to use the tablet-based training while sitting on a chair near the
robot. **d** Home-based scenario from the first field trials with Pepper motivating the person with
dementia to interact (Companion mode)

EU-funded projects, such as the AAL-JP project M3W, have developed seri-
ous games for dementia sufferers to measure and support their mental well-being.
Other relevant AAL-JP projects are Aladdin for the application in smart homes, and
Safemove for the improvement of mobility in the environment near the home.

13.3.3 Theratainment App

The theratainment app amicasa was developed in the Austrian national project Aktiv-
Daheim[1] in the style of a serious game (Fig. 13.3) that playfully supports a multi-

---

[1] www.aktivdaheim.at.

(a) **(b)**

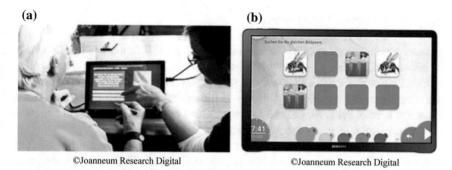

©Joanneum Research Digital ©Joanneum Research Digital

Fig. 13.3 **a** Playful multimodal training using a tablet PC with an eye-tracking device for executive functions diagnostics. **b** Graphical User Interface of the theratainment app for multimodal playful training

modal training scheme, with the purpose of supporting people living at home or at care centers. It includes cognitive and physical training, and can be extended with additional modalities such as odor and music perception. Studies demonstrate that physical training slows down the progress of cognitive impairment and can extend cognitive reserves (Christofoletti et al., 2008).

Sensor data are collected via a tablet PC that is equipped with a recommender engine that enables the user to perform personalized training units as a playful experience. The game with its sensor and diagnostic toolbox enables an innovative potential for entertainment, and the application of measurement and analysis methodologies based on behavior parameters (Paletta et al., 2018a).

13.3.4 Measures for the Analysis of Motivation for Stimulating Training

The results from measured outcomes of weekly sessions of training, as well as data received from trained caregivers, questionnaires, attention, emotion and motion features are collected for various objectives in data analysis. One relevant purpose is to focus multimodal therapeutic measures for optimization of the healthcare approach. A further purpose is appropriate adaptation of the training strategies. For the adjustment of weekly playful training programs, comprehensive medical, psychological, and care-oriented measures are usually considered.

Cognitive performance is analyzed for the estimation of dementia diagnostic features. Emotional aspects are investigated for motivational analysis and related to cognitive performance in order to analyze and adjust communication as well as training strategies.

13.3.4.1 Cognitive Performance

Eye movement features can be successfully used to provide indicators for Alzheimer diagnostics, considering indicative field studies that showed the discriminative power to classify into dementia and non-dementia participants exclusively from gaze data. The theratainment app-based approach to the evaluation of eye movement-based cognitive control analysis, i.e., the antisaccade test procedure (Paletta et al., 2018b), is characterized by a pervasive measurement paradigm.

The person with dementia is able and motivated to perform the training and serious game units at home, i.e., not in a laboratory environment. Consequently, the input data have to be filtered in order to gain the maximum quality for further processing and evaluation. Various eye movement features are extracted from the data. Areas of interest in the display are designed with respect to prosaccade and antisaccade behavior. Errors are determined from the violation of the antisaccade condition, i.e., turning attention to the opposite site of the visual stimulus (Paletta et al., 2018a).

In the test context, these features have demonstrated certain discriminative levels that would enable an indicative classification of dementia. These estimates may serve to initiate early warnings for professional support and intensification of health care in the case of an exceptional decrease of cognitive performance. Further evidence for cognitive performance will be collected from videogames played in the theratainment app.

13.3.4.2 Measurement of Emotion

Emotion plays a central role in the parametrization of human behavior, such as in the interplay between emotion and cognitive control via the dual competition framework (Pessoa, 2009). Emotion and motivation have an amplifying or inhibiting effect on the performance of human behavior. Mental health technologies have recently increasingly emerged from innovations in novel affordances in human-computer interaction as well as in the measurement technologies (Gregg & Tarrier, 2007).

Emotion plays a major role in process- or outcome-focused motivation (Toure-Tillery & Fishbach, 2014) which is an essential parameter in healthcare-oriented training processes. Emotion-oriented care can be more effective than standard care with regard to positive emotion in nursing home residents with mild to moderate dementia (Finnema, Dröes, Ettema, & van Tilburg, 2005).

Within the scope of AMIGO, we will investigate several strategies in terms of queried and manual as well as unobtrusive and unaware provision of emotion-relevant information. A principal service function provided by Pepper is to judge the user's emotion by means of sensors. Microphones allow Pepper to analyze the lexical field and the tone of voice to judge the emotional context (Softbank Robotics, 2015). In addition, Pepper processes images with shape recognition software to identify objects, individual faces, and their emotional states of the faces around him.

The validity of the emotion assessment by Pepper will be researched in detail. Furthermore, Pepper enables the selection of emoticons on its tablet PC in terms of an interaction request to the user. One study (Alberts, Vastenburg, & Desmet, 2013) demonstrated that Pick-A-Mood as an interactive query to particularly elderly users is significantly preferred to an emoticon-based display. A further investigation will be applied to facial emotion recognition from video input of the Pepper head camera, using state-of-the-art software (OpenFace, Baltrusaitis, Robinson, & Morency, 2016; AFFDEX, McDuff, El Kaliouby, Cohn, & Picard, 2015).

13.3.4.3 Measurement of Activation

The activation and engagement of the user during physical training activities will provide a further cue to determine the current motivational status. Since it is not possible to attach additional sensors to Pepper, its own sensors, such as the video cameras and the 3D sensor attached to the head will be used to extract cues about the activity status of the person with dementia. A first investigation of the feasibility of these information sources for activity recognition is described in Sect. 13.4.2.

13.3.4.4 Motivation Analysis and Training Adjustment

A notable increase in the efficiency of training is expected by means of an increase in the interaction frequency with the training app, as well as an increase in the quality of life.

Motivation analysis will be on the one hand measured by activation during physical training units. It will be applied by relating the average activation— i.e., described by the energy of the motion— of the person with dementia to the current measurement of activation. On the other hand, emotion analysis (Sect. 13.3.4.2) will demonstrate the pleasure of the user at the end of training units. Another input will be the measurement of performance in the playful training, such as the number of attempts needed to finish a puzzle in a memory game. The estimate of motivation will finally be used to adjust the current training strategy, considering that training that is too difficult will demotivate the user. Furthermore, animation procedures performed by Pepper can be adjusted to increase the interest of the person with dementia in continuing with training for further stimulation and with positive affect.

AMIGO expects even more significant results than when using only the tablet PC, since the social robot is supposed to engage even more. Finally, improvements in lifestyle, engagement, and motivation should be reflected in the overall diagnostic measurements.

13.4 First Results of the AMIGO Study

We are performing a mixed-method study in three main phases with the aim of refining and testing the social robot Pepper for use by people with dementia living at home. The following section describes the initial results of the first (qualitative study) and second (first field trial) phases.

13.4.1 Qualitative Study

The aim of the qualitative study was to explore the attitude, knowledge, and needs of potential users of socially assistive robots, like Pepper, in the field of dementia.

We conducted 23 individual interviews with people with dementia and 12 focus group interviews including 57 relatives, caregivers, dementia trainers, and (care) managers. The participants were from home care, nursing homes, hospitals, and day care centers. The interviews were held by professionals with experience in conducting interviews (clinical health psychologist, psychotherapist, nursing scientist, project assistant) in a quiet room of the different mentioned settings or in a quiet room of a social healthcare non-profit organization in the community. An interview guide had been developed based on current literature and expert knowledge, including the following questions:

- What kind of feelings does Pepper, as well as other robots, evoke in you?
- What (general) experiences have you had with robots?
- What ethical concerns do you have regarding the use of a robot like Pepper?
- I will show you 13 pages with sample pictures of different ADL. I would like to hear from you how a robot, like Pepper, might help you in your ADL/or your relative/client with dementia in their ADL. In our study, we used the following ADL-categories by Dijkstra (2017): eating and drinking, learning ability, mobility, body posture, daily activities, getting un(dressed), body temperature, hygiene, avoiding of danger, communication, contact with others, sense of rules and values, day/night pattern, continence, recreational activities.
- What kind of support would you or your relative/client with dementia need by a robot like Pepper?
- Imagine you or your relative/client with dementia is offered to test the robot Pepper for free? What do you think about it?
- Did I properly summarize the interview? Did I forget something?
- Would you like to tell me something else that is important to you, which has not been mentioned in our interview yet?

All interviews were recorded, transcribed, and analyzed by a nursing scientist using qualitative content analysis.

The results show that, with regard to attitudes, eight out of the 23 interviewees with dementia would not want to test the robot Pepper because they have a negative perception about robots in general. All other people with dementia were either positive toward or unsure about testing a robot.

> Yes, positive. That I might get to know him (laughs), yes. (Person with dementia)
> I really do not know (Person with dementia). (Person with dementia)

The interviewed relatives, professional caregivers, dementia trainers, and (care) managers had a mainly positive attitude toward testing Pepper. Often, feelings such as interest, curiosity, and fascination were described.

Yes I do, because I'm technically more like that, that I like such things and that I am interested in them. (…) so, I am very open towards novelties. (Relative)

In general, only a few of the people with dementia had knowledge of or experience with any kind of robots. Among the other participants, the (care) managers had the most knowledge about or experience with different types of robots. The best-known examples among all participants were robotic vacuum cleaners, robotic lawn mowers, humanoid robots (but mainly from media), industrial robots, and communication bots like Alexa.

So, personally, I have an Alexa at home. I have a cleaning robot … vacuums and mop … a lawnmower robot (laughs). (Nursing care manager)

With regard to the needs aspect, all participants perceived the support possibilities of a robot mainly in the areas of *avoiding danger* (e.g., falls, summoning help), *communication and contacts with others* (e.g., support with calls, use of voice commands), *daily activities* (e.g., regular reminders about appointments and household activities), *eating and drinking* (e.g., shopping lists, orders, reminders), *mobility and body posture* (e.g., motivation and instructions for activity training) as well as *recreational activities (e.g., music, dance, games, telling stories)* and *learning abilities* (e.g., reminders and support of cognitive training).

Below is a statement of a nurse with regard to the topic "*avoiding danger*," which was the most discussed one during the interviews.

…That would be best, if the robot could understand, ok, there's a patient lying on the ground, and he, for example, could initiate an emergency call. (nurse)

In view of these results, the functions of the robot Pepper were adapted for the field trials as shown in Fig. 13.4. For the first field trial, only the SOS call function and the calendar function were not implemented. Both functions will be scheduled for a second field trial.

13.4.2 First Field Trial

This first field trial is a follow-up study on the results of the qualitative study described in Sect. 13.4.1. The refined functions based on the qualitative results are shown in Fig. 13.4. For the first field trial, only the SOS call function and the calendar function had not been implemented. Both functions are scheduled for a second field trial.

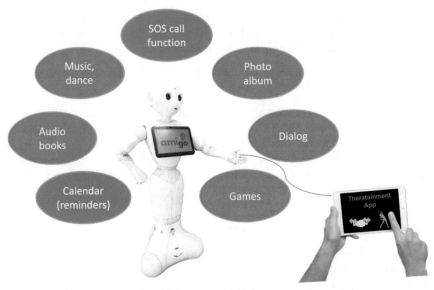

Fig. 13.4 Functions of Pepper for assisted health care and entertainment, adapted on the basis of results of the qualitative study presented in Sect. 13.4.1

13.4.2.1 Methods

The first field trial was a prototype test of the refined robot Pepper with the aim of exploring the usability and acceptance of Pepper by users in the home care setting.

In total, three people with dementia, three relatives, three dementia trainers, and three professional caregivers were included in the study. Data on usability and acceptance were collected using the Technology Usage Inventory by Kothgassner et al. (2012), as well as by means of observation and qualitative interviews. Movement data were collected by means of Pepper's eye camera. Ethical approval from the Medical University of Graz, Austria, was obtained and written informed consent was given by the people with dementia or their legal representative.

The test period was one week (Monday–Friday) per household. On Monday, Pepper was set up by a trained project member, and the people with dementia as well as their relatives were shown how to use the robot. Furthermore, data about the basic characteristics of the sample were collected by means of questionnaires during this day. On Tuesday and Thursday, a dementia trainer visited the participants to show how to use the robot. This visit included a one-hour training session for the person with dementia using the theratainment app for cognitive and physical training.

With the onset of the physical exercises, short videos were automatically captured by Pepper's head camera, which were analyzed with respect to the presence and activation of clients.

On Wednesday, a nurse visited the person with dementia for one hour and observed them in their use of the robot. Apart from these appointments, the people with dementia and their relatives were free to use the robot with all his functions as often as they liked. During the test week, a hotline was installed to enable the people with dementia or their relatives to contact project members if they had questions or in case of (technical) problems. On Friday, Pepper was removed, and interviews and questionnaires were conducted by a trained project member.

13.4.2.2 Results of the Multisensor Engagement Analysis

In order to measure the engagement, and from this further conclude about the current motivational status of the user, Pepper's own sensor results were interpreted for cues about the activity status of the person with dementia when they were, for example, engaged with a physical training unit. In a first feasibility study, the video stream of Pepper's top camera was investigated. This camera provides an output stream of 640 × 480 pixels at 30 fps or 2560 × 1920 pixels 1 fps. The video stream was analyzed for signs of human presence and motion. For this purpose, a state-of-the-art single frame computer vision analysis framework (Cao, Simon, Wei, & Sheikh, 2017) was applied in order to extract a skeleton representation including nodes and geometrical relations between the human extremities.

The first results gained with this methodology (Figs. 13.5 and 13.6) were very promising: from the tracking of nodes between video frames we expect to robustly extract motion features that will enable us to determine the energy and impulse of the user during physical training, and from this to support conclusions about the motivational status. Based on the measured user engagement, AMIGO's coaching approach would empower users and motivate them to engage in more intensive and beneficial training units.

Figure 13.6 provides particularly interesting results concerning the visibility of semantic nodes in video-based person detection. From these results, we conclude that the upper part of the body was very reliably detected. Central and lower parts of the body were often occluded or not visible due to the use of a chair to stabilize the stand of the client. However, the very high detection rate from videos grabbed directly in the home of the clients demonstrates that the methodology is very robust.

13.4.2.3 Results of the Usability and Acceptance Analysis

Initial results from the qualitative data show that most of the participants had a positive attitude toward continuing to use Pepper as additional support at home. The participants described positive feelings such as curiosity, interest, and surprise during the use of Pepper.

Benefits of support by Pepper during the test week were primarily recognized for the areas of *communication* (e.g., people with dementia enjoyed communicating with Pepper; they liked that Pepper himself started a communication with them and

(a)   **(b)**

©Joanneum Research Digital   ©Joanneum Research Digital

Fig. 13.5 Automated video-based analysis of activation of users. Images captured from the video of Pepper's head camera oriented in attention mode toward the user. The images are overlaid by annotations resulting from the automated person and human body component detections

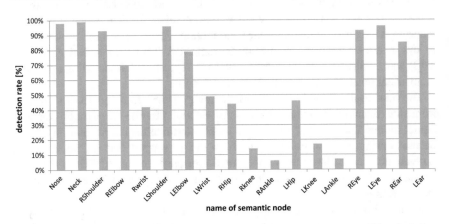

Fig. 13.6 Visibility of semantic nodes in video-based person detection. From these results, we conclude that the upper part of the body was very reliably detected. Central and lower parts of the body were often occluded or not visible due to the use of a chair to stabilize the stand of the client

they felt motivated to ask Pepper many questions), *social contacts* (e.g., people with dementia became more lively during the week that Pepper was in their household; Pepper promoted social contact with grandchildren; people with dementia took care of Pepper), *recreational activities* (e.g., people with dementia showed a strong interest in the music functions and the dance performance of Pepper), *learning ability* (e.g., people with dementia liked the cognitive training supported and motivated by Pepper; they had a feeling of learning new things) and *mobility* (e.g., the dance performance of Pepper motivated people with dementia to dance with him; apart from the special physical training included in the theratainment app, this was a motivation to be more physically active).

Regarding technical problems, the participants described connection problems between Pepper and the external tablet PC, touch screen problems of the external tablet PC (the touch screen was not sensitive enough), speech/communication problems with Pepper (e.g., Pepper had problems understanding quiet voices; if too many people were around Pepper, he had problems focusing on the right person). Some feature sequences were perceived as too fast (e.g., the change of pictures in the photo album, or physical exercises in the videos of the theratainment app). It was also reported that the functions of Pepper still need to be expanded (e.g., dialog options) and that the procedure to start Pepper and his functions should be easier to use for people with dementia.

13.5 Future Work

On the basis of the results of the qualitative study and the first field trial, Pepper will now be further refined. Notably, a calendar will be developed and implemented in Pepper to enable individual reminders (e.g., to take medication, household activities) for people with dementia. An SOS call function will be implemented as well. Additional music files and audiobooks will be included (taking into account the individual wishes of the participants). The image feature for the individual photo albums will be expanded. More dialog options as well as more motions and gestures will be developed for Pepper so that he can better communicate with the participants and is able to better support cognitive and physical training with the theratainment app.

The refined robot will then be further tested in a randomized controlled trial (second field trial) in a home environment. This study will start in May 2019 and will aim to examine the effect of the robot Pepper, combined with the theratainment app, on motivation (primary outcome) in persons with dementia compared to training with only the theratainment app. Secondary outcomes will include usability/acceptance, quality of life, cognition, mobility, care dependency, depression, behavioral problems, and caregiver burden.

Acknowledgements The research leading to these results has received funding from the Austrian BMVIT/FFG (No. 862051) by project AMIGO and project PLAYTIME of the AAL Programme of the European Union, by the Austrian BMVIT/FFG (No. 857334).

References

Adbi, J., Al-Hindawi, A., Ng, T., & Vizcaychipi, M. P. (2018). Scoping review on the use of socially assistive robot technology in elderly care. *BMJ Open, 12;8*(2), e018815. https://doi.org/10.1136/bmjopen-2017-018815.

ADI. (2013). *World Alzheimer report 2013: Journey of caring: An analysis of long-term care for dementia*. London, UK: ADI.

ADI (World Alzheimer Report). (2018). *The state of the art of dementia research: New frontiers.* London, UK: Alzheimer's Disease International.

Alberts, J. W., Vastenburg, M. H., & Desmet, P. M. A. (2013). Mood expression by seniors in digital communication; evaluative comparison of four mood-reporting instruments with elderly users. In *Proceedings of 5th International Congress of International Association of Societies of Design Research, Tokyo, Japan, 26–30 Aug 2013.*

Alzheimer's Association. (2018). Alzheimer's disease facts and figures. *Alzheimer's Dement, 14*(3), 367–429.

Baltrusaitis, T., Robinson, P., Morency, L. (2016). OpenFace: An open source facial behavior analysis tool. In *Proceedings of Workshop on Applications of Computer Vision (WACV).* IEEE.

Bedaf, S., Gelderblom, G. J., Syrdal, D. S., Lehmann, H., Michel, H., Hewson, D., ... de Witte L. (2013). Which activities threaten independent living of elderly when becoming problematic: Inspiration for meaningful service robot functionality. *Disability and Rehabilitation Assistive Technology, 9*(6), 445–452.

Bedarf, S., Draper, H., Gelderblom, G. J., Sorell, T., & De Witte, L. (2016). Can a service robot which supports independent living of older people disobey a command? The views of older people, informal carers and professional caregivers on the acceptability of robots. *International Journal of Social Robotics, 8*(3), 409–420.

Bedarf, S., Gelderblom, G. J., & De Witte, L. (2015). Overview and categorization of robots supporting independent living of elderly people: What activities do they support and how far have they developed. *Assistive Technology, 27*(2), 88–100.

Bekey, G., Ambrose, R., Kumar, V., Sanderson, A., Wilcox, B., & Zheng, Y. (2006). *International assessment of research and development in robotics.* Baltimore, Maryland: World Technology Evaluation Center (WTEC).

Bioethics Commission. (2018). Roboter in der Betreuung alter Menschen - Stellungnahme der Bioethikkommission [Robots in the care of older people—statement of the Bioethics Commission]. Wien: Geschäftsstelle der Bioethikkommission.

Boman, I. L., Lundberg, S., Starkhammar, S., & Nygård, L. (2014). Exploring the usability of a videophone mock-up for persons with dementia and their significant others. *BMC Geriatrics, 14,* 49.

Braunseis, F., Deutsch, T., Frese, T., & Sandholzer, H. (2012). The risk for nursing home admission did not change in ten years-a prospective cohort study with five-year follow-up. *Archives of Gerontology and Geriatrics, 54,* e63–e67.

Buhtz, C., Paulicke, D., Hirt, J., Schwarz, K., Stoevesandt, D., Meyer, G., & Jahn, P. (2018). Robotic systems for care at home: A scoping review (article in German). Z. Evid. Fortbild. Qual. Gesundh. wesen. ISSN 1865-9217.

Cao, Z., Simon, T., Wei, S.-E., Sheikh, Y. (2017). Realtime Multi-person 2D pose estimation using part affinity fields. In: *Proceedings of CVPR 2017.*

Christofoletti, G., Oliani, M. M., Gobbi, S., Stella, F., Bucken Gobbi, L. T., & Renato Canineu, P. (2008). A controlled clinical trial on the effects of motor intervention on balance and cognition in institutionalized elderly patients with dementia. *Clinical Rehabilitation, 22*(7), 618–626.

Dijkstra, A. (2017). Care dependency. In S. Schüssler & C. Lohrmann (Eds.), *Dementia in nursing homes.* Basel, Switzerland: Springer.

Feil-Seifer, D., & Matari, M. J. (2005). Defining socially assistive robotic. In *Proceedings of the 2005 IEEE 9th International Conference on Rehabilitation Robotics, ICORR, June 28–July 1, Chicago, USA* (pp. 465–468). https://doi.org/10.1109/icorr.2005.1501143.

Finn, M., & McDonald, S. (2011). Computerised cognitive training for older persons with mild cognitive impairment: A pilot study using a randomised controlled trial design. *Brain Impairment, 12*(3), 187–199.

Finnema, E., Dröes, R.-M., Ettema, T. P., & van Tilburg, W. (2005). The effect of integrated emotion-oriented care versus usual care on elderly persons with dementia in the nursing home and on nursing assistants: A randomized clinical trial. *International Journal of Geriatric Psychiatry, 20*(4), 330–343.

Forstmeier, S., & Maercker, A. (2015). Motivational processes in mild cognitive impairment and Alzheimer's disease: Results from the Motivational Reserve in Alzheimer's (MoReA) study. *BMC Psychiatry, 15,* 293.

Gregg, L., & Tarrier, N. (2007). Virtual reality in mental health. A review of the literature. *Social Psychiatry and Psychiatric Epidemiology, 42*(5), 343–354.

Ienca, M., Jotterand, F., Elger, B., Caon, M., Scoccia Pappagallo, A., Kressig, R.W., & Wangmo, T. (2017). Intelligent assistive technology for Alzheimer's disease and other dementias: A systematic review. *Journal of Alzheimer's Disease, 56*(4), 1301–1340. https://doi.org/10.3233/jad-161037.

IFR. (2018). *Executive summary world robotics 2018 service robots.* IFR.

Korczyn, A. D., Peretz, C., Aharonson, V., et al. (2007). Computer based cognitive training with CogniFit improved cognitive performance above the effect of classic computer games: prospective, randomized, double blind intervention study in the elderly. *Alzheimer's & Dementia: The Journal of the Alzheimer's Association, 3*(3), S171.

Kothgassner, O.D., Felnhofer, A., Hauk, N., Kasthofer, E., Gomm, J. & Kryspin-Exner, I. (2012). *TUI: Technology Usage Inventory. Fragebogen und Manual.* Wien: FFG

Lauriks, S., Reinersmann, A., Van der Roest, H. G., Meiland, F. J., Davies, R. J., Moelaert, F., … Dröes, R. M. (2007). Review of ICT-based services for identified unmet needs in people with dementia. *Ageing Res Review, 6,* 223–246.

Maercker, A., & Forstmeier, S. (2011). Healthy brain aging: The new concept of motivational reserve. *Psychiatrist, 2011*(35), 175–178.

Mann, J. A., McDonald, B. A., Kuo, I. H., Li, X., & Broadbent, E. (2015). People respond better to robots than computer tablets delivering healthcare instructions. *Computers in Human Behavior, 43,* 112–117.

Mao, H. F., Chang, L. H., Yao, G., Chen, W. Y., & Huang, W. N. (2015). Indicators of perceived useful dementia care assistive technology: Caregivers' perspectives. *Geriatrics & Gerontology International, 15*(8), 1049–1057.

McCallum, S., Boletsis, C. (2013). Dementia games: A literature review of dementia-related serious games. In *Serious games development and applications.* Lecture Notes in Computer Science (Vol. 8101, pp. 15–27). Springer.

McDuff, D., El Kaliouby, R., Cohn, J. F., & Picard, R. W. (2015). Predicting ad liking and purchase intent: Large-scale analysis of facial responses to ads. *Affective Computing, IEEE Transactions, 6*(3), 223–235.

Montero-Odasso, M., Bherer, L., Studenski, S., Gopaul, K., Oteng-Amoako, A., Woolmore-Goodwin, S., et al. (2015). Mobility and cognition in seniors. Report from the 2008 institute of aging (CIHR) mobility and cognition workshop. *Canadian Geriatrics Journal, 18*(3), 159–167.

Mordoch, E., Osterreicher, A., Guse, L., Roger, K., & Thompson, G. (2013). Use of social commitment robots in the care of elderly people with dementia: a literature review. *Maturitas, 74,*(1), 14–20.

Nordheim, J., Hamm, S., Kuhlmey, A., & Suhr, R. (2015). Tablet-PC und ihr Nutzen für demenzerkrankte Heimbewohner: Ergebnisse einer qualitativen Pilotstudie [Tablet computers and their benefits for nursing home residents with dementia: Results of a qualitative pilot study]. *Zeitschrift für Gerontologie und Geriatrie, 48,* 543–549.

OECD. (2015). *Addressing Dementia—the OECD response.* Paris, France: OECD Publishing.

Paletta, L., Lerch, A., Kemp, C., Pittino, L., Steiner, J., Panagl, M., et al. (2018a). Playful Multimodal training for persons with dementia with executive function based diagnostic tool. In *Proceedings of Pervasive Technologies Related to Assistive Environments (PETRA), Corfu, Greece, 26–29 June 2018.* ACM Press.

Paletta, L., Pszeida, M., Panagl, M. (2018b). Towards playful monitoring of executive functions: Deficits in inhibition control as indicator for cognitive impairment in first stages of Alzheimer. In: *Proceedings of 9th International Conference on Applied Human Factors and Ergonomics (AHFE 2018), Orlando, FL, 21–25 July 2018.* Springer.

Pessoa, L. (2009). How do emotion and motivation direct executive control? *Trends in Cognitive Sciences, 13,* 160–166.

Pino, M., Boulay, M., Jouen, F., & Rigaud, A. S. (2015). Are we ready for robots that care for us?" Attitudes and opinions of older adults toward socially assistive robots. *Frontiers in Aging Neuroscience, 23*(7), 141. https://doi.org/10.3389/fnagi.2015.00141.

Pripfl, J., Körtner, T., Batko-Klein, D., Hebesberger, D., Weninger, M., & Gisinger, C. (2016). Social service robots to support independent living: Experiences from a field trial. *Zeitschrift fur Gerontologie und Geriatrie, 49*(4), 282–287. https://doi.org/10.1007/s00391-016-1067-4.

Robert Koch Institut. (2015). *Gesundheit in Deutschland [Health in Germany], Gesundheits-berichterstattung des Bundes - Gemeinsam getragen von RKI und Destatis.* Berlin, Germany: RKI.

Robert, P. H., Mulin, E., Malle, P., & David, R. (2010). Apathy diagnosis, assessment, and treatment in Alzheimer's disease. In *CNS neuroscience & therapeutics.* Blackwell Publishing Ltd.

Rösler, U., Schmidt, K., Merda, M., & Melzer, M. (2018). *Digitalisierung in der Pflege (Digital-ization in nursing). Geschäftsstelle der Initiative Neue Qualität der Arbeit.* Berlin, Germany: Bundesanstalt für Arbeitsschutz und Arbeitsmedizin.

Rosseter, R. (2017). *Fact sheet: Nursing shortage.* Washington, DC, USA: AACN.

Schüssler S. (2015). Care dependency and nursing care problems in nursing home residents with and without dementia (Doctoral thesis, Medical University of Graz, Austria).

Shatil, E., Metzer, A., Horvitz, O., & Miller, A. (2010). Home-based personalized cognitive training in MS patients: A study of adherence and cognitive performance. *Neuro Rehabilitation, 26*(2), 143–153.

Shipstead, Z., Hicks, K., & Engle, R. W. (2012). Cogmed working memory training: Does the evidence support the claims? *Journal of Applied Research in Memory and Cognition, 1*(3), 185–193.

Smarr, C-A., Prakash, A., Beer, M., Mitzner, T. L., Kemp, C. C., & Rogers, W. A. (2012). Older adults' preferences for and acceptance of robot assistance for everyday living tasks. In *Proceedings of the Human Factors and Ergonomics Society. Annual Meeting* (Vol. 56, No. 1, pp. 153–157). https://doi.org/10.1177/1071181312561009.

Softbank Robotics. (2015). Pepper—The world's first personal robot that reads emotions. Softbank Rob. http://bit.ly/2oUMtvj.

Span, M., Hettinga, M., Vernooij-Dassen, M., Eefsting, J., & Smits, C. (2013). Involving people with dementia in the development of supportive IT applications: A systematic review. *Ageing Research Review, 12*(2), 535–551.

Spero, I. (Ed.) (2017). Neighbourhoods of the future—better homes for older adults—improving health, care, design and technology. In *Agile ageing alliance.* UK: McCarthy & Stone, CSL/RockCouture Productions Ltd.

Starkstein, S. E., Jorge, R., Mizrahi, R. (2006). The prevalence, clinical correlates and treatment of apathy in Alzheimer's disease. *The European Journal of Psychiatry, 20*(2), 96–106.

Sugihara, T., Fujinami, T., Phaal, R., & Ikawa, Y. (2013). A technology roadmap of assistive technologies for dementia care in Japan. *Dementia (London), 14*(1), 80–103.

Toure-Tillery, M., & Fishbach, A. (2014). How to measure motivation: A guide for the experimental social psychologist. *Social and Personality Psychology Compass.*

Wallenfels, M. (2016). Die Zukunft der Pflege durch Roboter. *ProCare, 8,* 42–45.

Wang, R. H., Sudhama, A., Begum, M., Huq, R., & Mihailidis, A. (2016). Robots to assist daily activities: views of older adults with Alzheimer's disease and their caregivers. *International Psychogeriatrics, 29*(1), 67–79.

WHO. (2007). Fact sheet, Investing in the health workforce enables stronger health systems. WHO.

Wilson, R. S., Schneider, J. A., Arnold, S. E., Bienias, J. L., & Bennett, D. A. (2007). Conscientiousness and the incidence of Alzheimer's disease and mild cognitive impairment. *Archives of General Psychiatry, 2007*(64), 1204–1212.

Wu, Y. H., Wrobel, J., Cornuet, M., Kerhervé, H., Damnée, S., & Rigaud, A. S. (2014). Acceptance of an assistive robot in older adults: A mixed-method study of human-robot interaction over a 1-month period in the Living Lab setting. *Clinical Interventions in Aging, 8*(9), 801–811.

Index

© Springer Nature Switzerland AG 2019
O. Korn (ed.), *Social Robots: Technological, Societal and Ethical Aspects of Human-Robot Interaction*, Human–Computer Interaction Series,
https://doi.org/10.1007/978-3-030-17107-0

Printed in the United States
By Bookmasters